Springer Proceedings in Mathematics & Statistics

Volume 436

This book series features volumes composed of selected contributions from workshops and conferences in all areas of current research in mathematics and statistics, including data science, operations research and optimization. In addition to an overall evaluation of the interest, scientific quality, and timeliness of each proposal at the hands of the publisher, individual contributions are all refereed to the high quality standards of leading journals in the field. Thus, this series provides the research community with well-edited, authoritative reports on developments in the most exciting areas of mathematical and statistical research today.

A. A. Ambily · V. B. Kiran Kumar
Editors

Semigroups, Algebras and Operator Theory

ICSAOT 2022, CUSAT, India, March 28–31

Springer

Editors
A. A. Ambily
Department of Mathematics
Cochin University of Science
and Technology
Kochi, Kerala, India

V. B. Kiran Kumar
Department of Mathematics
Cochin University of Science
and Technology
Kochi, Kerala, India

ISSN 2194-1009 ISSN 2194-1017 (electronic)
Springer Proceedings in Mathematics & Statistics
ISBN 978-981-99-6348-5 ISBN 978-981-99-6349-2 (eBook)
https://doi.org/10.1007/978-981-99-6349-2

Mathematics Subject Classification: 15B30, 18B40, 18D70, 47A10, 47L80, 60B15, 65D07

This Springer imprint is published by the registered company Springer Nature Singapore Pte Ltd.
The registered company address is: 152 Beach Road, #21-01/04 Gateway East, Singapore 189721, Singapore

Paper in this product is recyclable.

To
Prof. P. G. Romeo
Department of Mathematics, CUSAT.

Preface

The subject matter of these proceedings is connected with several branches of mathematics, including mathematical physics. Firstly, it reveals the importance of semigroup theory in various topics. These include the semigroup structure of regular probability measures. The algebra section discusses various themes, which display the subject's diverse nature and prominent applications. This section's interplay between geometric algebra and quantum mechanics is a noteworthy part. We have both theoretical and numerical approximation-type results in the operator theory part. The topics vary from C^* algebraic models to numerical solutions to some partial differential equations.

The presentation here allows readers with diverse mathematical backgrounds and topical interests to understand and appreciate these structures. The treatment would benefit active researchers and beginning graduate students in these areas. Some specific topics covered in the proceedings are explained now.

An inverse Clifford semigroup is a semilattice of groups. It is one of the earliest studied classes of semigroups. A Clifford semigroup's structural aspects are discussed here from a cross-connection perspective. In one of the articles, we see a generalization of classical Stone duality, referred to as commutative Stone duality. In another article, the notion of group Lie algebra is introduced. The article discussing the algebraic regularity of regular probability measures on a locally compact semigroup exhibits the interplay between abstract algebraic concepts and hardcore analysis tools from harmonic and functional analysis. An operator theory-flavored article represents resolvent algebra inside the full Toeplitz algebra over the Fock-Bargmann space in infinitely many variables. An expository article on applying geometric algebra to Koga's work on quantum mechanics will attract a broad audience from various branches of science. Articles discussing condition spectrum and numerical solutions of partial differential equations are also attractive to the pure and applied mathematics community.

Some articles are expository articles appearing here for the first time. These proceedings introduce beginning researchers and researchers in other areas to explore the scope of semigroups, algebras and operator theory so that they might become prepared and subsequently inspired to join the game.

Kochi, India

A. A. Ambily
V. B. Kiran Kumar

Acknowledgements

This volume is an outcome of the International Conference on Semigroups, Algebras and Operator Theory (ICSAOT-2022), held at the Department of Mathematics, Cochin University of Science and Technology, Kerala, March 28–31, 2022. This conference is the fourth in a series on the same theme. Professor P. G. Romeo was central in organizing the series' previous conferences (ICSAOT-2014, ICSAA-2015 and ICSAA-2019). This time, ICSAOT-2022 was organized to honor Prof. P. G. Romeo at his retirement. Due to the pandemic, we had to restrict ourselves from holding the conference in the online mode. We thank all the speakers and participants of the conference. The Cochin University of Science and Technology gave administrative and logistic support for the workshop. We gratefully acknowledge this support. We are also grateful to the faculty, administrative staff and research scholars of Cochin University of Science and Technology who helped us to organize the event and the technical assistance during the preparation of the volume for publication. We thank Dr. Neeraj Kumar, IIT Hyderabad, for helping us organize the event. We are thankful to the Department of Mathematics, Cochin University of Science and Technology research scholars for their support during the conference. Finally, we thank all the authors for contributing to the volume and the reviewers for helping us with the review process.

Contents

Operator Theory

Contributors

Madeline Al-Tahan Department of Mathematics and Statistics, Abu Dhabi University, Abu Dhabi, UAE

S. N. Arjun Department of Mathematics, Cochin University of Science and Technology, Kochi, KL, India

Noufal Asharaf Cochin University of Science and Technology, Kochi, India

Judy Augustine Department of Mathematics, University of Kerala, Thiruvananthapuram, Kerala, India

P. A. Azeef Muhammed Centre for Research in Mathematics and Data Science, Western Sydney University, Penrith, NSW, Australia

Athira Babu Cochin University of Science and Technology, Kochi, India

Wolfram Bauer Institut für Analysis, Hannover, Germany

Vadiraja Bhatta Department of Mathematics, Manipal Institute of Technology, Manipal Academy of Higher Education, Manipal, KA, India

K. V. Didimos Department of Mathematics, Sacred Heart College, Thevara, Kochi, KL, India

I. M. Evseev Lomonosov Moscow State University, Moscow, Russia;
Moscow Center of Fundamental and Applied Mathematics, Moscow, Russia

Robert Fulsche Institut für Analysis, Hannover, Germany

A. E. Guterman Lomonosov Moscow State University, Moscow, Russia;
Moscow Center of Fundamental and Applied Mathematics, Moscow, Russia;
Technion - Israel Institute of Technology, Haifa, Israel

Babushri Srinivas Kedukodi Department of Mathematics, Manipal Institute of Technology, Manipal Academy of Higher Education, Manipal, KA, India

Nikita V. Kitov Institute of Natural Sciences and Mathematics, Ural Federal University, Ekaterinburg, Russia

G. Krishna Kumar Department of Mathematics, University of Kerala, Thiruvananthapuram, Kerala, India

Syam Prasad Kuncham Department of Mathematics, Manipal Institute of Technology, Manipal Academy of Higher Education, Manipal, KA, India

Mark V. Lawson Department of Mathematics and the Maxwell Institute for Mathematical Sciences, Heriot-Watt University, Edinburgh, UK

M. N. N. Namboodiri IMRT, Thiruvananthapuram, India;
Formerly with Mathematics Department, CUSAT, Kochi, India

Harikrishnan Panackal Department of Mathematics, Manipal Institute of Technology, Manipal Academy of Higher Education, Manipal, KA, India

Pallavi Panjarike Department of Mathematics, Manipal Institute of Technology, Manipal Academy of Higher Education, Manipal, KA, India

C. S. Preenu Department of Mathematics, University College Thiruvananthapuram, Thiruvananthapuram, Kerala, India

A. R. Rajan Department of Mathematics, Institute of Mathematics Research and Training (IMRT), University of Kerala, Thiruvananthapuram, India

P. G. Romeo Department of Mathematics, Cochin University of Science and Technology, Kochi, Kerala, India

Rajani Salvankar Department of Mathematics, Manipal Institute of Technology, Manipal Academy of Higher Education, Manipal, KA, India

Kuncham Syam Prasad Department of Mathematics, Manipal Institute of Technology, Manipal Academy of Higher Education, Manipal, KA, India

Alanka Thomas Department of Mathematics, Cochin University of Science and Technology, Kochi, Kerala, India

Mikhail V. Volkov Institute of Natural Sciences and Mathematics, Ural Federal University, Ekaterinburg, Russia

Semigroups

Semigroups is a branch of algebra, which has applications in the areas biology, partial differential equation, formal languages, etc. The first part of this volume deals with semigroups and related areas.

Cross-Connections in Clifford Semigroups

P. A. Azeef Muhammed and C. S. Preenu

Abstract An inverse Clifford semigroup (often referred to as just a Clifford semigroup) is a semilattice of groups. It is an inverse semigroup and in fact, one of the earliest studied classes of semigroups (Clifford in Ann Math 42(4):1037–1049 (1941), [6]). In this short note, we discuss various structural aspects of a Clifford semigroup from a cross-connection perspective. In particular, given a Clifford semigroup S, we show that the semigroup $T\mathbb{L}(S)$ of normal cones is isomorphic to the original semigroup S, even when S is not a monoid. Hence, we see that cross-connection description degenerates in Clifford semigroups. Further, we specialise the discussion to provide the description of the cross-connection structure in an arbitrary semilattice, also.

Keywords Clifford semigroups · Inverse semigroups · Semilattices · Normal categories · Cross-connections

1 Introduction

Grillet [7] introduced cross-connections as a pair of functions to describe the interrelationship between the posets of principal left and right ideals of a regular semigroup. This construction involved building two intermediary semigroups and further identifying a fundamental image of the semigroup as a subdirect product, using the cross-connection functions. But since isomorphic posets give rise to isomorphic cross-connections, this construction could capture only fundamental regular semi-

P. A. Azeef Muhammed (✉)
Centre for Research in Mathematics and Data Science, Western Sydney University, Penrith, NSW 2751, Australia
e-mail: azeefp@gmail.com; A.ParayilAjmal@westernsydney.edu.au

C. S. Preenu
Department of Mathematics, University College Thiruvananthapuram, Thiruvananthapuram 695034, Kerala, India
e-mail: cspreenu@universitycollege.ac.in

A. A. Ambily and V. B. Kiran Kumar (eds.), *Semigroups, Algebras and Operator Theory*, Springer Proceedings in Mathematics & Statistics 436,
https://doi.org/10.1007/978-981-99-6349-2_1

groups. So, Nambooripad replaced posets with certain small categories to overcome this limitation. Hence, using the categorical theory of cross-connections, Nambooripad constructed arbitrary regular semigroups from their ideal structure.

Starting from a regular semigroup S, Nambooripad identified two small categories: $\mathbb{L}(S)$ and $\mathbb{R}(S)$ which abstract the principal left and right ideal structures of the semigroup, respectively. He showed that these categories are interconnected using a pair of functors. It can be seen that the object maps of these functors coincide with Grillet's cross-connection functions and so Nambooripad called his functors also cross-connections. Further, Nambooripad showed that this correspondence can be extended to an explicit category equivalence between the category of regular semigroups and the category of cross-connections. Hence the ideal structure of a regular semigroup can be completely captured using these 'cross-connected' categories.

Being a rather technical construction, it is instructive to work out the simplifications that arise in various special classes of semigroups. There has been several works in this direction [1–5, 11] and in this short article, we propose to outline how the construction simplifies in a couple of very natural classes of regular semigroups: namely, Clifford semigroups and semilattices. As the reader may see, this exercise also provides some useful illustrations to several cross-connection related subtleties.

In fact, Clifford semigroups are one of the first classes of inverse semigroups whose structure was studied. It was originally defined [6] as a union of groups in which idempotents commute. It may be noted that some authors refer to general union of groups also as Clifford semigroups but we shall follow Howie's [9] seminal treatise on semigroups and refer to a semilattice of groups as a Clifford semigroup. The following characterizations of Clifford semigroups will be useful in the sequel.

Theorem 1 ([9, Theorem 4.2.1], [8, Theorem 1.3.11]) *Let S be a semigroup. Then the following statements are equivalent.*

1. *S is a Clifford semigroup;*
2. *S is a semilattice of groups;*
3. *S is a strong semilattice of groups;*
4. *S is regular and the idempotents of S are central;*
5. *Every $\mathscr{D}$ class of S has a unique idempotent.*

In Nambooripad's cross-connection description, starting from a regular semigroup S, two small categories of principal left and right ideals (denoted by $\mathbb{L}(S)$ and $\mathbb{R}(S)$, respectively in the sequel) are defined. Then their inter-relationship is abstracted as a pair of functors called cross-connections. Conversely, given an abstractly defined pair of cross-connected categories (with some special properties), one can construct a regular semigroup. This correspondence between regular semigroups and cross-connections is proved to be a category equivalence.

The construction of the regular semigroup from the category happens in several layers and the crucial object here is the intermediary regular semigroup $T\mathbb{L}(S)$ of normal cones from the given category $\mathbb{L}(S)$. We shall see that when the semigroup S is Clifford, then the semigroup $T\mathbb{L}(S)$ is isomorphic to S. It is known that this is not true in general [5] even for an inverse semigroup S. The fact that the semigroup

$T\mathbb{L}(S)$ is isomorphic to S leads to a major degeneration of the cross-connection analysis in Clifford semigroups. This is discussed in the next section. In the last section, we specialise our discussion to an arbitrary semilattice and describe the cross-connections therein.

As mentioned above, since the article can be seen as a part of a continuing project of studying the various classes of regular semigroups within the cross-connection framework, we refer the reader to [1, 5, 11] for the preliminary notions and formal definitions in Nambooripad's cross-connection theory. We also refer the reader to [10] for the original treatise on cross-connections.

2 Normal Categories and Cross-Connections in Clifford Semigroups

Recall from [10] that given a regular semigroup S, the normal category $\mathbb{L}(S)$ of principal left ideals are defined as follows. The object set

$$v\mathbb{L}(S) := \{Se : e \in E(S)\}$$

and the morphisms in $\mathbb{L}(S)$ are partial right translations. In fact, the set of all morphisms between two objects Se and Sf may be characterised as the set $\{\rho(e, u, f) : u \in eSf\}$, where the map $\rho(e, u, f)$ sends $x \in Se$ to $xu \in Sf$.

First, we proceed to discuss some special properties of the category $\mathbb{L}(S)$ when S is a Clifford semigroup. This will lead us to the characterisation of the semigroup $T\mathbb{L}(S)$ of all normal cones in S.

Proposition 1 *Let S be a Clifford semigroup. Two objects in $\mathbb{L}(S)$ are isomorphic if and only if they are identical.*

Proof Clearly, identical objects are always isomorphic. Conversely, suppose Se and Sf are two isomorphic objects in $\mathbb{L}(S)$. Then by [10, Proposition III.13(c)], we have $e \mathscr{D} f$. Recall that in a Clifford semigroup, Green's relations $\mathscr{L}$, $\mathscr{R}$ and $\mathscr{D}$ are identical. Therefore $e\mathscr{L}f$ and hence $Se = Sf$. □

Given the normal category $\mathbb{L}(S)$ of principal left ideals of a regular semigroup S, it is known that two morphisms are equal, i.e. $\rho(e, u, f) = \rho(g, v, h)$ if and only if $e \mathscr{L} g$, $f \mathscr{L} h$, and $v = gu$. Also, given two morphisms $\rho(e, u, f)$ and $\rho(g, v, h)$, they are composable if and only $Sf = Sg$ so that $\rho(e, u, f)\rho(g, v, h) = \rho(e, uv, h)$. Now, we see that the equality of morphisms simplify when S is a Clifford semigroup.

Proposition 2 *Let S be a Clifford semigroup, then $\rho(e, u, f) = \rho(g, v, h)$ in the category $\mathbb{L}(S)$ if and only if $e = g$, $u = v$ and $f = h$.*

Proof Suppose that two morphisms $\rho(e, u, f) = \rho(g, v, h)$ are equal. Recall that by [10, Lemma II.12], this implies that $Se = Sg$, $Sf = Sh$ and $v = gu$. That is, the elements e and g are two $\mathscr{L}$ related idempotents. But since $\mathscr{L}$ and $\mathscr{H}$ are identical in a Clifford semigroup S and since a $\mathscr{H}$-class can contain at most one idempotent, we have $e = g$. Similarly we get $f = h$. So, the sets $eSf = gSh$ are equal and also $v = gu = eu = u$. □

Now, we proceed to characterise the building blocks of the cross-connection construction, namely the normal cones in the category $\mathbb{L}(S)$. An 'order-respecting' collection of morphisms in a normal category is defined as a normal cone.

Definition 1 Let $\mathbb{L}(S)$ be the normal category of principal left ideals in a regular semigroup S and $Sd \in v\mathbb{L}(S)$. A *normal cone with apex* Sd is a function $\gamma : v\mathbb{L}(S) \to \mathbb{L}(S)$ such that:

1. for each $Se \in v\mathbb{L}(S)$, one has $\gamma(Se) \in \mathbb{L}(S)(Se, Sd)$;
2. $\iota(Sf, Sg)\gamma(Sg) = \gamma(Sf)$ whenever $Sf \subseteq Sg$;
3. $\gamma(Sm)$ is an isomorphism for some $Sm \in v\mathbb{L}(S)$.

Now, for each $a \in S$, we can define a function $\rho^a : v\mathbb{L}(S) \to \mathbb{L}(S)$ as follows:

$$\rho^a(Se) := \rho(e, ea, f) \text{ where } f \mathscr{L} a. \tag{1}$$

It is easy to verify that the map ρ^a is a well-defined normal cone with apex Sf in the sense of Definition 1, see [10, Lemma III.15].

In the sequel, the normal cone ρ^a is called the *principal cone* determined by the element a. In particular, observe that, for an idempotent $e \in E(S)$, we have a principal cone ρ^e such that $\rho^e(Se) = \rho(e, e, e) = 1_{Se}$. This leads us to the most crucial proposition of this article.

Proposition 3 *In a Clifford semigroup S, every normal cone in* $\mathbb{L}(S)$ *is a principal cone.*

Proof Suppose γ is a normal cone in $\mathbb{L}(S)$ with vertex Se so that $\gamma(Se) = \rho(e, u, e)$ for some $u \in S$. Then for any $Sf \in v\mathbb{L}(S)$, we shall show that $\gamma(Sf) = \rho(f, fu, e) = \rho^u$ for some $u \in S$. To this end, first observe that since idempotents commute in S, we have $Sef = Sfe \subseteq Se$. So, by (2) of Definition 1, we have $\gamma(Sef) = \rho(ef, ef, e)\gamma(Se)$. Then,

$$\begin{aligned}
\gamma(Sef) &= \rho(ef, ef, e)\gamma(Se) && \text{since } Sef \subseteq Se\\
&= \rho(ef, ef, e)\rho(e, u, e)\\
&= \rho(ef, efu, e)\\
&= \rho(ef, fu, e) && \text{since } u \in eSe \text{ and } ef = fe.
\end{aligned}$$

Now for any $Sf \in v\mathbb{L}(S)$, let $\gamma(Sf) = \rho(f, v, e)$ for some $v \in fSe$. Then since $Sef \subseteq Sf$ also, we have

$$\begin{aligned}\gamma(Sef) &= \rho(ef, ef, f)\gamma(Sf)\\ &= \rho(ef, ef, f)\rho(f, v, e)\\ &= \rho(ef, efv, e)\\ &= \rho(ef, ev, e) && \text{since } fv = v\\ &= \rho(ef, ve, e) && \text{since idempotents are central in } S\\ &= \rho(ef, v, e) && \text{since } v \in fSe.\end{aligned}$$

Now, from the discussion above, we see that $\rho(ef, fu, e) = \rho(ef, v, e)$. Then, using Proposition 2, this implies that $v = fu$. Hence for all $Sf \in v\mathbb{L}(S)$, we have $\gamma(Sf) = \rho(f, fu, e)$ and so $\gamma = \rho^u$. □

In general, for an arbitrary regular semigroup S, two distinct principal cones ρ^a and ρ^b may be equal in $\mathbb{L}(S)$ even when $a \neq b$. But when S is Clifford, we proceed to show that it is not the case.

Proposition 4 *Let S be a Clifford semigroup. Given two principal cones ρ^a and ρ^b, we have $\rho^a = \rho^b$ if and only if $a = b$.*

Proof Clearly when $a = b$, then $\rho^a = \rho^b$. Conversely suppose $\rho^a = \rho^b$. Then their vertices coincide and so, we have $Sa = Sb$, then since S is Clifford and Green's relations coincide, we have $a\mathscr{H}b$. Now let e be the idempotent in $H_a = H_b$, the Green's $\mathscr{H}$ class containing a and b. Then $\rho^a(Se) = \rho(e, ea, e) = \rho(e, a, e)$. Similarly, we get $\rho^b(Se) = \rho(e, b, e)$. Since the cones are equal, the corresponding morphism components at each vertex coincide. Hence using Proposition 2, we have $a = b$. □

Recall from [10, Sect. III.1] that the set of all normal cones in a normal category forms a regular semigroup, under a natural binary operation. So, in particular, given the normal category $\mathbb{L}(S)$, the set $T\mathbb{L}(S)$ of all normal cones in the category $\mathbb{L}(S)$ is a regular semigroup. Now, we proceed to characterise this semigroup when S is Clifford.

Theorem 2 *Let S be a Clifford semigroup. Then the semigroup $T\mathbb{L}(S)$ of all normal cones in $\mathbb{L}(S)$ is isomorphic to the semigroup S.*

Proof Recall from [10, Sect. III.3.2] that the map $\bar{\rho}\colon a \mapsto \rho^a$ from a regular semigroup S to the semigroup $T\mathbb{L}(S)$ is a homomorphism. Now, when S is Clifford, by Proposition 3, we have seen that every normal cone in $T\mathbb{L}(S)$ is principal and hence the map $\bar{\rho}$ is surjective. Also, by Proposition 4, we see that the map $\bar{\rho}$ is surjective. Hence the map $\bar{\rho}$ is an isomorphism from the Clifford semigroup S to the semigroup $T\mathbb{L}(S)$. □

The above theorem characterises the semigroup of $T\mathbb{L}(S)$ of all normal cones in $\mathbb{L}(S)$; this naturally leads us to the complete description of the cross-connection structure of the semigroup S as follows.

Recall from [10, Sect. III.4] that given a normal category, it has an associated dual category whose objects are certain set-valued functors and morphisms are natural transformations. Now we proceed to characterise the normal dual $N^*\mathbb{L}(S)$ of the normal category $\mathbb{L}(S)$ of principal left ideals of a Clifford semigroup S.

Theorem 3 *Let S be a Clifford semigroup. Then the normal dual $N^*\mathbb{L}(S)$ of the normal category $\mathbb{L}(S)$ of principal left ideals in S is isomorphic to the normal category $\mathbb{R}(S)$ of principal right ideals in S.*

Proof It is known that [10, Theorem III.25] the normal dual $N^*\mathbb{L}(S)$ of the normal category $\mathbb{L}(S)$ is isomorphic to the normal category $\mathbb{R}(T\mathbb{L}(S))$ of principal right ideals of the regular semigroup $T\mathbb{L}(S)$. When S is a Clifford semigroup, by Theorem 2, we see that the semigroup $T\mathbb{L}(S)$ is isomorphic to S and hence $N^*\mathbb{L}(S)$ is isomorphic to $\mathbb{R}(S)$ as normal categories. □

Dually, we can easily prove the following results:

Theorem 4 *Let S be a Clifford semigroup. The semigroup $T\mathbb{R}(S)$ of all normal cones in the category $\mathbb{R}(S)$ of all principal right ideals in S is anti-isomorphic to the semigroup S. The normal dual $N^*\mathbb{R}(S)$ of the normal category $\mathbb{R}(S)$ is isomorphic to the normal category $\mathbb{L}(S)$.*

Recall from [10, Theorem IV.1] that the cross-connection of a regular semigroup S is defined as a quadruplet $(\mathbb{L}(S), \mathbb{R}(S), \Gamma, \Delta)$ such that $\Gamma\colon \mathbb{R}(S) \to N^*\mathbb{L}(S)$ and $\Delta\colon \mathbb{L}(S) \to N^*\mathbb{R}(S)$ are functors satisfying certain properties. Now, using Theorems 3 and 4, we see that both the functors Γ and Δ are in fact isomorphisms. And hence the cross-connection structure degenerates to isomorphisms of the associated normal categories in a Clifford semigroup S.

3 Cross-Connections of a Semilattice

Clearly, a semilattice is a Clifford semigroup. So, we specialise our discussion to a semilattice using the results in the previous section. The following theorem follows from Theorem 2.

Theorem 5 *Let S be a semilattice. Then the semigroup $T\mathbb{L}(S)$ of all normal cones in $\mathbb{L}(S)$ is isomorphic to the semilattice S.*

Theorems 3 and 4 when applied to a semilattice can be unified as follows:

Theorem 6 *Let S be a semilattice. Then the normal dual $N^*\mathbb{L}(S)$ of the normal category $\mathbb{L}(S)$ of principal left ideals in S is isomorphic to the normal category $\mathbb{R}(S)$ of principal right ideals in S. The normal dual $N^*\mathbb{R}(S)$ of the normal category $\mathbb{R}(S)$ is isomorphic to the normal category $\mathbb{L}(S)$.*

Hence, we see that both the cross-connections functors Γ and Δ are isomorphisms, when we have a semilattice also. So, the cross-connection structure degenerates to isomorphisms of the associated normal categories in a semilattice, too.

References

1. Azeef Muhammed, P.A.: Cross-connections and variants of the full transformation semigroup. Acta Sci. Math. (Szeged) **84**, 377–399 (2018)
2. Azeef Muhammed, P.A.: Cross-connections of linear transformation semigroups. Semigroup Forum **97**(3), 457–470 (2018)
3. Azeef Muhammed, P.A., Rajan, A.R.: Cross-connections of completely simple semigroups. Asian-European J. Math. **09**(03), 1650053 (2016)
4. Azeef Muhammed, P.A., Rajan, A.R.: Cross-connections of the singular transformation semigroup. J. Algebra Appl. **17**(3), 1850047 (2018)
5. Azeef Muhammed, P.A., Volkov, M.V., Auinger, K.: Cross-connection structure of locally inverse semigroups. Internat. J. Algebra Comput. **33**(1), 123–159 (2023)
6. Clifford, A.H.: Semigroups admitting relative inverses. Ann. Math. **42**(4), 1037–1049 (1941)
7. Grillet, P.A.: Structure of regular semigroups: 1. A representation, 2. Cross-connections, 3. The reduced case. Semigroup Forum **8**, 177–183, 254–259, 260–265 (1974)
8. Higgins, P.M.: Techniques of semigroup theory. Clarendon Press (1992)
9. Howie, J.M.: Fundamentals of Semigroup Theory, vol. 12. Oxford University Press (1995)
10. Nambooripad, K.S.S.: Theory of Cross-connections. Centre for Mathematical Sciences (1994)
11. Rajan, A.R., Preenu, C.S., Zeenath, K.S.: Cross connections for normal bands and partial order on morphisms. Semigroup Forum **104**, 125–147 (2022)

Non-commutative Stone Duality

Mark V. Lawson

Abstract I show explicitly that Boolean inverse semigroups are in duality with Boolean groupoids, a class of étale topological groupoids; this is what we mean by the term 'non-commutative Stone duality'. This generalizes classical Stone duality, which we refer to as 'commutative Stone duality'.

Keywords Stone duality · Inverse semigroups · Boolean inverse monoids

1 Introduction

The theory of what we term *non-commutative Stone duality* grew out of the work of a number of authors [13, 14, 18, 31, 35, 42, 45, 46]. In this chapter, I shall concentrate primarily on one aspect of that duality: namely, how Boolean inverse semigroups are in duality with a class of étale groupoids called Boolean groupoids. Specifically, we shall prove the following theorem and discuss some special cases (proved as Theorem 7).

Theorem 1 (Non-commutative Stone duality) *The category of Boolean inverse semigroups and callitic morphisms is dually equivalent to the category of Boolean groupoids and coherent, continuous, and covering functors.*

This theorem generalizes what you will find in [24] since we shall not assume that our topological groupoids are Hausdorff. Although this theorem can be gleaned from our papers [24–27, 30], what I describe here has not been reported in one place before.

You might think that this duality is of merely parochial interest. It isn't. The work of Matui [38, 39] deals with étale groupoids of just the kind that figure in our duality theorem. In addition, Matui refers constantly to 'compact-open $\mathcal{G}$-sets'.

M. V. Lawson (✉)
Department of Mathematics and the Maxwell Institute for Mathematical Sciences,
Heriot-Watt University, Riccarton, Edinburgh EH14 4AS, UK
e-mail: m.v.lawson@hw.ac.uk

A. A. Ambily and V. B. Kiran Kumar (eds.), *Semigroups, Algebras and Operator Theory*,
Springer Proceedings in Mathematics & Statistics 436,
https://doi.org/10.1007/978-981-99-6349-2_2

These are precisely what we call 'compact-open local bisections' and are elements of the inverse semigroup associated with the étale groupoid. Thus, even though Matui is not explicitly interested in Boolean inverse semigroups, they are there implicitly.

Sections 3–7 are devoted to proving the above theorem, whereas in Sect. 2 classical Stone duality is described since this sets the scene for our generalization. Thus, the reader familiar with classical Stone duality can start reading at Sect. 3. In Sect. 3, we describe the theory of Boolean inverse semigroups needed: these should be regarded as the non-commutative generalized Boolean algebras. In Sect. 4, we describe Boolean groupoids; these are a class of étale topological groupoids and play the role of 'non-commutative topological spaces'. In Sect. 5, we show how to pass from Boolean groupoids to Boolean inverse semigroups, where a key role is played by the compact-open local bisections of a Boolean groupoid. This is a comparatively easy construction. In Sect. 6, we show how to pass from Boolean inverse semigroups to Boolean groupoids by using ultrafilters. This is quite technical. In Sect. 7, we show, among other things, that the above two constructions are the inverse of each other. We also bring on board morphisms and establish our main duality theorem which generalizes the classical theory described in Sect. 2. Section 8 is dedicated to special cases: for example, those Boolean inverse semigroups that have all binary meets correspond to Hausdorff Boolean groupoids. There is one new application, in Sect. 9, which shows how our theory can be used to give an account of *unitization* first described in [58, Definition 6.6.1]. In Sect. 10, I shall show how the above theory can be derived within a more general framework using pseudogroups and arbitrary étale groupoids.

I shall assume you are conversant with the theory of inverse semigroups [20]; I will simply remind the reader of notation and terminology as we go along. Observe that whenever I refer to an order on an inverse semigroup, I mean the *natural partial order* defined on every inverse semigroup. I shall develop the theory of Boolean inverse semigroups from scratch, but this is no substitute for a proper development of that theory as in [58].

We shall also need some terminology from the theory of posets throughout. Let P be a poset with a minimum (or bottom) element denoted by zero 0. In this context, singleton sets such as $\{x\}$ will be written simply as x. If $X \subseteq P$, define

$$X^{\downarrow} = \{y \in P : y \leq x \text{ for some } x \in X\} \text{ and } X^{\uparrow} = \{y \in P : x \leq y \text{ for some } x \in X\}.$$

If $X = X^{\downarrow}$, we say that X is an *order-ideal*. An order-ideal of the form $a^{\downarrow}$ is said to be *principal*. If for any $x, y \in X$ there exists $z \in X$ such that $z \leq x, y$, we say that X is *downwardly directed*. If $X = X^{\uparrow}$, we say that X is *upwardly closed*.

Let P and Q be posets. A function $\theta \colon P \to Q$ is said to be *order-preserving* if $p_1 \leq p_2$ in P implies that $\theta(p_1) \leq \theta(p_2)$ in Q. An *order-isomorphism* is a bijection which is order-preserving and whose inverse is order-preserving. Let F be a *non-empty*[1] subset of P. We say that it is a *filter* if it is downwardly directed and upwardly

[1] Throughout this chapter, filters will be assumed non-empty.

closed. We say that it is *proper* if it does not contain the zero. A maximal proper filter is called an *ultrafilter*. Ultrafilters play an important role in this chapter.

A *meet semilattice* is a poset in which each pair of elements has a greatest lower bound (or meet); we write $x \wedge y$ for the meet of x and y. The following was proved as [6, Lemma 12.3] and is very useful.

Lemma 1 *Let P be a meet semilattice with bottom element 0, and let A be a proper filter in P. Then A is an ultrafilter if and only if the following holds: if $x \wedge y \neq 0$ for all $x \in A$ then $y \in A$.*

Let Y be a meet semilattice with bottom element 0. Let $X \subseteq Y$ be any subset. Define $X^{\wedge}$ to be the set of all finite meets of elements of X; we say that X has the *finite intersection property* if $0 \notin X^{\wedge}$.

Our references for topology are [52, 59]. The following contains the main results we need in this chapter that deal with topology. For the proof of (1), see [52, Sect. 23, Theorem A]; for (2), see [52, Sect. 26, Theorem D]; for (3), see [52, Sect. 21, Theorem A]; for (4), see [52, Sect. 21, Theorem D]; for (5), see [59, Theorem 18.2].

Lemma 2 *1. The product of any non-empty set of compact spaces is compact.*
2. Every compact subspace of a Hausdorff space is closed.
3. Every closed subspace of a compact space is compact.
4. A topological space is compact if and only if every set of closed sets with the finite intersection property has a non-empty intersection.
5. A Hausdorff space is locally compact if for every point x there is a compact set U such that $x \in U^{\circ}$, the union of all the open sets contained in U.

Let $\mathscr{B}$ be any set of subsets of a set X. We say that $\mathscr{B}$ is a *base* if the union of all elements in $\mathscr{B}$ is X and if $x \in B \cap C$, where $B, C \in \mathscr{B}$, then there is a $D \in \mathscr{B}$ such that $x \in D \subseteq B \cap C$. Given a base $\mathscr{B}$, we may define as a topology all those sets which are unions of subsets of $\mathscr{B}$; where the empty set is the union of the empty subset of $\mathscr{B}$. See [59, Theorem 5.3]. A space with a countable base is said to be *second countable*. A topological space is *discrete* if every subset is open. A subset of a topological space is said to be *clopen* if it is both open and closed. A topological space is said to be 0-*dimensional* if it has a base of clopen sets. If X is a topological space, we denote its set of open subsets by $\Omega(X)$.

2 Classical Stone Duality

In this section, we shall describe classical Stone duality which relates generalized Boolean algebras to locally compact Hausdorff 0-dimensional spaces. There are no new results in this section—the theory is classical—but it provides essential motivation for what we do when we come to study Boolean inverse semigroups. In Sect. 2.1, we shall recall the definition and first properties of Boolean algebras; in Sect. 2.2, we shall describe the structure of finite Boolean algebras; in Sect. 2.3, we describe classical Stone duality which deals with the relationship between Boolean algebras and compact Boolean spaces—this is the version of Stone duality that you will find well represented in the textbooks; finally, in Sect. 2.4, we describe the extension of classical Stone duality to generalized Boolean algebras and locally compact Boolean spaces.

2.1 Boolean Algebras

Boolean algebras may not rank highly in the pantheon of algebraic structures but they are, in fact, both mathematically interesting and remarkably useful; for example, they form the basis of measure theory.

Formally, a *Boolean algebra* is a 6-tuple $\left(B, \vee, \wedge, {}' , 0, 1\right)$ consisting of a set B, two binary operations $\vee$, called *join*, and $\wedge$, called *meet*, one unary operation $a \mapsto a'$, called *complementation*, and two constants 0 and 1 satisfying the following ten axioms:

(B1) $(x \vee y) \vee z = x \vee (y \vee z)$. (B6) $x \wedge 1 = x$.
(B2) $x \vee y = y \vee x$. (B7) $x \wedge (y \vee z) = (x \wedge y) \vee (x \wedge z)$.
(B3) $x \vee 0 = x$. (B8) $x \vee (y \wedge z) = (x \vee y) \wedge (x \vee z)$.
(B4) $(x \wedge y) \wedge z = x \wedge (y \wedge z)$. (B9) $x \vee x' = 1$.
(B5) $x \wedge y = y \wedge x$. (B10) $x \wedge x' = 0$.

The element 0 is called the *bottom* and 1 is called the *top*. A function $\theta \colon B \to C$ between two Boolean algebras is called a *homomorphism* if it preserves the two binary operations and the unary operation and maps constants to corresponding constants. Boolean algebras and their homomorphisms form a category. Observe that for a function to be a homomorphism of Boolean algebras, it is enough that it preserves the constants and maps meets and joins to meets and joins, respectively; this is because the complement of x is the unique element y such that $1 = x \vee y$ and $0 = x \wedge y$. The following lemma summarizes some important properties of Boolean algebras that readily follow from these axioms.

Lemma 3 *In a Boolean algebra B, the following hold for all $x, y \in B$:*

(1) $x \vee x = x$ *and* $x \wedge x = x$. (5) $x \vee y = x \vee \left(y \wedge x'\right)$.
(2) $x \wedge 0 = 0$. (6) $x'' = x$.
(3) $1 \vee x = 1$. (7) $(x \vee y)' = x' \wedge y'$.
(4) $x = x \vee (x \wedge y)$. (8) $(x \wedge y)' = x' \vee y'$.

The theory of Boolean algebras is described in an elementary fashion in [9] and from a more advanced standpoint in [15, 50]. The first two chapters of [12] approach the subject of Stone duality from the perspective of frame theory.

Examples of Boolean Algebras

1. The basic example of a Boolean algebra is the *power set Boolean algebra* which consists of the set of all subsets, $\mathsf{P}(X)$, of the non-empty set X with the operations $\cup$, $\cap$, and $\overline{A} = X \setminus A$ and the two constants $\varnothing$, X.
2. We denote by $\mathbb{B}$ the unique two-element Boolean algebra.
3. Let A be any finite non-empty set. We call A an *alphabet*. Denote the free monoid on A by A^*; observe that A^* is countably infinite. Elements of the free monoid are called *strings*. The *length* of a string x is denoted by $|x|$. The empty string is denoted by ε. A subset of A^* is called a *language* over A. Recall that a language over an alphabet A is said to be *recognizable* if there is a finite-state automaton that accepts it. By Kleene's theorem [21], the set of recognizable languages over A is equal to the set of regular languages over A. Denote the set of regular languages over A by $\mathrm{Reg}(A)$. This set is a Boolean algebra (with extra operations). This Boolean structure can be exploited to provide a sophisticated way of studying families of regular languages [8, 44].

Boolean algebras have their roots in the work of George Boole, though the definition of Boolean algebras seems to have been inspired by his work rather than originating there [1, 11]. Until the 1930s, research on Boolean algebras was essentially about axiomatics. However, with Stone's paper [54], stability in the definition of Boolean algebras emerges because he showed that each Boolean algebra could be regarded as a (unital) ring in which each element was idempotent; rings such as this are called *Boolean rings*. In the language of category theory, his result shows that the category of unital Boolean algebras is isomorphic to the category of unital Boolean rings (where I have stressed the fact that the Boolean algebra has a top element and the ring has an identity). The following result describes how the correspondence between Boolean algebras and Boolean rings works at the algebraic level.

Theorem 2 *1. Let B be a Boolean algebra. Define $a + b = \left(a \wedge b'\right) \vee \left(a' \wedge b\right)$ (the symmetric difference) and $a \cdot b = a \wedge b$. Then $(B, +, \cdot, 1)$ is a Boolean ring.*
2. Let $(R, +, \cdot, 1)$ be a Boolean ring. Define $a \vee b = a + b + a \cdot b$, $a \wedge b = a \cdot b$, and $a' = 1 - a$. Then $\left(R, \vee, \wedge, ', 0, 1\right)$ is a Boolean algebra.
3. The constructions (1) and (2) above are mutually inverse.

The above result is satisfying since the definition of Boolean rings could hardly be simpler but also raises the interesting question of why Marshall H. Stone (1903–1989), a functional analyst, should have been interested in Boolean algebras in the first place. The reason is that Stone worked on the spectral theory of symmetric operators, and this led to an interest in algebras of commuting projections. Such algebras are naturally Boolean algebras. The connection between Boolean inverse semigroups and C^*-algebras continues this tradition. The following theorem [7] puts this connection in a slightly wider context. If R is a ring, denote its set of idempotents by $\mathsf{E}(R)$.

Theorem 3 *Let R be a unital commutative ring. Then the set $\mathsf{E}(R)$ is a Boolean algebra when we define $e \vee f = e + f - ef$, $e \wedge f = e \cdot f$, and $e' = 1 - e$.*

By Theorems 2 and 3, it is immediate that each Boolean algebra arises as the Boolean algebra of idempotents of a commutative ring.

So far we have viewed Boolean algebras as purely algebraic objects, but in fact they come equipped with a partial order that underpins this algebraic structure. Let B be a Boolean algebra. For $x, y \in B$, define $x \leq y$ if and only if $x = x \wedge y$; we say that y *lies above* x. The proofs of the following are routine.

Lemma 4 *With the above definition, we have the following:*

1. *$\leq$ is a partial order on B.*
2. *$x \leq y$ if and only if $y = x \vee y$.*
3. *$a \wedge b = \mathrm{glb}\{a, b\}$ and $a \vee b = \mathrm{lub}\{a, b\}$.*

2.2 Finite Boolean Algebras

In this section, we describe the structure of all finite Boolean algebras. It is only included to provide motivation for the sections that follow. The crucial idea is that of an atom. A non-zero element $x \in B$ of a Boolean algebra is called an *atom* if $y \leq x$ implies that either $x = y$ or $y = 0$. We denote the set of atoms of the Boolean algebra B by $\mathsf{at}(B)$. The proof of the following is immediate from the definition.

Lemma 5 *Let x and y be atoms. Then either $x = y$ or $x \wedge y = 0$.*

The proof of the following is immediate.

Lemma 6 *Let B be a finite Boolean algebra. Then every non-zero element lies above an atom.*

The following lemma will be useful.

Lemma 7 *Let B be a Boolean algebra. If $a \not\leq b$ then $a \wedge b' \neq 0$.*

Proof Suppose that $a \wedge b' = 0$. Then $a = a \wedge 1 = a \wedge \left(b \vee b'\right) = a \wedge b$ from which we get that $a \leq b$, which is a contradiction. □

Let B be a finite Boolean algebra. For each $a \in B$, denote by $\mathscr{U}_a$ the set of all atoms in B below a. By Lemma 6, it follows that $a \neq 0$ implies that $\mathscr{U}_a \neq \varnothing$. The important properties of the sets $\mathscr{U}_a$ we shall need are listed below.

Lemma 8 *Let B be a finite Boolean algebra.*

1. *$\mathscr{U}_0 = \varnothing$.*
2. *$\mathscr{U}_1 = \mathsf{at}(B)$.*
3. *$\mathscr{U}_a \cap \mathscr{U}_b = \mathscr{U}_{a \wedge b}$.*
4. *$\mathscr{U}_a \cup \mathscr{U}_b = \mathscr{U}_{a \vee b}$.*
5. *$\mathscr{U}_{a'} = \overline{\mathscr{U}_a}$.*

6. $\mathscr{U}_a \subseteq \mathscr{U}_b$ *if and only if* $a \leq b$.
7. $\mathscr{U}_a = \mathscr{U}_b$ *if and only if* $a = b$.

Proof (1) It is immediate.

(2) It is immediate.

(3) Let $x \in \mathscr{U}_a \cap \mathscr{U}_b$. Then $x \leq a$ and $x \leq b$. It follows that $x \leq a \wedge b$ and so $x \in \mathscr{U}_{a \wedge b}$. Now suppose that $x \in \mathscr{U}_{a \wedge b}$. Then $x \leq a \wedge b$. But $a \wedge b \leq a, b$. It follows that $x \in \mathscr{U}_a \cap \mathscr{U}_b$.

(4) Let $x \in \mathscr{U}_a \cup \mathscr{U}_b$. Then $x \leq a$ or $x \leq b$. In either case, $x \leq a \vee b$. It follows that $x \in \mathscr{U}_{a \vee b}$. Conversely, suppose that $x \in \mathscr{U}_{a \vee b}$. Then $x \leq a \vee b$. It follows that $x = (a \wedge x) \vee (b \wedge x)$. But x is an atom. So, either $a \wedge x = x$ or $b \wedge x = x$; that is either $x \leq a$ or $x \leq b$, whence $x \in \mathscr{U}_a \cup \mathscr{U}_b$.

(5) Suppose that $x \in \overline{\mathscr{U}_a}$. Then $x \nleq a$. It follows that $x \wedge a' \neq 0$ by Lemma 7. But x is an atom, and so $x \leq a'$. The proof of the reverse inclusion follows by part (6) of Lemma 3.

(6) Suppose that $\mathscr{U}_a \subseteq \mathscr{U}_b$. If $a \nleq b$ then $a \wedge b' \neq 0$ by Lemma 7. By Lemma 6, there is an atom $x \leq a \wedge b'$. It follows that $x \leq a$ and $x \leq b'$ and so $x \nleq b$ by part (5) above. This contradicts our assumption. It follows that $a \leq b$. The proof of the converse is immediate.

(7) This is immediate by part (6). □

Proposition 1 *Every finite Boolean algebra is isomorphic to the Boolean algebra of subsets of a finite set.*

Proof Let B be a finite Boolean algebra. As our set, we take $\mathsf{at}(B)$, the set of atoms of B. Define a function $B \to \mathsf{P}(\mathsf{at}(B))$, the set of all subsets of $\mathsf{at}(B)$, by $a \mapsto \mathscr{U}_a$. By part (7) of Lemma 8, this is an injective homomorphism of Boolean algebras. It remains only to prove that it is surjective. Let $A = \{x_1, \ldots, x_n\}$ be any non-empty set of atoms. Put $a = \bigvee_{i=1}^{n} x_i$. We shall prove that $\mathscr{U}_a = A$. Let x be any atom such that $x \leq a$. Then $x = (x \wedge x_1) \vee \cdots \vee (x \wedge x_n)$. By Lemma 5, it follows that $x = x_i$ for some i. □

We now place the above result in its proper categorical context.

Theorem 4 (Stone duality for finite Boolean algebras) *The category of finite Boolean algebras and homomorphisms between them is dually equivalent to the category of finite sets and functions between them.*

Proof Let $\theta \colon B \to C$ be a homomorphism between finite Boolean algebras. Define $\theta^\sharp \colon \mathsf{at}(C) \to \mathsf{at}(B)$ by $\theta^\sharp(f) = e$ if $f \leq \theta(e)$ and e is an atom. We shall prove that $\theta^\sharp$ really is a function. Suppose that e and e' are distinct atoms such that $\theta(e) \geq f$ and $\theta(e') \geq f$. Then $\theta(e \wedge e') \geq f$ but $e \wedge e' = 0$, since e and e are distinct atoms, which implies that $f = 0$. This is a contradiction. Therefore, $\theta^\sharp$ is a partial function. We have that $1 = \bigvee_{e \in \mathsf{at}(B)} \theta(e)$ inside B. Thus $1 = \theta(1) = \bigvee_{\theta(e) \in \mathsf{at}(B)} \theta(e)$. But $1 \geq f$. It follows that $f = \bigvee_{\theta(e) \in \mathsf{at}(B)} \theta(e) \wedge f$. But f is an atom, so that either $\theta(e) \wedge f = 0$ or $\theta(e) \wedge f = f$. In the latter case, $f \leq \theta(e)$. We have therefore proved that $\theta^\sharp$ is a function.

Let $\alpha\colon \mathsf{at}(C) \to \mathsf{at}(B)$ be any function. Define $\alpha^\flat\colon B \to C$ by

$$\alpha^\flat(e) = \bigvee_{\alpha(f)\leq e} f.$$

We prove that $\alpha^\flat$ is a homomorphism of Boolean algebras. The join of the empty set is 0 and so $\alpha^\flat(0) = 0$. Similarly, $\alpha^\flat(1) = 1$. The fact that $\alpha^\flat$ preserves joins follows from Lemma 8. The fact that $\alpha^\flat$ preserves meets is straightforward. It follows that $\alpha^\flat$ is a homomorphism of Boolean algebras.

Let $\theta\colon B \to C$ be a homomorphism of Boolean algebras. Then $\theta^\sharp\colon \mathsf{at}(C) \to \mathsf{at}(B)$ is a well-defined function. We calculate the effect of $\left(\theta^\sharp\right)^\flat$ on atoms. Let e be an atom of B. Then by definition

$$\left(\theta^\sharp\right)^\flat(e) = \bigvee_{\theta^\sharp(f)\leq e} f.$$

But e is an atom and so $\theta^\sharp(f) = e$. It follows that $f \leq \theta(e)$. We are therefore looking at the join of all atoms below $\theta(e)$, which is exactly $\theta(e)$. We have therefore proved that $\left(\theta^\sharp\right)^\flat = \theta$ on atoms. It follows that $\left(\theta^\sharp\right)^\flat = \theta$ as functions.

Let $\alpha\colon \mathsf{at}(C) \to \mathsf{at}(B)$ be any function. Let f be any atom of C. Then $\left(\alpha^\flat\right)^\sharp(f) = e$ if and only if $f \leq \alpha^\flat(e)$ and e is an atom. By definition

$$\alpha^\flat(e) = \bigvee_{\alpha(i)\leq e} i = \bigvee_{\alpha(i)=e} i.$$

It follows that $f = i$ for some i. Thus $\alpha(f) = e$. It follows that $\left(\alpha^\flat\right)^\sharp = \alpha$.

It is now routine to check that we have defined a dual equivalence of categories. □

2.3 *Arbitrary Boolean Algebras*

In the light of Proposition 1, it is tempting to conjecture that every Boolean algebra is isomorphic to a powerset Boolean algebra. However, this turns out to be false; powerset Boolean algebras always have atoms, but there are Boolean algebras that have no atoms at all (the *atomless* Boolean algebras. See [9, p. 118, Chap. 4]. The Lindenbaum–Tarski Boolean algebra constructed from classical propositional logic is another example.). To describe arbitrary Boolean algebras, we have to adopt a different approach, and this was just what Stone did [55]. You will find classical Stone duality described in the following references [2, 9, 12, 15]. The approach is symbolized below and can be viewed as a generalization of the finite case described in the previous section:

$$\begin{array}{ccc} \text{atom} & \xrightarrow{\text{replaced by}} & \text{ultrafilter} \\ \text{powerset} & \xrightarrow{\text{replaced by}} & \text{topological space} \\ \text{atom } x \leq a & \xrightarrow{\text{replaced by}} & a \in F \text{ ultrafilter.} \end{array}$$

Let B be an arbitrary Boolean algebra. There is no reason for B to have atoms, so we have to find 'atom substitutes' that do always exist. This is the role of the ultrafilters. A non-empty subset F of a Boolean algebra B is called a *filter* if it is closed under meets and upwardly closed. The filter F is said to be *proper* if $0 \notin F$. Let $a \in B$. Then $a^{\uparrow}$ is a filter called the *principal filter generated by* a. A proper filter F is said to be *prime* if $a \vee b \in F$ implies that $a \in F$ or $b \in F$. A maximal proper filter is called an *ultrafilter*.

Lemma 9 *The following are equivalent for a proper filter F in a Boolean algebra B:*

1. *F is an ultrafilter.*
2. *For each non-zero $a \in B$, either $a \in F$ or $a' \in F$.*
3. *F is a prime filter.*

Proof (1)⇒(2). Let F be an ultrafilter. Suppose that $a \notin F$. Then by Lemma 1, there exists $b \in F$ such that $a \wedge b = 0$. Now $1 = a \vee a'$. Thus $b = (b \wedge a) \vee (b \wedge a')$. So, $b = b \wedge a'$. It follows that $b \leq a'$ giving $a' \in F$, as required.

(2)⇒(3). We prove that F is a prime filter. Let $a \vee b \in F$. Suppose that $a \notin F$ and $b \notin F$. Then, by assumption, $a' \in F$ and $b' \in F$ so that $a' \wedge b' \in F$. Thus $(a \vee b)' \in F$ which is a contradiction.

(3)⇒(1). Let F be a prime filter. We prove that F is an ultrafilter. We shall use Lemma 1. Assume that $a \in B$ is such that $a \wedge b \neq 0$ for all $b \in F$. We shall prove that $a \in A$. Suppose for the sake of argument that $a \notin F$. Now, $1 = a \vee a' \in F$ and so either $a \in F$ or $a' \in F$. It follows that $a' \in F$ but $a \wedge a' = 0$ which contradicts our assumption about a. Thus $a \in F$. We have therefore proved that F is an ultrafilter. □

In the light of the above result, the terms 'prime filter' and 'ultrafilter' are interchangeable in a Boolean algebra.

Ultrafilters are connected with homomorphisms to the two-element Boolean algebra, $\mathbb{B}$. The following is the ultrafilter version of [12, Proposition 2.2].

Lemma 10 *Let B be a Boolean algebra.*

1. *Let $\theta \colon B \to \mathbb{B}$ be a homomorphism of Boolean algebras. Then $F = \theta^{-1}(1)$ is an ultrafilter.*
2. *Let F be an ultrafilter. Then the characteristic function $\chi_F \colon B \to \mathbb{B}$ is a homomorphism of Boolean algebras.*

Proof These are both straightforward to prove using Lemma 9. □

The above lemma actually establishes a bijection between ultrafilters in B and homomorphisms from B to the two-element Boolean algebra $\mathbb{B}$. We can connect atoms with special kinds of ultrafilters; this enables us to link what we are doing in this section with what we did previously.

Lemma 11 *Let B be a Boolean algebra. The principal filter $F = a^{\uparrow}$ is a prime filter if and only if a is an atom.*

Proof Let a be an atom. Then $a^{\uparrow}$ is a filter. We shall prove that it is a prime filter. Suppose that $b \vee c \in F$. Then $a \leq b \vee c$. Thus $a = (a \wedge b) \vee (a \wedge c)$. It cannot happen that both $a \wedge b = 0$ and $a \wedge c = 0$. Also $a \wedge b \leq a$ and $a \wedge c \leq a$. But a is an atom. If $a \wedge b = a$, then $a \leq b$ and $b \in F$; if $a \wedge b = 0$ then $a \wedge c = a$ implying that $a \leq c$ and so $c \in F$. This proves that $a^{\uparrow}$ is a prime filter. We now prove the converse. Suppose that F is an ultrafilter. We prove that a is an atom. Suppose not. Then there is $0 < b < a$. Then $b^{\uparrow}$ is a filter and $F \subset b^{\uparrow}$. But this contradicts the assumption that F is an ultrafilter. It follows that a must be an atom. □

The above lemma is only interesting in the light of the following result. The routine proof uses Zorn's Lemma[2] or see [15, Chap. 1, Proposition 2.16].

Lemma 12 (Boolean Prime Ideal Theorem) *A non-empty subset of a Boolean algebra is contained in an ultrafilter if and only if it has the finite intersection property.*

The first corollary below is the analogue of the result for finite Boolean algebras that every non-zero element is above an atom.

Corollary 1 *Every non-zero element of a Boolean algebra is contained in an ultrafilter.*

The second corollary says that there are enough ultrafilters to separate points; this is the analogue of the result that says in a finite Boolean algebra each element is a join of the atoms below it.

Corollary 2 *Let a and b be distinct non-zero elements of a Boolean algebra. Then there is an ultrafilter that contains one of the elements and omits the other.*

Proof Since $a \neq b$ then either $a \nleq b$ or $b \nleq a$. Suppose that $a \nleq b$. Then $a \wedge b' \neq 0$ by Lemma 7. Thus by Corollary 1, there is an ultrafilter F that contains $a \wedge b'$. It follows that $a \in F$ and $b \notin F$. □

Ultrafilters are the first step in generalizing the theory of finite Boolean algebras to arbitrary Boolean algebras. The second is to introduce topological spaces to replace powersets. A compact Hausdorff space which is 0-dimensional is called a *Boolean space*; for emphasis, these will sometimes also be referred to in this chapter as *compact Boolean spaces*.

---

[2] The fairy godmother of mathematics.

Lemma 13 *The clopen subsets of a Boolean space form a Boolean algebra.*

Proof Let X be a Boolean space and denote by $\mathsf{B}(X)$ the set of all clopen subsets of X. Observe that $\varnothing, X \in \mathsf{B}(X)$. If $A, B \in \mathsf{B}(X)$, then $A \cap B, A \cup B \in \mathsf{B}(X)$. Finally, if $A \in \mathsf{B}(X)$ then $\overline{A} \in \mathsf{B}(X)$. □

Let B be a Boolean algebra. Define $\mathsf{X}(B)$ to be the set of ultrafilters on B. If $a \in B$, denote by $\mathscr{U}_a$ the set of ultrafilters containing a. By Lemma 11, this notation is consistent with that introduced in Sect. 2.2.

Lemma 14 *Let B be a Boolean algebra.*

1. $\mathscr{U}_0 = \varnothing$.
2. $\mathscr{U}_1 = \mathsf{X}(B)$.
3. $\mathscr{U}_a \cap \mathscr{U}_b = \mathscr{U}_{a \wedge b}$.
4. $\mathscr{U}_a \cup \mathscr{U}_b = \mathscr{U}_{a \vee b}$.
5. $\mathscr{U}_{a'} = \overline{\mathscr{U}_a}$.

Proof The proofs of (1) and (2) are straightforward. The proof of (3) follows from the fact that filters are closed under meets. The proof of (4) follows from the fact that ultrafilters are prime filters. The proof of (5) follows from the fact that $a \wedge a' = 0$ and the fact that ultrafilters are proper filters, and part (2) of Lemma 9. □

The above lemma tells us that the collection of sets $\mathscr{U}_a$, where $a \in B$, forms the base for a topology on $\mathsf{X}(B)$. We shall first of all determine the salient properties of this topological space.

Lemma 15 *For each Boolean algebra B, the topological space $\mathsf{X}(B)$ is Boolean.*

Proof The base of the topology consists of sets of the form $\mathscr{U}_a$. These are open by fiat. But by part (5) of Lemma 14, they are also closed. It follows that $\mathsf{X}(B)$ is 0-dimensional. We prove that this space is Hausdorff. Let A and B be distinct ultrafilters. Then there exists $a \in A \setminus B$, and so $a \notin B$. We now use Lemma 9 to deduce that $a' \in B$. It follows that $A \in \mathscr{U}_a$ and $B \in \mathscr{U}_{a'}$. By part (3) of Lemma 14, we have that $\mathscr{U}_a \cap \mathscr{U}_{a'} = \varnothing$. Thus, the space $\mathsf{X}(B)$ is Hausdorff. Finally, we prove that the space $\mathsf{X}(B)$ is compact. Let $\mathscr{C} = \{\mathscr{U}_a : a \in I\}$ be a cover of $\mathsf{X}(B)$. Suppose that no finite subset of $\mathscr{C}$ covers $\mathsf{X}(B)$. Then for any $a_1, \ldots, a_n \in I$ we have that $\mathscr{U}_{a_1} \cup \cdots \cup \mathscr{U}_{a_n} \neq \mathsf{X}(B)$. It follows that $a_1 \vee \cdots \vee a_n \neq 1$ and so $a_1' \wedge \cdots \wedge a_n' \neq 0$. Thus the set $I' = \{a' : a \in I\}$ has the finite intersection property. By Lemma 12, there is an ultrafilter F such that $I' \subseteq F$. By assumption, $F \in \mathscr{U}_a$ for some $a \in I$ and so $a, a' \in F$, which is a contradiction. □

The topological space $\mathsf{X}(B)$ is called the *Stone space* of the Boolean algebra B. We can now assemble Lemmas 13 and 15 into the first main result.

Proposition 2 *1. Let B be a Boolean algebra. Then $B \cong \mathsf{B}(\mathsf{X}(B))$, where here $\cong$ means an isomorphism of Boolean algebras.*
2. Let X be a Boolean space. Then $X \cong \mathsf{X}(\mathsf{B}(X))$, where here $\cong$ means a homeomorphism of topological spaces.

Proof (1) Define $\alpha \colon B \to \mathsf{B}(\mathsf{X}(B))$ by $a \mapsto \mathscr{U}_a$. By Lemma 14 this is a homomorphism of Boolean algebras. It is injective by Corollary 2. We prove that it is surjective. An element of $\mathsf{B}(\mathsf{X}(B))$ is a clopen subset of $\mathsf{X}(B)$. Since it is open, it is a union of open sets of the form $\mathscr{U}_a$, but closed subsets of compact spaces are compact by part (3) of Lemma 2. It follows that it is a union of a finite number of sets of the form $\mathscr{U}_a$ and so must itself be of that form.

(2) Let $x \in X$. Define O_x to be the set of all clopen sets that contain x. It is easy to check that this is a prime filter in $\mathsf{B}(X)$ and so $O_x \in \mathsf{X}(\mathsf{B}(X))$. Define $\beta \colon X \to \mathsf{X}(\mathsf{B}(X))$ by $x \mapsto O_x$. Since both domain and codomain spaces are compact and Hausdorff, to prove that β is a homeomorphism it is enough to prove that it is bijective and continuous. Suppose that $O_x = O_y$. If $x \neq y$ then by the fact that X is Hausdorff, we can find disjoint open sets U and V such that $x \in U$ and $y \in V$. But X is 0-dimensional and so we can assume, without loss of generality, that U and V are clopen from which we deduce that $O_x \neq O_y$. It follows that β is injective. Next, let F be any ultrafilter in $\mathsf{B}(X)$. Then this is an ultrafilter consisting of clopen subsets of a compact space. Since F is a filter, it has the finite intersection property. By part (4) of Lemma 2, it follows that there is an element x in the intersection of all the elements of F. Thus $F \subseteq O_x$. But F is an ultrafilter and so $F = O_x$. We have therefore proved our function is a bijection. Finally, we prove continuity. Let U be an open subset of $\mathsf{X}(\mathsf{B}(X))$. Then U is a union of the basic open sets which are clopen. These have the form $\mathscr{U}_A$ where A is a clopen subset of X. Thus it is enough to calculate $\beta^{-1}(\mathscr{U}_A)$. But $O_x \in \mathscr{U}_A$ if and only if $x \in A$. Thus $\beta^{-1}(\mathscr{U}_A) = A$. □

We can extend the above result to maps to obtain the following:

Theorem 5 (Classical Stone duality I) *The category of Boolean algebras and their homomorphisms is dually equivalent to the category of Boolean spaces and continuous functions between them.*

Proof In Lemma 10, we proved that there is a bijective map between the ultrafilters in B and the Boolean algebra homomorphisms from B to $\mathbb{B}$, the 2-element Boolean algebra. This bijection associates with the ultrafilter F its characteristic function χ_F. Let $\theta \colon B_1 \to B_2$ be a homomorphism between Boolean algebras. Let $F \in \mathsf{X}(B_2)$ be an ultrafilter. Then $\chi_F \theta$ is the characteristic function of an ultrafilter in B_1. In this way, we can map homomorphisms $B_1 \to B_2$ to continuous functions $\mathsf{X}(B_1) \leftarrow \mathsf{X}(B_2)$ with a consequent reversal of arrows. In the other direction, let $\phi \colon X_1 \to X_2$ be a continuous function. Then ϕ^{-1} maps clopen sets to clopen sets. In this way, we can map continuous functions $X_1 \to X_2$ to homomorphisms $\mathsf{B}(X_1) \leftarrow \mathsf{B}(X_2)$. The result now follows from Proposition 2. □

Examples of Classical Stone Duality

1. Let B be a finite Boolean algebra. By Lemma 11, the ultrafilters of B are in bijective correspondence with the atoms of B. We may therefore identify the elements of $\mathsf{X}(B)$ with the set of atoms of B. Let $a \in B$. We describe the set $\mathscr{U}_a$ in terms of atoms. The ultrafilter $b^{\uparrow} \in \mathscr{U}_a$ if and only if $b \leq a$. So, the set $\mathscr{U}_a$ is in bijective correspondence with the set of atoms below a. It follows that the Boolean space $\mathsf{X}(B)$ is homeomorphic with the discrete space of atoms of B. Let $\theta \colon B \to C$ be a homomorphism of finite Boolean algebras. If $c \in C$ is an atom then $c^{\uparrow}$ is an ultrafilter in C and so $\theta^{-1}\left(c^{\uparrow}\right)$ is an ultrafilter in B. It follows that $\theta^{-1}\left(c^{\uparrow}\right) = b^{\uparrow}$, where b is an atom in B. Thus $x \geq b$ if and only if $\theta(x) \geq c$. We therefore have that $\theta(b) \geq c$. But b is the only atom of B which has this property. In this way, the classical theory of finite Boolean algebras can be derived from Stone duality; that is, Theorem 4 is a special case of Theorem 5.
2. Tarski proved that any two atomless, countably infinite Boolean algebras are isomorphic [9, Chap. 16, Theorem 10]. It makes sense, therefore, to define the *Tarski algebra*[3] to be an atomless, countably infinite Boolean algebra. We describe the Stone space of the Tarski algebra. An element x of a topological space is said to be *isolated* if $\{x\}$ is open. Suppose that a is an atom of the Boolean algebra B. By definition $\mathscr{U}_a$ is the set of all ultrafilters that contain a. But $a^{\uparrow}$ is an ultrafilter containing a by Lemma 11 and, evidently, the only one. Thus $\mathscr{U}_a$ is an open set containing one point and so the point $a^{\uparrow}$ is isolated. Suppose that $\mathscr{U}_a$ contains exactly one point F. Then F is the only ultrafilter containing a. Suppose that a were not an atom. Then we could find $0 \neq b < a$. Thus $a = b \vee \left(a \wedge b'\right)$. By Corollary 1, there is an ultrafiler F_1 containing b, and there is an ultrafilter F_2 containing $a \wedge b'$. Then $F_1 \neq F_2$ but both contain a. This is a contradiction. It follows that a is an atom. We deduce that the atoms of the Boolean algebra determine the isolated points of the associated Stone space. It follows that a Boolean algebra has the property that every element is above an atom (that is, it is *atomic*) if and only if the isolated points in its Stone space form a dense subset. We deduce that the Stone space associated with an atomless Boolean algebra has no isolated points. If B is countable then its Stone space is second-countable since B is isomorphic to the set of all clopen subsets of the Stone space of B. The Stone space of the Tarski algebra is therefore a 0-dimensional, second-countable, compact, Hausdorff space with no isolated points. Observe by [57, Theorem 9.5.10] that such a space is metrizable. It follows by Brouwer's theorem, [59, Theorem 30.3], that the Stone space of the Tarski algebra is the *Cantor space*.
3. The Cantor space, described in (2) above, often appears in disguise. Let A be any finite set with at least two elements. Denote by $X = A^{\omega}$ the set of all right-infinite strings of elements over A. We can regard this set as the product space $A^{\mathbb{N}}$ which is compact since A is finite by part (1) of Lemma 2. For each finite string $x \in A^*$ denote by xX the subset of X that consists of all elements of X that begin with the finite string x. This is an open set of X. If $a \in A$ denote $\hat{a} = A \setminus \{a\}$. Let

[3] Not an established term.

$x = x_1 \ldots x_n$ have length $n \geq 1$. Then $\overline{xX} = \hat{x}_1 X \cup x_1\hat{x}_2 X \cup \cdots x_1 \ldots x_{n-1}\hat{x}_n X$. It follows that if xX is open then $\overline{xX}$ is open. Thus the sets xX are clopen. The set A^* is countably infinite and so the number of clopen subsets is countably infinite. If $x, y \in A^*$, then there are a few possibilities. If neither x nor y is the prefix of the other, then $xX \cap yX = \varnothing$. Now, suppose that $x = yu$. Then $xX = yuX \subseteq yX$ from which it follows that $xX \cap yX = xX$. An open subset U of X has the form $U = X_1 \ldots X_n X$, where the X_i are subsets of A. This is a union of sets of the form xX where $x \in X_1 \ldots X_n$. It follows that the sets $\varnothing$ and xX, where $x \in A^*$, form a clopen base for the topology on X. If w and w' are distinct elements of X, then they will differ in the nth position and so belong to disjoint sets of the form xX. It follows that X is a second-countable Boolean space. This space cannot have any isolated points: if $\{w\}$ is an open subset, then it must be a union of sets of the form xX but this is impossible. It follows that X is the Cantor space. We refer the reader to [22, 23] and [32, Sect. 5] for more on this topological space.

4. We construct the Stone spaces of the powerset Boolean algebras $\mathsf{P}(X)$. The isolated points of the Stone space of $\mathsf{P}(X)$ form a dense subset of the Stone space which is homeomorphic to the discrete space X. Thus the Stone space of $\mathsf{P}(X)$ is a compact Hausdorff space that contains a copy of the discrete space X. In fact, the Stone–Čech compactification of the discrete space X is precisely the Stone space of $\mathsf{P}(X)$. See [52, Sect. 30, Theorem A] and [52, Sect. 75].
5. Let A be any finite alphabet. We shall use notation from the theory of regular expressions so that $L + M$ means $L \cup M$ and x can mean $\{x\}$ (but also the string x in a different context). A language L over A is said to be *definite*[4] if $L = X + YA^*$ where $X, Y \subseteq A^*$ are finite languages. Denote the set of definite languages by $\mathscr{D}$. This forms a Boolean algebra. The Stone space of the Boolean algebra $\mathscr{D}$ can be described as follows. Put $\mathscr{X} = A^* + A^\omega$. If x and y are distinct elements of $\mathscr{X}$, define $x \wedge y$ to be the largest common prefix of x and y. Define $d(x, y) = 0$ if $x = y$ else $d(x, y) = \exp(-|x \wedge y|)$. Then d is an ultrametric on $\mathscr{X}$, and $\mathscr{X}$ is a complete metric space with respect to this ultrametric. The open balls are of the form $\{x\}$ or $xA^* + xA^\omega$ where $x \in A^*$ and form a base for the topology. Thus $\mathscr{X}$ is a Boolean space. It can be proved that the Stone space of $\mathscr{D}$ is this ultrametric space $\mathscr{X}$. See [43] for more on this example.

2.4 Generalized Boolean Algebras

There is a generalization of classical Stone duality that relates what are termed generalized Boolean algebras to locally compact Boolean spaces. At the level of objects, this was described in [55, Theorem 4] and at the level of homomorphisms in [4].

---

[4] Strictly speaking, 'reverse definite'.

In elementary work [12], Boolean algebras are usually defined with a top element and globally defined complements. However, this is too restrictive for the applications we have in mind; it corresponds in topological language to only looking at compact spaces even though many mathematically interesting spaces are locally compact. In this section, I shall study what are termed 'generalized Boolean algebras' or, to adapt terminology current in ring theory, *non-unital Boolean algebras*. Similarly, in this section, a *distributive lattice* will always have a bottom but not necessarily a top. Let D be a distributive lattice equipped with a binary operation $\setminus$ such that for all $x, y \in D$ we have that $0 = y \wedge (x \setminus y)$ and $x = (x \wedge y) \vee (x \setminus y)$. I say that such a distributive lattice is a *generalized Boolean algebra*. Let B be a generalized Boolean algebra. A subset $C \subseteq B$ is said to be a *subalgebra* if it contains the bottom element of B, is closed under meets, and joins and is closed under the operation $\setminus$. Such a C is a generalized Boolean algebra in its own right.

If $b \leq a$ in a lattice then $[b, a]$ denotes the set of all elements x of the lattice such that $b \leq x \leq a$. We call the set $[b, a]$ an *interval*. If $c \in [b, a]$ then a *complement* of c is an element $d \in [b, a]$ such that $c \wedge d = b$ and $c \vee d = a$. We say that $[b, a]$ is *complemented* if every element has a complement. Let D be a distributive lattice. We say it is *relatively complemented* if for every pair $b \leq a$, the interval $[b, a]$ is complemented.

Lemma 16 *Let D be a distributive lattice. Then the following are equivalent:*

1. *D is a generalized Boolean algebra.*
2. *Each non-zero principal order-ideal of D is a unital Boolean algebra.*
3. *D is relatively complemented.*

Proof (1)⇒(2). Let $a \in B$ be non-zero. Then $a^{\downarrow}$ is a distributive lattice with bottom element 0 and top element a. Let $b \leq a$. Then $b \wedge (a \setminus b) = 0$ and $b \vee (a \setminus b) = a$. It follows that within $a^{\downarrow}$, we should define $b' = a \setminus b$. Thus each non-zero principal order-ideal is a unital Boolean algebra.

(2)⇒(1). Let $x, y \in D$ where $x \neq 0$. Then $x \wedge y \leq x$. Define $x \setminus y = (x \wedge y)'$ where $(x \wedge y)'$ is the complement of $x \wedge y$ in the Boolean algebra $x^{\downarrow}$. By definition $(x \wedge y) \vee (x \setminus y) = x$ and $y \wedge (x \setminus y) = y \wedge (x \setminus y) \wedge x = (x \wedge y) \wedge (x \setminus y) = 0$. If $x = 0$ then define $0 = (0 \setminus 0)$.

(1)⇒(3). Let $b \leq a$. Let $x \in [b, a]$. Put $y = (a \setminus x) \vee b$. Then $x \wedge y = b$ and $x \vee y = a$. We have proved that in each interval, every element has a complement.

(3)⇒(1). It is immediate. □

In the light of the above result, we shall regard generalized Boolean algebras as distributive lattices with zero in which each non-zero principal order-ideal is a Boolean algebra. For example, let B be the set of all finite subsets of $\mathbb{N}$. Then B is a generalized Boolean algebra but not a (unital) Boolean algebra.

We may define ultrafilters and prime filters in generalized Boolean algebras just as we defined them in unital Boolean algebras. Let B be a generalized Boolean algebra. Define $\mathsf{X}(B)$ to be the set of ultrafilters on B. If $a \in B$, denote by $\mathscr{U}_a$ the set of ultrafilters containing a.

Lemma 17 *Let B be a generalized Boolean algebra and let a be any non-zero element. Then there is an order-isomorphism between the filters in the Boolean algebra $a^{\downarrow}$ and the filters in B that contain a. Under this order-isomorphism, proper filters correspond to proper filters, and ultrafilters to ultrafilters.*

Proof Let F be a filter of B that contains a. Put $F_{\downarrow} = F \cap a^{\downarrow}$. Then $F_{\downarrow}$ is non-empty and it is straightforward to show that it is a filter. Observe that if $F_1 \subseteq F_2$ are filters of B that contain a then $(F_1)_{\downarrow} \subseteq (F_2)_{\downarrow}$.

Let G be a filter of $a^{\downarrow}$. The proof that $G^{\uparrow}$, taken in B, is a filter of B that contains a is straightforward. Observe that if $G_1 \subseteq G_2$ are both filters of $a^{\downarrow}$ then $G_1^{\uparrow} \subseteq G_2^{\uparrow}$.

It is now routine to check that $F = \left(F_{\downarrow}\right)^{\uparrow}$ and $G = \left(G^{\uparrow}\right)_{\downarrow}$. We have therefore established our order-isomorphism. Since B and $a^{\downarrow}$ have the same bottom element, it is routine to check that proper filters in $a^{\downarrow}$ are mapped to proper filters in B, and that ultrafilters in $a^{\downarrow}$ are mapped to ultrafilters in B. □

Part (1) below was proved as [56, Theorem 3] and part (2) below was proved as [30, Proposition 1.6] and uses Lemma 17.

Lemma 18 *1. In a distributive lattice, every ultrafilter is a prime filter.*
2. A distributive lattice is a generalized Boolean algebra if and only if every prime filter is an ultrafilter.

It follows that in a generalized Boolean algebra, prime filters and ultrafilters are the same.

Let X be a Hausdorff space. Then X is *locally compact* if each point of X is contained in the interior of a compact subset [59, Theorem 18.2]. If V is a set, then its *interior* is denoted by V°; this is the union of all the open sets contained in V.

Lemma 19 *Let X be a Hausdorff space. Then the following are equivalent.*

1. *X is locally compact and 0-dimensional.*
2. *X has a base of compact-open sets.*

Proof (1)⇒(2). Let U be a clopen set (since the space is 0-dimensional). Let $x \in U$. Since X is locally compact, there exists a compact set V such that $x \in V^{\circ}$. Now $x \in U \cap V^{\circ}$ is open and X has a basis of clopen sets. In particular, we can find a clopen set W such that $x \in W \subseteq U \cap V^{\circ} \subseteq V$. By part (3) of Lemma 2, W is a closed subset of the compact set V and so W is compact. It follows that $x \in W \subseteq U$ where W is compact-open. Thus X has a base of compact-open sets.

(2)⇒(1). By part (2) of Lemma 2, every compact subset of a Hausdorff space is closed. It follows that X has a basis of clopen subset. It is immediate that the space is locally compact. □

We define a *locally compact Boolean space* to be a 0-dimensional, locally compact Hausdorff space. Let X be a locally compact Boolean space. Denote by $\mathsf{B}(X)$ the set of all compact-open subsets of X. The proof of the following is straightforward using Lemma 2.

Lemma 20 *Let X be a locally compact Boolean space. Then under the usual operations of union and intersection, the poset $\mathsf{B}(X)$ is a generalized Boolean algebra.*

The proof of the following lemma is routine, once you recall that ultrafilters and prime filters are the same thing in generalized Boolean algebras.

Lemma 21 *Let B be a generalized Boolean algebra.*

1. $\mathscr{U}_0 = \varnothing$.
2. $\mathscr{U}_a \cap \mathscr{U}_b = \mathscr{U}_{a \wedge b}$.
3. $\mathscr{U}_a \cup \mathscr{U}_b = \mathscr{U}_{a \vee b}$.

The above lemma tells us that the collection of all sets of the form $\mathscr{U}_a$, where $a \in B$, is the base for a topology on $\mathsf{X}(B)$. We shall first of all determine the salient properties of the topological space $\mathsf{X}(B)$.

Lemma 22 *For each generalized Boolean algebra B, the topological space $\mathsf{X}(B)$ is a locally compact Boolean space.*

Proof Let A and B be distinct ultrafilters. Let $a \in A \setminus B$; such an element exists since we cannot have that A is a proper subset of B since both are ultrafilters. By Lemma 1, there exists $b \in B$ such that $a \wedge b = 0$. Observe that $\mathscr{U}_a \cap \mathscr{U}_b = \varnothing$ and $A \in \mathscr{U}_a$ and $B \in \mathscr{U}_b$. We have proved that $\mathsf{X}(B)$ is Hausdorff. It only remains to prove that each of the sets $\mathscr{U}_a$ is compact. Suppose that $\mathscr{U}_a \subseteq \bigcup_{i \in I} \mathscr{U}_{b_i}$. Observe that $\mathscr{U}_a = \bigcup_{i \in I} \mathscr{U}_{a \wedge b_i}$. So, without loss of generality, we can assume that $b_i \leq a$. Thus, we are given that $\mathscr{U}_a = \bigcup_{i \in I} \mathscr{U}_{b_i}$ where $b_i \leq a$. By assumption, $a^{\downarrow}$ is a unital Boolean algebra. The result now follows by Lemmas 17 and 15. □

Just as before, the topological space $\mathsf{X}(B)$ is called the *Stone space* of the generalized Boolean algebra B. We can now assemble Lemmas 20 and 22 into the following result.

Proposition 3 *1. Let B be a generalized Boolean algebra. Then $B \cong \mathsf{B}(\mathsf{X}(B))$, where $\cong$ means an isomorphism of generalized Boolean algebras.*
2. Let X be a locally compact Boolean space. Then $X \cong \mathsf{X}(\mathsf{B}(X))$, where $\cong$ means a homeomorphism of topological spaces.

Proof (1) Define $\alpha \colon B \to \mathsf{B}(\mathsf{X}(B))$ by $a \mapsto \mathscr{U}_a$. Let a and b be elements of B. Suppose that $\mathscr{U}_a = \mathscr{U}_b$. Then both a and b are in the order-ideal $(a \vee b)^{\downarrow}$. It follows by Lemma 17 and Proposition 2 that $a = b$. By Lemma 21, the bottom element is mapped to the bottom element and binary meets and binary joins are preserved. It remains to show that it is surjective. Let U be a compact-open set of $\mathsf{B}(\mathsf{X}(B))$. Since it is open, it is a union of basic open sets and since it is compact it is a union of a finite number of basic open sets. But this implies that U is a basic open set and so $U = \mathscr{U}_a$ for some $a \in B$.

(2) For each $x \in X$, denote by O_x the set of all compact-open sets that contain x. It is easy to check that O_x is a prime filter in $\mathsf{B}(X)$ and so we have defined a

map from X to $\mathsf{X}(\mathsf{B}(X))$. This map is injective because locally compact Boolean spaces are Hausdorff. Let F be an arbitrary ultrafilter in $\mathsf{B}(X)$. This is therefore an ultrafilter whose elements are compact-open. Let $V \in F$. We now use part (2) of Lemma 2: compact subsets of Hausdorff spaces are closed. Since F is an ultrafilter, all intersections of elements of F with V are non-empty and the set of sets so formed has the finite intersection property. It follows that there is a point x that belongs to them all. Thus x belongs to every element of F. Thus $F \subseteq O_x$ from which we get equality since we are dealing with ultrafilters. We have therefore established that we have a bijection. In particular, every ultrafilter in $\mathsf{B}(X)$ is of the form O_x. We prove that this function and its inverse are continuous. Let V be a compact-open set in X. Then the image of V under our map is the set $S = \{O_x : x \in V\}$. But V is an element of $\mathsf{B}(X)$. The set S is just $\mathscr{U}_V$. We now show that the inverse function is continuous. An element of a base for the topology on $\mathsf{B}(X)$ is of the form $\mathscr{U}_a$ where $a \in \mathsf{B}(X)$. Let $a = V$ be a compact-open subset of X. A typical element of $\mathscr{U}_V$ is O_x where $x \in V$. It follows that the inverse image of $\mathscr{U}_V$ is V. □

Let $\theta \colon B \to C$ be a homomorphism of generalized Boolean algebras. We say it is *proper* if $C = \mathrm{im}(\theta)^{\downarrow}$; in other words, each element of C is below an element of the image. A continuous map between topological spaces is said to be *proper* if the inverse images of compact sets are compact.

Theorem 6 (Commutative Stone duality II) *The category of generalized Boolean algebras and proper homomorphisms is dually equivalent to the category of locally compact Boolean spaces and proper continuous homomorphisms.*

Proof Let $\theta \colon B_1 \to B_2$ be a proper homomorphism between generalized Boolean algebras and let F be an ultrafilter in B_2. Then $\theta^{-1}(F)$ is non-empty because the homomorphism is proper and it is an ultrafilter in B_1. We therefore have a map $\theta^{-1} \colon \mathsf{X}(B_2) \to \mathsf{X}(B_1)$. Let $\phi \colon X_1 \to X_2$ be a proper continuous map between locally compact Boolean spaces and let U be a compact-open subset of X_2. Then $\phi^{-1}(U)$ is also compact-open. We therefore have a map $\phi^{-1} \colon \mathsf{B}(X_2) \to \mathsf{B}(X_1)$. It is now routine to check using Proposition 3 that we have a duality between categories. □

The above theorem generalizes Theorem 5 since homomorphisms between unital Boolean algebras are automatically proper, and in a Hausdorff space compact sets are closed and closed subsets of compact spaces are themselves compact by Lemma 2 and so continuous maps between Boolean spaces are automatically proper. We shall now generalize the above theorem in the sections that follow.

Every locally compact Hausdorff space admits a one-point compactification [52, Sect. 37] in which the resulting space is compact Hausdorff. If the original space is 0-dimensional, so, too, is its one-point compactification; see [9, Exercise 43.19] and [55, p. 387].

Lemma 23 *The one-point compactification of a locally compact Boolean space is a compact Boolean space.*

We shall return to this lemma later in Sect. 9.

3 Boolean Inverse Semigroups

Let S be an inverse semigroup. We denote its semilattice of idempotents by $\mathsf{E}(S)$. If $X \subseteq S$, define $\mathsf{E}(X) = X \cap \mathsf{E}(S)$. If $a \in S$, write $\mathbf{d}(a) = a^{-1}a$ and $\mathbf{r}(a) = aa^{-1}$. If the product ab is only defined when $\mathbf{d}(a) = \mathbf{r}(b)$, then we talk about the *restricted product*. We say that $a, b \in S$ are *compatible*, written as $a \sim b$, if $a^{-1}b$ and ab^{-1} are both idempotents. A pair of elements being compatible is a necessary condition for them to have an upper bound. A subset is said to be *compatible* if every pair of elements in that subset is compatible. The following was proved in [20, Lemmas 1.4.11, 1.4.12].

Lemma 24 *1.* $s \sim t$ *if and only if* $s \wedge t$ *exists and* $\mathbf{d}(s \wedge t) = \mathbf{d}(s) \wedge \mathbf{d}(t)$ *and* $\mathbf{r}(s \wedge t) = \mathbf{r}(s) \wedge \mathbf{r}(t)$.
2. If $a \sim b$ *then* $a \wedge b = ab^{-1}b = bb^{-1}a = aa^{-1}b = ba^{-1}a$.

We now suppose our inverse semigroup contains a zero. If e and f are idempotents then we say they are *orthogonal*, written as $e \perp f$, if $ef = 0$. If a and b are elements of an inverse semigroup with zero we say that they are *orthogonal*, written as $a \perp b$, if $\mathbf{d}(a) \perp \mathbf{d}(b)$ and $\mathbf{r}(a) \perp \mathbf{r}(b)$. If $a \perp b$ then $a \sim b$; in this case, if $a \vee b$ exists we often write $a \oplus b$ and talk about *orthogonal joins*. This terminology can be extended to any finite set.

An inverse semigroup with zero is said to be *distributive* if it has all binary compatible joins and multiplication distributes over such joins. The semilattice of idempotents of a distributive inverse semigroup is a distributive lattice.

An inverse semigroup is a *meet-semigroup* if it has all binary meets. Let S be an arbitrary inverse semigroup. A function $\phi \colon S \to \mathsf{E}(S)$ is called a *fixed-point operator* if for each $a \in S$ the element $\phi(a)$ is the largest idempotent less than or equal to a. The proofs of the following can be found in [34] or follow from the definition.

Lemma 25 *Let S be an inverse semigroup.*

1. *S is a meet-semigroup if and only if it has a fixed-point operator ϕ.*
2. *If S is a meet-semigroup, then we may define ϕ by* $\phi(a) = a \wedge \mathbf{d}(a) \; (= a \wedge \mathbf{r}(a))$.
3. *If ϕ is a fixed-point operator, then* $\phi(ae) = \phi(a)e$ *and* $\phi(ea) = e\phi(a)$ *for all* $e \in \mathsf{E}(S)$.
4. *If ϕ is a fixed-point operator, then* $a \wedge b = \phi\left(ab^{-1}\right)b$.

It is important to be able to manipulate meets and joins in a distributive inverse semigroup. The following result tells us how. Part (1) was proved as [20, Proposition 1.4.17], part (2) was proved as [20, Proposition 1.4.9], and parts (3), (4), and (5) were proved as [24, Lemma 2.5].

Lemma 26 *In a distributive inverse semigroup, the following hold.*

1. *In a distributive inverse monoid, if $a \vee b$ exists then*

$$\mathbf{d}(a \vee b) = \mathbf{d}(a) \vee \mathbf{d}(b) \text{ and } \mathbf{r}(a \vee b) = \mathbf{r}(a) \vee \mathbf{r}(b).$$

2. *If $a \wedge b$ exists then $ac \wedge bc$ exists and $(a \wedge b)c = ac \wedge bc$, and dually.*
3. *Suppose that $\bigvee_{i=1}^{m} a_i$ and $c \wedge \left(\bigvee_{i=1}^{m} a_i\right)$ exist. Then all the meets $c \wedge a_i$ exist as does the join $\bigvee_{i=1}^{m} c \wedge a_i$, and we have that*

$$c \wedge \left(\bigvee_{i=1}^{m} a_i\right) = \bigvee_{i=1}^{m} c \wedge a_i.$$

4. *Suppose that a and $b = \bigvee_{j=1}^{n} b_j$ are such that all the meets $a \wedge b_j$ exist. Then $a \wedge b$ exists and is equal to $\bigvee_j a \wedge b_j$.*
5. *Let $a = \bigvee_{i=1}^{m} a_i$ and $b = \bigvee_{j=1}^{n} b_j$ and suppose that all meets $a_i \wedge b_j$ exist. Then $\bigvee_{i,j} a_i \wedge b_j$ exists as does $a \wedge b$ and we have that $a \wedge b = \bigvee_{i,j} a_i \wedge b_j$.*

A distributive inverse semigroup is said to be *Boolean* if its semilattice of idempotents is a generalized Boolean algebra.

Examples of Boolean Inverse Semigroups

1. Let X be an infinite set. Denote by $\mathcal{I}^{\text{fin}}(X)$ the set of all partial bijections of the set X with finite domains. Then this is a Boolean inverse semigroup that is not a Boolean inverse monoid.
2. Symmetric inverse monoids $\mathcal{I}(X)$ are Boolean inverse monoids. Its Boolean algebra of idempotents is isomorphic to the power set Boolean algebra $\mathsf{P}(X)$. If X is finite with n elements, then we denote the symmetric inverse monoid on an n-element set by $\mathcal{I}_n$.
3. Groups with zero adjoined, denoted by G^0. These may look like chimeras but they are honest-to-goodness Boolean inverse monoids whose Boolean algebra of idempotents is isomorphic to the two-element Boolean algebra.
4. R_n is the set of all $n \times n$ rook matrices [51]. These are all $n \times n$ matrices over the numbers 0 and 1 such that each row and each column contains at most one non-zero entry. In fact, R_n is isomorphic to $\mathcal{I}_n$.
5. $R_n\left(G^0\right)$ is the set of all $n \times n$ rook matrices over a group with zero [36]. These are all $n \times n$ matrices over the group with zero G^0 in which each row and each column contains at most one non-zero entry.
6. Let S be a Boolean inverse semigroup. Then the set $M_\omega(S)$ of all $\omega \times \omega$ generalized rook matrices over S is a Boolean inverse semigroup. See [16].

Lemma 27 *Let S be a Boolean inverse semigroup. Let $b \leq a$. Then there is a unique element, denoted by $(a \setminus b)$, such that the following properties hold: $(a \setminus b) \leq a$, $b \perp (a \setminus b)$, and $a = b \vee (a \setminus b)$.*

Proof Observe that there is an order-isomorphism from $a^\downarrow$ to $\mathbf{d}(a)^\downarrow$ under the map $x \mapsto \mathbf{d}(x)$. If $b \leq a$ then define $(a \setminus b)$ to be the unique element below a that corresponds to $\mathbf{d}(a) \setminus \mathbf{d}(b)$ under the above order-isomorphism. □

Let S be a Boolean inverse semigroup. If $X \subseteq S$ is any non-empty subset, then $X^\vee$ denotes the set of all binary joins of compatible pairs of elements of X. Clearly,

$X \subseteq X^{\vee}$. A subset I of S is said to be an *additive ideal* if I is a *semigroup ideal* of S (that is, $SI \subseteq I$ and $IS \subseteq I$) and I is closed under binary compatible joins. If $\theta\colon S \to T$ is a morphism of Boolean inverse semigroups, then the set $K = \{s \in S\colon \theta(s) = 0\}$ is called the *kernel* of θ and is clearly an additive ideal of S. We say that S is *0-simplifying* if $S \neq \{0\}$, and the only additive ideals are $\{0\}$ and S itself. Let e and f be any idempotents. We say that a non-empty finite set $X = \{x_1, \ldots, x_n\}$ is a *pencil from e to f* if $e = \bigvee_{i=1}^{n} \mathbf{d}(x_i)$ and $\mathbf{r}(x_i) \leq f$. Suppose that $e = 0$. Then all of the $x_i = 0$. It follows that there is always a pencil from 0 to any idempotent f. On the other hand if $f = 0$ then $\mathbf{r}(x_i) = 0$, and so $x_i = 0$ and it follows that $e = 0$. It is easy to check that if I is an additive ideal and $f \in I$, where f is an idempotent, and there is a pencil from e to f, where e is an idempotent, then $e \in I$.

Lemma 28 *Let S be a Boolean inverse semigroup not equal to zero. Then S is 0-simplifying if and only if whenever e and f are non-zero idempotents there is a pencil from e to f.*

Proof Suppose that S is 0-simplifying, and let e and f be non-zero idempotents. Then $(SfS)^{\vee}$ is a non-zero additive ideal of S. By assumption $S = (SfS)^{\vee}$. It follows that $e \in (SfS)^{\vee}$. It is now routine to check that there is a pencil from e to f. We now prove the converse. Let I be a non-zero additive ideal of S and let $s \in S$ be arbitrary and non-zero. Let $f \in I$ be any non-zero idempotent. Then there is a pencil from $s^{-1}s$ to f. It follows that $s^{-1}s \in I$. But I is a semigroup ideal and so $s \in I$. We have proved that $I = S$. □

Let X be a pencil from e to f in a Boolean inverse semigroup. We can always assume, for any distinct $x, y \in X$, that $\mathbf{d}(x) \perp \mathbf{d}(y)$.

Let T be a Boolean inverse semigroup. An inverse subsemigroup S of T is said to be a *subalgebra* if S is closed under binary compatible joins taken in T and if $e, f \in \mathsf{E}(S)$ then $e \setminus f \in \mathsf{E}(S)$. Observe that S is a Boolean inverse semigroup for the induced operations; observe that $\mathsf{E}(S)$ is a subalgebra of the generalized Boolean algebra $\mathsf{E}(T)$.

We should note that Wehrung [58, Definition 3.1.17] uses the term *additive inverse subsemigroup* for what we have termed a subalgebra.

Our perspective is that Boolean inverse semigroups are non-commutative generalizations of generalized Boolean algebras.

4 Boolean Groupoids

We shall assume that the reader is familiar with the basic ideas and definitions of category theory as described in the first few chapters of Mac Lane [40], for example. Our goal is just to present the perspective on categories needed in this chapter.

A category is usually regarded as a ‘category of structures’ of some kind, such as the category of sets or the category of groups. A (small) category can, however, also be regarded as an algebraic structure, that is, as a set equipped with a partially defined binary operation satisfying certain axioms. We shall need both perspectives in this chapter, but the latter perspective will be foremost. This algebraic approach to categories was an important ingredient in Ehresmann's work [5] and was applied by Philip Higgins to prove some basic results in group theory [10].

To define the algebraic notion of a category, we begin with a set C equipped with a partial binary operation which we denote by concatenation. We write $\exists ab$ to mean that the product ab is defined. An *identity* in such a structure is an element e such that if $\exists ae$ then $ae = a$ and if $\exists ea$ then $ea = a$. A *category* is a set equipped with a partial binary operation satisfying the following axioms:

(C1) $\exists a\,(bc)$ if and only if $\exists\,(ab)\,c$ and when one is defined so is the other and they are equal.
(C2) $\exists abc$ if and only if $\exists ab$ and $\exists bc$.
(C3) For each a there is an identity e, perforce unique, such that $\exists ae$, and there exists an identity f, perforce unique, such that $\exists fa$. I shall write $\mathbf{d}\,(a) = e$ and $\mathbf{r}\,(a) = f$ and draw the picture

$$f \xleftarrow{a} e.$$

The set of all elements from e to f is called a *hom-set* and denoted by $\hom(e, f)$.

You can check that $\exists ab$ if and only if $\mathbf{d}\,(a) = \mathbf{r}\,(b)$. A category with one identity is a monoid. Thus, viewed in this light, categories are 'monoids with many identities'. The morphisms between categories are called *functors*; we shall think of functors as generalizations of monoid homomorphisms.

Let C be a category. If $A, B \subseteq C$ then we may define AB to be that subset of C which consists of all products ab where $a \in A$, $b \in B$ and ab is defined in the category. We call this *subset multiplication*.

We now define groupoids. An element a of a category is said to be *invertible* if there exists an element b such that ab and ba are identities. If such an element b exists, it is unique and is called the *inverse* of a; we denote the inverse of a when it exists by a^{-1}. A category in which every element is invertible is called a *groupoid*.

Examples of Groupoids

1. A groupoid with one identity is a group. Thus groupoids are 'groups with many identities'.
2. A set can be regarded as a groupoid in which every element is an identity.
3. Equivalence relations can be regarded as groupoids. They correspond to *principal groupoids*; that is, those groupoids in which given any identities e and f there is at most one element g of the groupoid such that $f \xleftarrow{g} e$. A special case of such groupoids are the *pair groupoids*, $X \times X$, which correspond to equivalence relations having exactly one equivalence class.

If G is a groupoid and $A \subseteq G$, then A^{-1} is the set of all inverses of elements of A.

We shall need the following notation for the maps involved in defining a groupoid (not entirely standard). Define $\mathbf{d}(g) = g^{-1}g$ and $\mathbf{r}(g) = gg^{-1}$. If $g \in G$, define $\mathbf{i}(g) = g^{-1}$. Put

$$G * G = \{(g, h) \in G \times G : \mathbf{d}(g) = \mathbf{r}(h)\},$$

and define $\mathbf{m}: G * G \to G$ by $(g, h) \mapsto gh$. If $U, V \subseteq G$, define $U * V = (U \times V) \cap (G * G)$. The set of identities of G is denoted by G_o. If e is an identity in G then G_e is the set of all elements a such that $a^{-1}a = e = aa^{-1}$. We call this the *local group at* e. Put $\operatorname{Iso}(G) = \bigcup_{e \in G_o} G_e$. This is called the *isotropy groupoid* of G.

We now show how to construct all groupoids. Let G be a groupoid. We say that elements $g, h \in G$ are *connected*, denoted by $g \equiv h$, if there is an element $x \in G$ such that $\mathbf{d}(x) = \mathbf{d}(h)$ and $\mathbf{r}(x) = \mathbf{d}(g)$. The $\equiv$-equivalence classes are called the *connected components of the groupoid*. If $\exists gh$, then necessarily $g \equiv h$. It follows that $G = \coprod_{i \in I} G_i$ where the G_i are the connected components of G. Each G_i is a connected groupoid. So, it remains to describe the structure of all connected groupoids. Let X be a non-empty set and let H be a group. The set of triples $X \times H \times X$ becomes a groupoid when we define $(x, h, x')(x', h', x'') = (x, hh', x'')$ and $(x, h, y)^{-1} = (y, h^{-1}, x)$. It is easy to check that $X \times H \times X$ is a connected groupoid. Now let G be an arbitrary groupoid. Choose, and fix, an identity e in G. Denote the local group at e by H. For each identity f in G, choose an element x_f such that $\mathbf{d}(x_f) = e$ and $\mathbf{r}(x_f) = f$. Put $X = \{x_f : f \in G_o\}$. We prove that G is isomorphic to $X \times H \times X$. Let $g \in G$. Then $x_{\mathbf{r}(g)}^{-1} g x_{\mathbf{d}(g)} \in H$. Define a map from G to $X \times H \times X$ by $g \mapsto \left(x_{\mathbf{r}(g)}, x_{\mathbf{r}(g)}^{-1} g x_{\mathbf{d}(g)}, x_{\mathbf{d}(g)}\right)$. It is easy to show that this is a bijective functor.

We shall need some special kinds of functors in our duality theory called covering functors. These we define now. Let G be any groupoid and e any identity. The *star of* e, denoted by St_e, consists of all elements $g \in G$ such that $\mathbf{d}(g) = e$. Let $\theta : G \to H$ be a functor between groupoids. Then for each identity $e \in G$, the functor θ induces a function θ_e mapping St_e to $\mathrm{St}_{\theta(e)}$. If all these functions are injective (respectively, surjective), then we say that θ is *star-injective* (respectively, *star-surjective*). A *covering functor* is a functor which is star-bijective. The following was proved as [24, Lemma 2.26].

Lemma 29 *Let* $\theta : G \to H$ *be a covering functor between groupoids. Suppose that the product* ab *is defined in* H *and that* $\theta(x) = ab$. *Then there exist* $u, v \in G$ *such that* $x = uv$ *and* $\theta(u) = a$ *and* $\theta(v) = b$.

The key definition needed to relate groupoids and inverse semigroups in our non-commutative generalization of Stone duality is the following. A subset $A \subseteq G$ is called a *local bisection* if $A^{-1}A, AA^{-1} \subseteq G_o$.

Lemma 30 *A subset $A \subseteq G$ is a local bisection if and only if $a, b \in A$ and $\mathbf{d}(a) = \mathbf{d}(b)$ implies that $a = b$, and $\mathbf{r}(a) = \mathbf{r}(b)$ implies that $a = b$.*

Proof Suppose that A is a local bisection. Let $a, b \in A$ such that $\mathbf{d}(a) = \mathbf{d}(b)$. Then the product ab^{-1} exists and, by assumption, is an identity. It follows that $a = b$. A similar argument shows that if $a, b \in A$ are such that $\mathbf{r}(a) = \mathbf{r}(b)$ then $a = b$. We now prove the converse. We prove that $A^{-1}A \subseteq G_o$. Let $a, b \in A$ and suppose that $a^{-1}b$ exists. Then $\mathbf{r}(a) = \mathbf{r}(b)$. By assumption, $a = b$ and so $a^{-1}b$ is an identity. The proof that $AA^{-1} \subseteq G_o$ is similar. □

What follows is based on [46, 53].

Just as we can study topological groups, so we can study topological groupoids. A *topological groupoid* is a groupoid G equipped with a topology, and G_o is equipped with the subspace topology, such that the maps $\mathbf{d}, \mathbf{r}, \mathbf{m}, \mathbf{i}$ are all continuous functions where $\mathbf{d}, \mathbf{r} \colon G \to G$ and $\mathbf{m} \colon G * G \to G$. Clearly, it is just enough to require that $\mathbf{m}$ and $\mathbf{i}$ are continuous.

A topological groupoid is said to be *open* if the map $\mathbf{d}$ is an open map; it is said to be *étale* if the map $\mathbf{d}$ is a local homeomorphism. It is worth observing (see [53, p. 12]) that the definition of étale is based on the function $\mathbf{d} \colon G \to G$.

In this chapter, we shall focus on étale topological groupoids. The obvious question is why should étale groupoids be regarded as a nice class of topological groupoids? The following result due to Pedro Resende provides us with one reason. For the following, see [46, Exercises I.1.8].

Lemma 31 *Let G be a topological groupoid. Then G is étale if and only if $\Omega(G)$, the set of all open subsets of G, is a monoid under subset multiplication with G_o as the identity.*

We may paraphrase the above theorem by saying that étale groupoids are those topological groupoids that have an algebraic alter ego.

We say that an étale topological groupoid is *Boolean* if its space of identities is a locally compact Boolean space.

> *Our perspective is that Boolean groupoids are 'non-commutative' generalizations of locally compact Boolean spaces.*

Thinking of topological groupoids as non-commutative spaces in this way goes back to [18, 45].

5 From Boolean Groupoids to Boolean Inverse Semigroups

The goal of this section is to show how to construct Boolean inverse semigroups from Boolean groupoids.

Lemma 32 *The set of all local bisections of a discrete groupoid forms a Boolean inverse monoid under subset multiplication.*

Proof Let A and B be local bisections. We prove that AB is a local bisection. We calculate $(AB)^{-1}AB$. This is equal to $B^{-1}A^{-1}AB$. Now $A^{-1}A$ is a set of identities. Thus $B^{-1}A^{-1}AB \subseteq B^{-1}B$. But $B^{-1}B$ is a set of identities. It follows that $(AB)^{-1}AB$ is a set of identities. By a similar argument, we deduce that $AB(AB)^{-1}$ is a set of identities. We have therefore proved that the product of two local bisections is a local bisection. The proof of associativity is straightforward. Since G_o is a local bisection, we have proved that the set of local bisections is a monoid. Observe that if A is a local bisection, then $A = AA^{-1}A$ and $A = A^{-1}AA^{-1}$. Thus the semigroup is regular. Suppose that $A^2 = A$, where A is a local bisection. Then $a = bc$ where $b, c \in A$. But $\mathbf{d}(a) = \mathbf{d}(c)$, and so $a = c$, and $\mathbf{r}(a) = \mathbf{r}(b)$, and so $a = b$. It follows that $a = a^2$. But the only idempotents in groupoids are identities and so a is an identity. We have shown that if $A^2 = A$ then $A \subseteq G_o$. It is clear that if $A \subseteq G_o$ then $A^2 = A$. We have therefore proved that the idempotent local bisections are precisely the subsets of the set of identities. The product of any two such idempotents is simply their intersection and so idempotents commute with each other. It follows that our monoid is inverse. It is easy to check that $A \leq B$ in this inverse semigroup precisely when $A \subseteq B$. Now, the idempotents are the subsets of the set of identities and the natural partial order is subset inclusion. It follows that the set of identities is a Boolean algebra, since it is isomorphic to the Boolean algebra of all subsets of G_o. Suppose that A and B are local bisections such that $A \sim B$. Then it is easy to check that $A \cup B$ is a local bisection. Clearly, subset multiplication distributes over such unions. We have therefore proved that the monoid is a Boolean inverse monoid. □

A subset $A \subseteq G$ of a groupoid is called a *bisection* if

$$A^{-1}A, AA^{-1} = G_o.$$

The following is immediate by Proposition 32 and tells us that we may also construct groups from groupoids.

Corollary 3 *The set of bisections of a discrete groupoid is just the group of units of the inverse monoid of all local bisections of the discrete groupoid.*

Definition Let G be a Boolean groupoid. Denote by $\mathsf{KB}(G)$ the set of all compact-open local bisections of G.

Proposition 4 *Let G be a Boolean groupoid. Then $\mathsf{KB}(G)$ is a Boolean inverse semigroup.*

Proof Let U and V be two compact-open local bisections. Since the groupoid G is étale, the product UV is open by Lemma 31. The product of local bisections is a local bisection by the proof of Lemma 32. It remains to show that UV is compact. Let $UV \subseteq \bigcup_{i \in I} A_i$ where the A_i are open local bisections—since the open local bisections of étale groupoids form a base for the topology. Then $U^{-1}U \cap VV^{-1} = U^{-1}UVV^{-1} \subseteq \bigcup_{i \in I} U^{-1}A_iV^{-1}$. The sets $U^{-1}A_iV^{-1}$ are open local bisections. By assumption $U^{-1}U \cap VV^{-1}$ is compact-open; here we use the fact that the identity space of a Boolean groupoid is a locally compact Boolean space. Thus $U^{-1}U \cap VV^{-1} = U^{-1}UVV^{-1} \subseteq \bigcup_{i=1}^{n} U^{-1}A_iV^{-1}$, relabeling if necessary. It follows that $UV \subseteq \bigcup_{i=1}^{n} UU^{-1}A_iV^{-1}V \subseteq \bigcup_{i=1}^{n} A_i$ and so UV is compact. It follows that $\mathsf{KB}(G)$ is a semigroup. The proof that it is a Boolean inverse semigroup now follows easily from what we have done above and Lemma 32. □

6 From Boolean Inverse Semigroups to Boolean Groupoids

The goal of this section is to show how to construct Boolean groupoids from Boolean inverse semigroups.

Let S be an inverse semigroup. A non-empty subset $A \subseteq S$ is said to be a (respectively, *proper*) *filter* (respectively, if $0 \notin A$), if the set A is downwardly directed and upwardly closed. A maximal proper filter is called an *ultrafilter*. Let A be a proper filter in a distributive inverse semigroup S. We say it is *prime* if $a \vee b \in A$ implies that $a \in A$ or $b \in A$. The proof of the following is straightforward.

Lemma 33 *Let S be a Boolean inverse semigroup. Then A is an ultrafilter if and only if A^{-1} is an ultrafilter.*

A subset A of an inverse semigroup S is said to be a *coset* if $a, b, c \in A$ implies that $ab^{-1}c \in A$. The following extends [24, Lemma 2.6].

Lemma 34 *Every filter is a coset.*

Proof Let A be a filter and let $a, b, c \in A$. Let $d \in A$ where $d \leq a, b, c$. Then $d = dd^{-1}d \leq ab^{-1}c$. It follows that $ab^{-1}c \in A$. □

We define an *idempotent filter* to be a filter that contains an idempotent. The following was proved for cosets as [19, Proposition 1.5] and so follows by Lemma 34.

Lemma 35 *A filter is idempotent if and only if it is an inverse subsemigroup.*

We now relate idempotent filters in S to filters in $\mathsf{E}(S)$.

Lemma 36 *Let S be an inverse semigroup. There is an order-isomorphism between the idempotent filters in S and the filters in $\mathsf{E}(S)$, in which proper filters correspond to proper filters.*

Proof Let A be an idempotent filter in S. We prove first that $\mathsf{E}(A)$ is a filter in $\mathsf{E}(S)$. Let $e, f \in \mathsf{E}(A)$. By assumption, $e, f \in A$. Thus there is an element $i \in A$ such that $i \leq e, f$. But the set of idempotents of an inverse semigroup is an order-ideal. It follows that i is an idempotent. But $i \leq e, f$ and so $i \leq ef$. However, A is a filter and so since $i \in A$ we have that $ef \in A$ and so $ef \in \mathsf{E}(A)$. Let $e \in \mathsf{E}(A)$ and let $e \leq f$ where f is an idempotent. Then $f \in A$ and so $f \in \mathsf{E}(A)$. We have therefore proved that $\mathsf{E}(A)$ is a filter in $\mathsf{E}(S)$. It is clear that if $A \subseteq B$ then $\mathsf{E}(A) \subseteq \mathsf{E}(B)$. Now, let F be a filter in $\mathsf{E}(S)$; it is easy to check that $F^{\uparrow}$ is an idempotent filter in S. It is clear that if $F \subseteq G$ then $F^{\uparrow} \subseteq G^{\uparrow}$. Let A be an idempotent filter in S. We prove that $A = \mathsf{E}(A)^{\uparrow}$. It is clear that $\mathsf{E}(A)^{\uparrow} \subseteq A$. Let $a \in A$. By assumption $e \in A$ for some idempotent e. But $a, e \in A$. There exists $f \leq a, e$ which has to be an idempotent since the set of idempotents of an inverse semigroup is an order-ideal. It follows, in particular, that $f \leq a$, which proves the claim. Let F be a filter in $\mathsf{E}(S)$. It is now routine to check that $F = \mathsf{E}\left(F^{\uparrow}\right)$. We have therefore proved our order-isomorphism, and it is clear that proper filters map to proper filters. □

Lemma 37 *Let A be an idempotent filter in a distributive inverse semigroup S.*

1. *A is a prime filter in S if and only if $\mathsf{E}(A)$ is a prime filter in $\mathsf{E}(S)$.*
2. *A is an ultrafilter in S if and only if $\mathsf{E}(A)$ is an ultrafilter in $\mathsf{E}(S)$.*

Proof (1) Suppose first that A is a prime filter in S. We saw in Lemma 36 that $\mathsf{E}(A)$ is a proper filter in $\mathsf{E}(S)$. Suppose that $e \vee f \in \mathsf{E}(A)$. Then $e \vee f \in A$. It follows that $e \in A$ or $f \in A$. It is now immediate that $\mathsf{E}(A)$ is a prime filter in $\mathsf{E}(S)$. We now prove the converse. Suppose that $\mathsf{E}(A)$ is a prime filter in $\mathsf{E}(S)$. We prove that A is a prime filter in S. Let $a \vee b \in A$. Then $\mathbf{d}(a \vee b) = \mathbf{d}(a) \vee \mathbf{d}(b) \in \mathsf{E}(A)$. It follows that $\mathbf{d}(a) \in \mathsf{E}(A)$ or $\mathbf{d}(b) \in \mathsf{E}(A)$. Without loss of generality, we suppose the former. So, $a \vee b \in A$ and $\mathbf{d}(a) \in A$. But A is an inverse subsemigroup and so $a = (a \vee b)\,\mathbf{d}(a) \in A$. This proves that A is a prime filter in S.

(2) Suppose that A is an ultrafilter in S. We prove that $\mathsf{E}(A)$ is an ultrafilter in $\mathsf{E}(S)$. Suppose that F is a proper filter in $\mathsf{E}(S)$ such that $\mathsf{E}(A) \subseteq F$. We claim that $F^{\uparrow}$ is a proper filter in A. Let $a, b \in F^{\uparrow}$. The $e \leq a$ and $f \leq b$ where $e, f \in F$. But then $ef \leq a, b$ and $ef \in F$. The set $F^{\uparrow}$ is clearly upwardly closed and evidently proper. But $A \subseteq F^{\uparrow}$ and A is a maximal proper filter. We deduce that $A = F^{\uparrow}$. It is now immediate that $\mathsf{E}(A) = F$. We have therefore proved that $\mathsf{E}(A)$ is an ultrafilter in $\mathsf{E}(S)$. We now prove the converse. Suppose that $\mathsf{E}(A)$ is an ultrafilter in $\mathsf{E}(S)$. We prove that A is an ultrafilter in S. Suppose that $A \subseteq B$ where B is a proper filter in S. Since A is an idempotent filter so too is B. We therefore have that $A = \mathsf{E}(A)^{\uparrow} \subseteq \mathsf{E}(B)^{\uparrow} = B$. But $\mathsf{E}(A) \subseteq \mathsf{E}(B)$ and, by assumption, $\mathsf{E}(A)$ is an ultrafilter in $\mathsf{E}(S)$. It follows that $\mathsf{E}(A) = \mathsf{E}(B)$ from which we deduce that $A = B$. □

Lemma 38 *Let A be a (respectively, proper) filter. Then $\left(A^{-1}A\right)^{\uparrow}$ is an (respectively, proper) idempotent filter. Likewise, $\left(AA^{-1}\right)^{\uparrow}$ is an idempotent filter.*

Proof We prove that $\left(A^{-1}A\right)^{\uparrow}$ is a filter. Let $x, y \in \left(A^{-1}A\right)^{\uparrow}$. Then $a^{-1}b \leq x$ and $c^{-1}d \leq y$ where $a, b, c, d \in A$. Let $z \leq a, b, c, d$ where $z \in A$. Then $z^{-1}z \leq x, y$. It is clear that $\left(A^{-1}A\right)^{\uparrow}$ is upwardly closed. It is also clear that if A is a proper filter so too is $\left(A^{-1}A\right)^{\uparrow}$. The fact that $\left(A^{-1}A\right)^{\uparrow}$ is an idempotent filter is immediate. □

Let A be a (respectively, proper) filter. Then A^{-1} is a (respectively, proper) filter. If A is a (proper) filter, define $\mathbf{d}(A) = \left(A^{-1}A\right)^{\uparrow}$ and $\mathbf{r}(A) = \left(AA^{-1}\right)^{\uparrow}$. By Lemma 38, these are both (proper) idempotent filters. In fact, we have the following which is easy to prove using Lemma 35.

Lemma 39 *Let S be a Boolean inverse semigroup. If A is an idempotent ultrafilter then $\mathbf{d}(A) = A$ and $\mathbf{r}(A) = A$.*

We can now explain why we have used the term 'coset'.

Lemma 40 *Let A be a filter in an inverse semigroup. Then $A = (a\mathbf{d}(A))^{\uparrow}$ and $A = (\mathbf{r}(A)a)^{\uparrow}$, where $a \in A$.*

Proof We prove that $A = (a\mathbf{d}(A))^{\uparrow}$. Let $x \in A$. Then, since $a \in A$, there exists $b \in A$ such that $b \leq x, a$. Observe that $ab^{-1}b \leq xx^{-1}x = x$ and $ab^{-1}b \in a\mathbf{d}(A)$. Thus $A \subseteq (a\mathbf{d}(A))^{\uparrow}$. To prove the reverse inclusion, let $x \in (a\mathbf{d}(A))^{\uparrow}$. Then $ab^{-1}c \leq x$ where $a, b, c \in A$. But we proved above that A was a coset and so $ab^{-1}c \in A$. It follows that $x \in A$. □

Lemma 41 *Let A be a filter in a distributive inverse semigroup.*

1. *A is a prime filter if and only if $\mathbf{d}(A)$ is a prime filter, and dually.*
2. *A is an ultrafilter if and only if $\mathbf{d}(A)$ is an ultrafilter, and dually.*

Proof (1) Suppose that A is a prime filter. We prove that $\mathbf{d}(A)$ is a prime filter. Let $x \vee y \in \mathbf{d}(A)$. Then $a^{-1}b \leq x \vee y$. It follows that $aa^{-1}b \leq a(x \vee y)$. But $aa^{-1}b \in A$ since A is a coset. Thus $ax \vee ay \in A$. Without loss of generality, suppose that $ax \in A$. Then $a^{-1}(ax) \in A^{-1}A$ and so $x \in \mathbf{d}(A)$. We have therefore proved that $\mathbf{d}(A)$ is a prime filter. Suppose now that $\mathbf{d}(A)$ is a prime filter. We prove that A is a prime filter. Let $x \vee y \in A$. Then $\mathbf{d}(x \vee y) \in \mathbf{d}(A)$. Without loss of generality, suppose that $\mathbf{d}(x) \in \mathbf{d}(A)$. It follows that $a^{-1}b \leq \mathbf{d}(x)$ where $a, b \in A$. Thus $(x \vee y)a^{-1}b \leq (x \vee y)\mathbf{d}(x) = x$ where $(x \vee y)a^{-1}b \in A$ since A is a coset. We have proved that $x \in A$.

(2) Suppose that A is an ultrafilter. We prove that $\mathbf{d}(A)$ is an ultrafilter. Suppose that $\mathbf{d}(A) \subseteq B$ where B is a proper filter in S. Observe that B is an idempotent filter. Let $a \in A$. Then $A = (a\mathbf{d}(A))^{\uparrow}$. It follows that $A \subseteq (aB)^{\uparrow}$. By assumption $A = (aB)^{\uparrow}$. Thus $\mathbf{d}(A) = B$, as required. Suppose now that $\mathbf{d}(A)$ is an ultrafilter. We prove that A is an ultrafilter. Suppose that $A \subseteq B$ where B is a proper filter. Then $\mathbf{d}(A) \subseteq \mathbf{d}(B)$. By assumption $\mathbf{d}(A) = \mathbf{d}(B)$ from which it follows that $A = B$. □

The following result is important. In particular, prime filters are easier to work with than ultrafilters.

Lemma 42 *In a Boolean inverse semigroup, prime filters are the same as ultrafilters.*

Proof The proper filter A is prime if and only if $\mathbf{d}(A)$ is prime by Lemma 41. By Lemma 37, the proper filter $\mathbf{d}(A)$ is prime if and only if $\mathsf{E}(\mathbf{d}(A))$ is a prime filter in the generalized Boolean algebra $\mathsf{E}(S)$. But in a generalized Boolean algebra, by Lemma 18, prime filters are the same as ultrafilters. We can now work our way backwards to establish the claim. □

Definition Let A and B be ultrafilters. Define $A \cdot B$ precisely when $\mathbf{d}(A) = \mathbf{r}(B)$, in which case, $A \cdot B = (AB)^{\uparrow}$. Observe that $\mathbf{d}\left(A^{-1}\right) = \mathbf{r}(A)$. It follows that $A^{-1} \cdot A = \mathbf{d}(A)$ and $A \cdot A^{-1} = \mathbf{r}(A)$.

Lemma 43 *Let S be a Boolean inverse semigroup.*

1. *If A is an ultrafilter, then both $\mathbf{d}(A)$ and $\mathbf{r}(A)$ are ultrafilters.*
2. *If A and B are ultrafilters of S such that $A \cdot B$ is defined, then $A \cdot B$ is an ultrafilter of S such that $\mathbf{d}(A \cdot B) = \mathbf{d}(B)$ and $\mathbf{r}(A \cdot B) = \mathbf{r}(A)$.*
3. *If A is an ultrafilter in S, then $A \cdot \mathbf{d}(A)$ is defined and $A \cdot \mathbf{d}(A) = A$. Similarly, $\mathbf{r}(A) \cdot A = A$.*
4. *$(A \cdot B) \cdot C = A \cdot (B \cdot C)$ when the product is defined.*

Proof (1) This follows by Lemma 41.

(2) We prove first that if A and B are proper filters, then $A \cdot B$ is a proper filter. Let $x, y \in A \cdot B$. Then $ab \leq x$ and $cd \leq y$ where $a \in A$, $b \in B$, $c \in A$, and $d \in B$. Let $u \leq a, c$ and $u \in A$, and let $v \leq b, d$ and $v \in B$. Then $uv \leq ab \leq x$ and $uv \leq cd \leq y$. Observe that $uv \in AB$ and so we have proved that $(AB)^{\uparrow}$ is downwardly directed. It is clearly upwardly directed. Suppose that $0 \in (AB)^{\uparrow}$. Then $ab = 0$ for some $a \in A$ and $b \in B$. But $\left(A^{-1}A\right)^{\uparrow} = \left(BB^{-1}\right)^{\uparrow}$. It follows that $a^{-1}a \in \left(BB^{-1}\right)^{\uparrow}$. Thus $bb^{-1} \leq a^{-1}a$ where $b \in B$. It follows that $0 = ab = a\left(aa^{-1}\right)b \geq bb^{-1}b = b$. It follows that $b = 0$ which contradicts the assumption that B is a proper filter. It is routine to prove that $\mathbf{d}(A \cdot B) = \mathbf{d}(B)$. The result now follows by Lemma 41.

(3) This is straightforward on the basis of Lemma 34.

(4) This follows by (2) above and Lemma 40. □

Definition Let S be a Boolean inverse semigroup. Denote by $\mathsf{G}(S)$ the set of prime filters on S.

Using the identification between prime filters and ultrafilters proved in Lemma 42 together with Lemmas 43, 33, and 39, we have proved the following.

Lemma 44 *Let S be a Boolean inverse semigroup. Then $\mathsf{G}(S)$ is a groupoid with respect to the partially defined operation $\cdot$, where the identities of this groupoid are precisely the idempotent ultrafilters.*

Let S be a Boolean inverse semigroup. Let $a \in S$. Denote by $\mathscr{U}_a$ the set of all prime filters in S that contain a. Observe that if $a = 0$ then $\mathscr{U}_0 = \varnothing$.

Lemma 45 *Let S be a Boolean inverse semigroup.*

1. *Each non-zero element of S is contained in a prime filter. Thus $\mathscr{U}_a \neq \varnothing$ if and only if $a \neq 0$.*
2. *$(\mathscr{U}_a)^{-1} = \mathscr{U}_{a^{-1}}$.*
3. *Let $A \in \mathscr{U}_a \cap \mathscr{U}_b$. Then there exists $c \leq a, b$ such that $A \in \mathscr{U}_c \subseteq \mathscr{U}_a \cap \mathscr{U}_b$.*
4. *If $a \sim b$ then $\mathscr{U}_{a \vee b} = \mathscr{U}_a \cup \mathscr{U}_b$.*

Proof (1) Let $a \neq 0$. Then $\mathbf{d}(a) \neq 0$. Let F be any prime filter in $\mathsf{E}(S)$ that contains $\mathbf{d}(a)$. Then $F^{\uparrow}$ is an idempotent prime filter in S. Thus $A = \left(aF^{\uparrow}\right)^{\uparrow}$ is a prime filter in S that contains a.

(2) It is straightforward.

(3) Let $A \in \mathscr{U}_a \cap \mathscr{U}_b$. It follows that $a, b \in A$. But A is a filter, and so there is $c \in A$ such that $c \leq a, b$. It follows that $A \in \mathscr{U}_c$ and that $\mathscr{U}_c \subseteq \mathscr{U}_a \cap \mathscr{U}_b$.

(4) We suppose that $a \sim b$ and so $a \vee b$ exists. The inclusion $\mathscr{U}_a \cup \mathscr{U}_b \subseteq \mathscr{U}_{a \vee b}$ is immediate. The reverse inclusion follows from the fact that ultrafilters are the same as prime filters in a Boolean inverse semigroup by Lemma 42. □

By Lemma 45, the collection of all sets $\mathscr{U}_a$, where $a \in S$, forms a base for a topology on $\mathsf{G}(S)$. Denote by $\mathsf{F}(S)$ the set of all idempotent prime filters. This can be topologized by giving it the subspace topology. Thus the sets of the form $\mathscr{U}_a \cap \mathsf{F}(S)$ form a base for the topology on $\mathsf{F}(S)$. Observe that

$$\mathscr{U}_a \cap \mathsf{F}(S) = \bigcup_{f \leq a, f^2 = f} \mathscr{U}_f.$$

It follows by the above observation and Lemma 35 that the collection of sets $\mathscr{U}_e$, where e is an idempotent, forms a base for the subspace topology on $\mathsf{F}(S)$. By Lemmas 36 and 42, we see that there is a bijection between the set $\mathsf{F}(S)$ and the set of prime filters of $\mathsf{E}(S)$. If $e \in \mathsf{E}(S)$, then we denote the set of ultrafilters of $\mathsf{E}(S)$ that contain e by $\mathscr{V}_e$. The bijection above restricts to a bijection between $\mathscr{U}_e$ and $\mathscr{V}_e$ for each idempotent e. We have therefore proved the following.

Lemma 46 *Let S be a Boolean inverse semigroup. Then the topological space of idempotent prime filters is homeomorphic to the Stone space of $\mathsf{E}(S)$.*

Lemma 47 *Let A and B be filters such that $A \cap B \neq \varnothing$ and $\mathbf{d}(A) = \mathbf{d}(B)$ (respectively, $\mathbf{r}(A) = \mathbf{r}(B)$). Then $A = B$.*

Proof Let $a \in A \cap B$. Put $\mathbf{d}(A) = C$. Then $A = (aC)^{\uparrow} = B$. The proof of the other case is similar. □

Proposition 5 *Let S be a Boolean inverse semigroup. Then $\mathsf{G}(S)$ is a Boolean groupoid.*

Proof First we show that $\mathsf{G}(S)$ is a topological groupoid. By part (2) of Lemma 45, the inversion map is continuous. We observe that

$$\mathbf{m}^{-1}(\mathscr{U}_s) = \left(\bigcup_{0 \neq ab \leq s} \mathscr{U}_a \times \mathscr{U}_b \right) \cap (\mathsf{G}(S) * \mathsf{G}(S))$$

for all $s \in S$. The proof is straightforward and the same as Step 3 of the proof of [24, Proposition 2.22] and shows that $\mathbf{m}$ is a continuous function.

We show that $\mathsf{G}(S)$ is étale. It is enough to show that the map from $\mathscr{U}_a$ to $\mathscr{U}_{\mathbf{d}(a)}$ given by $A \mapsto \mathbf{d}(A)$ is a homeomorphism. The proof of this is the same as the proof of Step 4 of the proof of [24, Proposition 2.22].

The fact that the identity space of $\mathsf{G}(S)$ is homeomorphic to the Stone space of $\mathsf{E}(S)$ follows by Lemma 46. This tells us that our étale topological groupoid is a Boolean groupoid. □

Definition If S is a Boolean inverse semigroup, then we refer to $\mathsf{G}(S)$ as the *Stone groupoid* of S.

7 Non-commutative Stone Duality

In this section, we shall generalize Theorem 6 by replacing generalized Boolean algebras by Boolean inverse semigroups, and locally compact Boolean spaces by Boolean groupoids.

7.1 Properties of Prime Filters

Our first goal now is to prove that we have *enough* ultrafilters in a Boolean inverse semigroup. We adapt to our setting the proofs to be found in [12, Chap. I, Sect. 2]. Any proofs that are omitted can be found in [30]. Let S be a distributive inverse semigroup. An order-ideal of S closed under binary joins is called an *additive order-ideal*. If A is an order-ideal, then $A^{\vee}$ denotes the set of all binary joins of compatible pairs of elements of A. If A is an order-ideal then $A^{\vee}$ is an additive order-ideal containing A. We say that an additive order-ideal A is *prime* if $a^{\downarrow} \cap b^{\downarrow} \subseteq A$ implies that $a \in A$ or $b \in A$. The proof of the following is straightforward or can be found as [30, Lemma 3.10].

Lemma 48 *Let S be a distributive inverse semigroup. Then A is a prime filter if and only if $S \setminus A$ is a prime additive order-ideal.*

The following can be proved using Zorn's lemma.

Lemma 49 *Let S be a distributive inverse semigroup. Let I be an additive order-ideal of S, and let F be a filter disjoint from I. Then there is an additive order-ideal J maximal with respect to two properties: $I \subseteq J$ and $J \cap F = \varnothing$.*

The proof of the following is straightforward.

Lemma 50 *Let S be a distributive inverse semigroup. Let I be an additive order-ideal and let a be an arbitrary element of S. Then $I \cup a^{\downarrow}$ is an order-ideal and*

$$\left(I \cup a^{\downarrow}\right)^{\vee} = \{x \vee b \colon x \in I, b \leq a, x \sim b\}.$$

Lemma 51 *Let S be a distributive inverse semigroup. Let F be a filter in S and let J be an additive order-ideal maximal among all additive order-ideals disjoint from F. Then J is a prime additive order-ideal.*

The following lemma is crucial to our program.

Lemma 52 *Let S be a Boolean inverse semigroup. Let $a, b \in S$ such that $b \nleq a$. Then there is an ultrafilter that contains b but omits a.*

Proof By assumption, $b^{\uparrow} \cap a^{\downarrow} = \varnothing$. Now $b^{\uparrow}$ is a filter and $a^{\downarrow}$ is an additive order-ideal. By Lemma 49, there is an additive order-ideal J maximal with respect to two properties: $a^{\downarrow} \subseteq J$ and $J \cap b^{\uparrow} = \varnothing$. By Lemma 51, we have that J is prime. By Lemma 48, the complement of J in S is a prime filter containing b but omitting a. □

Lemma 53 *Let S be a Boolean inverse semigroup.*

1. *$\mathscr{U}_a \subseteq \mathscr{U}_b$ if and only if $a \leq b$.*
2. *$\mathscr{U}_a = \mathscr{U}_b$ if and only if $a = b$.*
3. *$\mathscr{U}_a \cdot \mathscr{U}_b = \mathscr{U}_{ab}$.*
4. *$\mathscr{U}_a$ contains only idempotent ultrafiters if and only if a is an idempotent.*
5. *$\mathscr{U}_a$ is compact.*
6. *$\mathscr{U}_a$ is an idempotent if and only if a is an idempotent.*
7. *$\mathscr{U}_a \sim \mathscr{U}_b$ if and only if $a \sim b$.*

Proof (1) Suppose that $\mathscr{U}_a \subseteq \mathscr{U}_b$. If $a \nleq b$ then by Lemma 52 there exists an ultrafilter that contains a and omits b, which contradicts our assumption. Thus $a \leq b$ as required.

(2) Immediate by (1) above.

(3) Let $A \in \mathscr{U}_a$ and $B \in \mathscr{U}_b$ such that $A \cdot B$ is defined. Then $A \cdot B \in \mathscr{U}_{ab}$. The proof of the reverse inclusion is the same as the proof of part (4) of [24].

(4) Only one direction needs proving. Observe that $\mathscr{U}_a \subseteq \mathscr{U}_{a^2}$ since any ultrafilter that contains a is an inverse subsemigroup and so must also contain a^2. We therefore have that $a \leq a^2$ by (1) above from which it follows that $a = a^2$.

(5) By Lemma 40, there is a bijection between the set $\mathscr{U}_a$ and the set $\mathscr{U}_{\mathbf{d}(a)}$ given by $A \mapsto \mathbf{d}(A)$. A base for the open sets of $\mathscr{U}_a$ is the collection $\mathscr{U}_c$ where $c \leq b$. By the above bijection, the set $\mathscr{U}_c$ is mapped to the set $\mathscr{U}_{\mathbf{d}(c)}$. It follows that there is a homeomorphism between $\mathscr{U}_a$ and the set $\mathscr{U}_{\mathbf{d}(a)}$. But by Lemma 46, the space $\mathscr{U}_{\mathbf{d}(a)}$ is homeomorphic with the space $\mathscr{V}_{\mathbf{d}(a)}$ of the prime filters in $\mathsf{E}(S)$ which contain $\mathbf{d}(a)$. But the sets $\mathscr{V}_{\mathbf{d}(a)}$ are compact by the proof of Lemma 22. It follows that $\mathscr{U}_a$ is compact.

(6) This follows by part (3) above and part (2).

(7) Only one direction needs proving. Suppose that $\mathscr{U}_a \sim \mathscr{U}_b$. Then by part (2) of Lemma 45 and part (3) above, both $\mathscr{U}_{a^{-1}b}$ and $\mathscr{U}_{ab^{-1}}$ are idempotents. It follows by part (6) above that $a \sim b$. □

7.2 Properties of Compact-Open Local Bisections

Let G be a Boolean groupoid and let $g \in G$. Define $\mathscr{F}_g$ to be the set of compact-open local bisections of G that contain g.

Lemma 54 *Let G be a Boolean groupoid.*

1. *$\mathscr{F}_g$ is a prime filter in $\mathsf{KB}(G)$.*
2. *Every prime filter in $\mathsf{KB}(G)$ is of the form $\mathscr{F}_g$ for some $g \in G$.*
3. *$\mathscr{F}_g \cdot \mathscr{F}_{g^{-1}} = \mathscr{F}_{\mathbf{r}(g)}$ and $\mathscr{F}_{g^{-1}} \cdot \mathscr{F}_g = \mathscr{F}_{\mathbf{d}(g)}$.*
4. *Suppose that gh is defined in G. Then $\mathbf{d}(\mathscr{F}_g) = \mathbf{r}(\mathscr{F}_h)$ and $\mathscr{F}_g \cdot \mathscr{F}_h = \mathscr{F}_{gh}$.*
5. *If $\mathscr{F}_g = \mathscr{F}_h$. Then $g = h$.*

Proof (1) Let $U, V \in \mathscr{F}_g$. Then $U \cap V$ is an open set containing g. We now use the fact that the compact-open local bisections of G form a base. There is therefore a compact-open local bisection W containing g such that $W \subseteq U \cap V$. It follows that $\mathscr{F}_g$ is downwardly directed. It is clearly closed upwards and doesn't contain the empty set. It is clearly a prime filter.

(2) The proof is the same as the proof of [24, part (5) of Lemma 2.19].

(3) We shall prove $\mathscr{F}_g \cdot \mathscr{F}_{g^{-1}} = \mathscr{F}_{\mathbf{r}(g)}$ since the proof of the other case is similar. We use the fact that in a Boolean groupoid, the product of compact-open local bisections is a compact-open local bisection by Proposition 4. Thus $\mathscr{F}_g \cdot \mathscr{F}_{g^{-1}} \subseteq \mathscr{F}_{\mathbf{r}(g)}$. But both left-hand side and right-hand side are prime filters in a Boolean inverse semigroup. It follows that both are ultrafilters and so must be equal.

(4) The proof is similar to the proof of (3).

(5) Let $g, h \in U \in \mathscr{F}_g = \mathscr{F}_h$. We have that $\mathbf{d}(\mathscr{F}_g) = \mathbf{d}(\mathscr{F}_h)$. By definition, $\mathbf{d}(\mathscr{F}_g) = \mathscr{F}_g^{-1} \cdot \mathscr{F}_g$ and this is equal to $\mathscr{F}_{\mathbf{d}(g)}$ by (3) above. Thus $\mathscr{F}_{\mathbf{d}(g)} = \mathscr{F}_{\mathbf{d}(h)}$. It follows that the compact-open subsets of G_o that contain $\mathbf{d}(g)$ are the same as the compact-open subsets of G_o that contain $\mathbf{d}(h)$. But the groupoid G is Boolean and so G_o is a locally compact Boolean space. It follows that $\mathbf{d}(g) = \mathbf{d}(h)$. But g and h certainly both belong to the same compact-open local bisection and so $g = h$. □

7.3 Proof of the First Part of the Main Theorem

Given a Boolean inverse semigroup S, then by Proposition 5 we have shown how to construct a Boolean groupoid $\mathsf{G}(S)$, and given a Boolean groupoid G, then by Proposition 4 we have shown how to construct a Boolean inverse semigroup $\mathsf{KB}(G)$. The following result tells us what happens when we iterate these two constructions.

Proposition 6 *1. Let S be a Boolean inverse semigroup. Define a function $\alpha : S \to \mathsf{KB}(\mathsf{G}(S))$ by $\alpha(a) = \mathscr{U}_a$. Then this is an isomorphism of semigroups.*

2. *Let G be a Boolean groupoid. Define a function $\beta : G \to \mathsf{G}(\mathsf{KB}(G))$ by $\beta(g) = \mathscr{F}_g$. Then β is an isomorphism of groupoids and a homeomorphism.*

Proof (1) By Lemma 47, the set $\mathscr{U}_a$ is a local bisection. It is open by definition of the topology. It is compact by part (5) of Lemma 53. Thus $\mathscr{U}_a$ is a compact-open local bisection. It follows that the function is well-defined. It is a semigroup homomorphism by part (3) of Lemma 53, since $\mathscr{U}_a \mathscr{U}_b = \mathscr{U}_{ab}$. It is injective by part (2) of Lemma 53. It remains to show that it is surjective. Let U be any compact-open local bisection of $\mathsf{G}(S)$. Since it is open it is a union of sets of the form $\mathscr{U}_a$ and since it is compact it is a union of a finite number of sets of this form. It follows that $U = \bigcup_{i=1}^{n} \mathscr{U}_{a_i}$. But $\mathscr{U}_{a_i} \subseteq U$ which is the natural partial order in the inverse semigroup $\mathsf{KB}(\mathsf{G}(S))$. It follows that the set of elements of the form $\mathscr{U}_{a_i}$ is compatible. By part (7) of Lemma 53, it follows that set $\{a_1, \ldots, a_m\}$ is compatible. Put $a = \bigvee_{i=1}^{n} a_i$. Then $\alpha(a) = U$. We have therefore proved that α is an isomorphism of semigroups.

(2) By Lemma 54, β is a bijective functor. It remains to show that it is a homeomorphism. Since G is a Boolean groupoid, a base for the topology on G is provided by the compact-open local bisections. Let U be a compact-open local bisection of G. Then $U \in \mathsf{KB}(G)$. We may therefore form the set $\mathscr{U}_U$ which is a typical element of the base for the topology on $\mathsf{G}(\mathsf{KB}(G))$. It is now easy to check (or see the proof of [24, Part (1) of Proposition 2.23]) that the bijection β restricts to a bijection between U and $\mathscr{U}_U$. □

7.4 *Proof of the Main Theorem*

Our goal now is to take account of appropriate morphisms in our constructions. We refer the reader to [17] for information about more general kinds of morphisms.

The statement and proof of [30, part (3), Lemma 3.11] *is incorrect. We now give the correct statement and proof.*

Lemma 55 *Let $\theta \colon S \to T$ be a morphism of distributive inverse semigroups. Then for each prime filter P we have that $\theta^{-1}(P)$ is non-empty if and only if each $t \in T$ can be written as $t = \bigvee_{i=1}^{n} t_i$ where each $t_i \leq \theta(s_i)$ for some $s_i \in S$.*

Proof We prove the easy direction first. Suppose that for each $t \in T$ we can write $t = \bigvee_{i=1}^{n} t_i$ where each $t_i \leq \theta(s_i)$ for some $s_i \in S$. Let P be any prime filter. By assumption it is non-empty. Let $t \in P$. By assumption, we can write $t = \bigvee_{i=1}^{n} t_i$ where each $t_i \leq \theta(s_i)$ for some $s_i \in S$. But P is a prime filter. Thus $t_i \in P$ for some i. It follows that $\theta(s_i) \in P$ for some $s_i \in S$, whence $\theta^{-1}(P)$ is non-empty. We now prove the converse. Suppose that $t \in S$ which cannot be written in the stated form. Then $t \notin \left(\mathrm{im}(\theta)^{\downarrow}\right)^{\vee}$. Put $I = \left(\mathrm{im}(\theta)^{\downarrow}\right)^{\vee}$. Then $t^{\uparrow} \cap I = \emptyset$. We now use Section 7.1 to deduce that there is a prime filter P that contains t and is disjoint from J. But this implies that $\theta^{-1}(P)$ is empty which is a contradiction. It follows that no such element t exists. □

A morphism $\theta \colon S \to T$ of Boolean inverse semigroups is said to be *weakly-meet-preserving* if given $t \leq \theta(a), \theta(b)$ there exists $c \leq a, b$ such that $t \leq \theta(c)$. The following is [58, Proposition 3-4.6].

Lemma 56 *Let S be a Boolean inverse semigroup. Let I be an additive ideal of S.*

1. *Define $(a, b) \in \varepsilon_I$ if and only if there exists $c \leq a, b$ such that $a \setminus c, b \setminus c \in I$. Then ε_I is an additive congruence with kernel I.*
2. *If σ is any additive congruence with kernel I then $\varepsilon_I \subseteq \sigma$.*

An additive congruence is *ideal-induced* if it equals ε_I for some additive ideal I. The following result is due to Ganna Kudryavtseva (private communication).

Proposition 7 *A morphism of Boolean inverse semigroups is weakly-meet-preserving if and only if its associated congruence is ideal-induced.*

Proof Let I be an additive ideal of S and let ε_I be its associated additive congruence on S. Denote by $\nu\colon S \to S/\varepsilon_I$ its associated natural morphism. We prove that ν is weakly-meet-preserving. Denote the ε_I-class containing s by $[s]$. Let $[t] \leq [a], [b]$. Then $[t] = [at^{-1}t]$ and $[t] = [bt^{-1}t]$. By definition, there exist $u, v \in S$ such that $u \leq t, at^{-1}t$ and $v \leq t, bt^{-1}t$ such that $t \setminus u, at^{-1}t \setminus u, t \setminus v, bt^{-1}t \setminus v \in I$. Now $[t] = [u] = [at^{-1}t]$ and $[t] = [v] = [bt^{-1}t]$. Since $u, v \leq t$, it follows that $u \sim v$ and so $u \wedge v$ exists. Clearly, $u \wedge v \leq a, b$. In addition $[t] = [u \wedge v]$. We have proved that ν is weakly-meet-preserving.

Conversely, let $\theta\colon S \to T$ be weakly-meet-preserving. We prove that it is determined by its kernel I. By part (2) of Lemma 56, it is enough to prove that if $\theta(a) = \theta(b)$ then we can find $c \leq a, b$ such that $a \setminus c, b \setminus c \in I$. Put $t = \theta(a) = \theta(b)$. Then there exists $c \leq a, b$ such that $t \leq \theta(c)$. It is easy to check that $\theta(a \setminus c) = 0 = \theta(b \setminus c)$. We have therefore proved that $a \setminus c, b \setminus c \in I$ and so $(a, b) \in \varepsilon_I$. $\square$

We now combine the above two properties. A morphism $\theta\colon S \to T$ of Boolean inverse semigroups is said to be *callitic*[5] if it satisfies two conditions:

1. We require that θ be *proper*. This means that for each $t \in T$, we can write $t = \bigvee_{i=1}^{n} t_i$ where each $t_i \leq \theta(s_i)$ for some $s_i \in S$.
2. We require that θ be *weakly-meet-preserving*.

A continuous function between topological spaces is said to be *coherent* if the inverse images of compact-open subsets are compact-open. You can easily check that the collection of Boolean inverse semigroups and callitic morphisms forms a category, as does the collection of Boolean groupoids and coherent, continuous covering functors. We can now state and prove the main theorem of this chapter.

Theorem 7 (Non-commutative Stone duality) *The category of Boolean inverse semigroups and callitic morphisms is dually equivalent to the category of Boolean groupoids and coherent, continuous covering functors.*

[5] I made this word up. It comes from the Greek word 'kallos' meaning beauty. I simply wanted to indicate that these maps were sufficiently 'nice'.

Proof Let $\theta\colon S \to T$ be a callitic morphism between Boolean inverse semigroups. Let B be a prime filter in T. We prove that $\theta^{-1}(B)$ is a prime filter in S. By Lemma 55 this set is non-empty. Let $x, y \in \theta^{-1}(B)$. Then $\theta(x), \theta(y) \in B$. But B is a filter, and so there is an element $b \in B$ such that $b \leq \theta(x), \theta(y)$. Since θ is weakly-meet-preserving, there is $s \in S$ such that $s \leq x, y$ and $b \leq \theta(s)$. But $b \in B$ and so $\theta(s) \in B$ and so $s \in \theta^{-1}(B)$. Let $x \in \theta^{-1}(B)$ and $x \leq y$. Then $\theta(x) \leq \theta(y)$ and $\theta(x) \in B$. It follows that $\theta(y) \in B$ and so $y \in \theta^{-1}(B)$. Let $x \vee y \in \theta^{-1}(B)$. Then $\theta(x \vee y) \in B$. Now we use the fact that θ is also a morphism to get that $\theta(x) \vee \theta(y) \in B$. But B is a prime filter. Without loss of generality, suppose that $\theta(x) \in B$ and so $x \in \theta^{-1}(B)$. Put $\theta^{\star} = \theta^{-1}$. We have therefore defined a function $\theta^{\star}\colon \mathsf{G}(T) \to \mathsf{G}(S)$. It remains to show that $\theta^{\star}$ is a coherent, continuous covering functor.

The bulk of the proof is taken up by showing that θ^{-1} is a functor. Let F be an idempotent prime filter in T. Thus by Lemma 35, this is an inverse subsemigroup of T. Then $\theta^{-1}(F)$ is a prime filter in S and the inverse image of an inverse subsemigroup is an inverse subsemigroup. It follows that $\theta^{-1}(F)$ is an idempotent prime filter. We have therefore shown that θ^{-1} maps identities to identities.

We next prove that if F and G are prime filters such that $F^{-1} \cdot F = G \cdot G^{-1}$ then

$$\left(\theta^{-1}(F)\theta^{-1}(G)\right)^{\uparrow} = \theta^{-1}\left((FG)^{\uparrow}\right).$$

We prove first that

$$\theta^{-1}(F)\theta^{-1}(G) \subseteq \theta^{-1}(FG).$$

Let $s \in \theta^{-1}(F)\theta^{-1}(G)$. Then $s = ab$ where $a \in \theta^{-1}(F)$ and $b \in \theta^{-1}(G)$. Thus $\theta(s) = \theta(a)\theta(b) \in FG$. It follows that $s \in \theta^{-1}(FG)$. Observe that $\theta^{-1}(X)^{\uparrow} \subseteq \theta^{-1}\left(X^{\uparrow}\right)$. It follows that

$$\left(\theta^{-1}(F)\theta^{-1}(G)\right)^{\uparrow} \subseteq \theta^{-1}\left((FG)^{\uparrow}\right).$$

We now prove the reverse inclusion. Let $s \in \theta^{-1}\left((FG)^{\uparrow}\right)$. Then $\theta(s) \in F \cdot G$ and so $fg \leq \theta(s)$ for some $f \in F$ and $g \in G$. The map θ is assumed proper, and so we may quickly deduce that there exists $v \in S$ such that $\theta(v) \in G$. Consider the product $\theta(s)\theta(v)^{-1}$. Since $\theta(s) \in F \cdot G$ and $\theta(v)^{-1} \in G^{-1}$ we have that $\theta(s)\theta(v)^{-1} \in F \cdot G \cdot G^{-1} = F \cdot F^{-1} \cdot F = F$. Thus $\theta\left(sv^{-1}\right) \in F$, and we were given $\theta(v) \in G$, and clearly $\left(sv^{-1}\right)v \leq s$. Put $a = sv^{-1}$ and $b = v$. Then $ab \leq s$ where $\theta(a) \in F$ and $\theta(b) \in G$. It follows that $s \in \left(\theta^{-1}(F)\theta^{-1}(G)\right)^{\uparrow}$.

We may now show that θ^{-1} is a functor. Let F be a prime filter. Observe that $\theta^{-1}(F)^{-1} = \theta^{-1}\left(F^{-1}\right)$. We have that

$$\left(\theta^{-1}\left(F^{-1}\right)\theta^{-1}(F)\right)^{\uparrow} = \left(\theta^{-1}(F)^{-1}\theta^{-1}(F)\right)^{\uparrow} = \mathbf{d}\left(\theta^{-1}(F)\right)$$

and

$$\theta^{-1}\left(\left(F^{-1}F\right)^{\uparrow}\right) = \theta^{-1}\left(\mathbf{d}\left(F\right)\right).$$

Hence by our result above

$$\theta^{-1}\left(\mathbf{d}\left(F\right)\right) = \mathbf{d}\left(\theta^{-1}\left(F\right)\right).$$

A dual result also holds and so θ^{-1} preserves the domain and codomain operations. Suppose that $\mathbf{d}\left(F\right) = \mathbf{r}\left(G\right)$ so that $F \cdot G$ is defined. By our calculation above, $\mathbf{d}\left(\theta^{-1}\left(F\right)\right) = \mathbf{r}\left(\theta^{-1}\left(G\right)\right)$ and so the product $\theta^{-1}\left(F\right) \cdot \theta^{-1}\left(G\right)$ is defined. By our main result above, we have that

$$\theta^{-1}\left(F \cdot G\right) = \theta^{-1}\left(F\right) \cdot \theta^{-1}\left(G\right),$$

as required. We have therefore shown that θ^{-1} is a functor. The proof that θ^{-1} is a covering functor follows the same lines as the proof of [24, Proposition 2.15]: the proof of star injectivity uses Lemma 47, and the proof of star surjectivity uses the same lemma and Lemma 41. To show that θ^{-1} is continuous, observe that a basic open set of $\mathsf{G}\left(S\right)$ has the form $\mathscr{U}_s$ for some $s \in S$. It is simple to check that this is pulled back under the map $\left(\phi^{-1}\right)^{-1}$ to the set $\mathscr{U}_{\theta(s)}$. We now prove coherence. Let X be a compact-open subset of $\mathsf{G}\left(S\right)$. Then X may be written as a union of compact-open local bisections. Thus, by compactness, we can write X as a union of a finite number of compact-open local bisections. We now use the previous result to deduce that the inverse image of X is a finite union of compact-open local bisections and so is itself a compact-open set.

Let $\phi \colon G \to H$ be a coherent, continuous covering functor. Let U be a compact-open local bisection of H. Then $\phi^{-1}\left(U\right)$ is compact-open since ϕ is coherent. Without loss of generality, we can assume that it is non-empty. We prove that it is a local bisection. Let $g, h \in \phi^{-1}\left(U\right)$ be such that $\mathbf{d}\left(g\right) = \mathbf{d}\left(h\right)$. Then $\theta\left(g\right), \theta\left(h\right) \in U$ and $\mathbf{d}\left(\theta\left(g\right)\right) = \mathbf{d}\left(\theta\left(h\right)\right)$. By assumption, U is a local bisection and so $\theta\left(g\right) = \theta\left(h\right)$. We now use the fact that ϕ is star-injective to deduce that $g = h$. A dual result proves that U is a local bisection. Put $\phi^{-1} = \phi_{\star}$. We have therefore defined a function $\phi_{\star} \colon \mathsf{KB}\left(H\right) \to \mathsf{KB}\left(G\right)$. It remains to show that $\phi_{\star}$ is a callitic morphism.

The proof that it is a semigroup homomorphism follows the same lines as the proof in [24, Proposition 2.17] where we use Lemma 29. We show that this map is proper and weakly-meet-preserving. We prove first that ϕ^{-1} is proper. Let $B \in \mathsf{KB}\left(G\right)$. Then B is a non-empty compact-open local bisection in G. Let $g \in B$. Then $\phi\left(g\right) \in H$. Clearly, H is an open set containing $\phi\left(g\right)$. Since H is étale, it follows that H is a union of compact-open local bisections and so $\phi\left(g\right) \in C_g$ for some compact-open local bisection C_g in H. Since ϕ is continuous and coherent, $g \in \phi^{-1}\left(C_g\right)$ is compact-open and because ϕ is a covering functor $\phi^{-1}\left(C_g\right)$ is a local bisection. It follows that $B \subseteq \bigcup_{g \in B} \phi^{-1}\left(C_g\right)$. Since B is compact, we may in fact write $B \subseteq \bigcup_{i=1}^{m} \phi^{-1}\left(C_{g_i}\right)$ for some finite set of elements $g_1, \ldots, g_m \in B$. Put $B_i = B \cap \phi^{-1}\left(C_{g_i}\right)$. This is clearly an open local bisection and $B = \bigcup_{i=1}^{m} B_i$ and each $B_i \subseteq \phi^{-1}\left(C_i\right)$ where $C_i = C_{g_i}$.

We prove that we may find compact-open local bisections D_i such that B is the union of the D_i and $D_i \subseteq \phi^{-1}(C_i)$. Since B_i is an open local bisection, it is a union of compact-open local bisections. Amalgamating these unions we have that B is a union of compact-open local bisections each of which is a subset of one of the $\phi^{-1}(C_i)$. It follows that B is a union of a finite number of such compact-open local bisections. Define D_i to be the union of those which are contained in $\phi^{-1}(C_i)$ and the result follows. We now prove that ϕ^{-1} is weakly-meet-preserving. Let $A, B \in \mathsf{KB}(H)$. Then A and B are compact-open local bisections of H. Let Y be any compact-open local bisection of G such that $Y \subseteq \phi^{-1}(A), \phi^{-1}(B)$. Clearly, $Y \subseteq \phi^{-1}(A \cap B)$. We can at least say that $A \cap B$ is an open local bisection and $\theta(Y) \subseteq A \cap B$. Since ϕ is continuous, we know that $\phi(Y)$ is compact. Now H has a base of compact-open local bisections. It follows that $A \cap B$ is a union of compact-open local bisections. But $\theta(Y)$ is compact and so $\theta(Y)$ is contained in a finite union of compact-open local bisections that is also contained in $A \cap B$. Thus $\theta(Y) \subseteq V = \bigcup_{i=1}^{m} V_i \subseteq A \cap B$. Now $A \cap B$ a local bisection implies that V is a local bisection. It is evident that V is a compact-open local bisection itself. We therefore have $\theta(Y) \subseteq V \subseteq A \cap B$. Hence $Y \subseteq \theta^{-1}(\theta(Y)) \subseteq \theta^{-1}(V) \subseteq \theta^{-1}(A \cap B)$.

Let $\theta \colon S \to T$ be a callitic morphism between two Boolean inverse semigroups. Define $\mathsf{G}(\theta) = \theta^{\star}$. Let $\phi \colon G \to H$ be a coherent continuous covering functor between Boolean groupoids. Define $\mathsf{KB}(\phi) = \phi_{\star}$. It is now routine to check that we have defined functors, where G takes us from Boolean inverse semigroups and callitic morphisms to the dual of the category of Boolean groupoids and coherent continuous covering functors; and where KB takes us from the category of Boolean groupoids and coherent continuous covering functors to the dual of the category of Boolean inverse semigroups and callitic morphisms.

Let $\theta \colon S \to T$ be a callitic morphism between two Boolean inverse semigroups. We shall compare this to the callitic morphism $(\theta^{\star})_{\star} : \mathsf{KB}(\mathsf{G}(S)) \to \mathsf{KB}(\mathsf{G}(T))$ using Proposition 6. We have to compute $(\theta^{\star})_{\star}(\mathscr{U}_s)$. It is routine to check that this is $\mathscr{U}_{\theta(s)}$. Let $\phi \colon G \to H$ be a coherent continuous covering functor between Boolean groupoids. We shall compare this to the coherent continuous covering functor $(\phi_{\star})^{\star} : \mathsf{G}(\mathsf{KB}(G)) \to \mathsf{G}(\mathsf{KB}(H))$ using Proposition 6. We have to compute $(\phi_{\star})^{\star}(\mathscr{F}_g)$. It is routine to check that this is $\mathscr{F}_{\phi(g)}$.

The functor $\mathsf{KB} \circ \mathsf{G}$ is now clearly naturally isomorphic with the identity functor on the category of Boolean inverse semigroups and callitic morphisms, whereas the functor $\mathsf{G} \circ \mathsf{KB}$ is now clearly naturally isomorphic with the identity functor on the category of Boolean groupoids and coherent continuous covering functors. □

8 Special Cases

In this section, we shall describe some special cases of Theorem 7. None of these is new, but they have not appeared altogether in this way before.

By the results of Sects. 2.3 and 2.4, a generalized Boolean algebra with an identity is nothing other than a Boolean algebra. It follows that the Boolean inverse semigroups with the property that the space of identities of their Stone groupoids are compact are precisely the Boolean inverse monoids.

The group of units of a Boolean inverse monoid is just the set of all compact-open bisections of the associated Stone groupoid; the compact-open bisections form what is known as the *topological full group* of the Boolean groupoid.

Suppose that S is a countable Boolean inverse semigroup. Then the Stone groupoid of S must have a countable base of compact-open local bisections. It follows that its Stone groupoid is second-countable.

In Sect. 2.3, we showed that the Stone space of the Tarski algebra is the Cantor space. Define a *Tarski monoid* to be a countably infinite Boolean inverse monoid whose semilattice of idempotents forms a Tarski algebra. Define a *Tarski groupoid* to be a second-countable Boolean groupoid whose space of identities is the Cantor space.

Following Wehrung [58], define a *semisimple* Boolean inverse semigroup to be one in which for each element a the principal order-ideal $a^{\downarrow}$ is finite.

Proposition 8 *Let S be a Boolean inverse semigroup. Then S is semisimple if and only if $\mathsf{G}(S)$ carries the discrete topology.*

Proof Let a be any element of such a semigroup. Then $\mathscr{U}_a = \bigcup_{i=1}^{n} \mathscr{U}_{a_i}$, where $a_1, \ldots, a_n$ are all the atoms below a; this is proved by observing that in a semisimple Boolean inverse semigroup, every non-zero element lies above an atom and so every non-zero element is a join of atoms. If a is an atom then $\mathscr{U}_a$ contains exactly one element. It follows that the sets $\mathscr{U}_a$, where a is an atom, form a base for the topology on $\mathsf{G}(S)$, and so this set is equipped with the discrete topology. Conversely, suppose that $\mathsf{G}(S)$ is equipped with the discrete topology. Then, for each $a \in S$, we have that $\mathscr{U}_a$ is finite since it is compact-open. But $b \leq a$ if and only if $\mathscr{U}_b \subseteq \mathscr{U}_a$. This proves that the Boolean inverse semigroup S is semisimple. □

If S is a semisimple Boolean inverse semigroup, then the Stone groupoid $\mathsf{G}(S)$ is isomorphic to the set of atoms of S equipped with the restricted product; see [29] for the structure of semisimple Boolean inverse semigroups. The significance of semisimple Boolean inverse semigroups is explained by the following result. We say that a Boolean inverse semigroup is *atomless* if it has no atoms.

Proposition 9 (Dichotomy theorem) *Let S be a 0-simplifying Boolean inverse semigroup. Then either S is semisimple or S is atomless.*

Proof Suppose that a is an atom. Then $\mathbf{d}(a)$ is an atom. We can therefore assume, without loss of generality, that there is at least one idempotent atom e. Let x be any non-zero element of S. We prove that $x^{\downarrow}$ is finite. It is enough to prove that $\mathbf{d}(x)^{\downarrow}$ is finite. Let $f \leq \mathbf{d}(x)$. Since the semigroup is assumed to be 0-simplifying, there is a pencil from f to e. There is therefore a finite set of elements $\{x_1, \ldots, x_n\}$ such that $f = \bigvee_{i=1}^{n} \mathbf{d}(x_i)$ and $\mathbf{r}(x_i) \leq e$. But e is an atom. Without loss of generality, we assume that all the x_i are non-zero. Thus $\mathbf{r}(x_i) = e$. It follows that each $\mathbf{d}(x_i)$ is an atom and we have proved that f is a join of a finite number of atoms. It follows that $\mathbf{d}(x)^{\downarrow}$ is a finite Boolean algebra. □

Our next result tells us when the Stone groupoid is Hausdorff. It is a good example of the connection between algebra and topology.

Proposition 10 *Let S be a Boolean inverse semigroup. Then S is a meet-semigroup if and only if $\mathsf{G}(S)$ is Hausdorff.*

Proof Let S be a Boolean inverse semigroup and suppose that it is a meet-semigroup. Let F and G be distinct ultrafilters in $\mathsf{G}(S)$; since neither ultrafilter can be a subset of the other, we can find elements $a \in F \setminus G$ and $b \in G \setminus F$. The element $a \wedge b$ exists by assumption and also by assumption $a \neq a \wedge b$ and $b \neq a \wedge b$. Observe that $F \in \mathscr{U}_{a \setminus (a \wedge b)}$ and $G \in \mathscr{U}_{b \setminus (a \wedge b)}$. Let $x \leq a \setminus (a \wedge b), b \setminus (a \wedge b)$. Then $x \leq a \wedge b$. It follows that $x = 0$. We have proved that $\mathscr{U}_{a \setminus (a \wedge b)} \cap \mathscr{U}_{b \setminus (a \wedge b)} = \varnothing$ and so we have proved that $\mathsf{G}(S)$ is Hausdorff. Conversely, suppose that $\mathsf{G}(S)$ is Hausdorff. Let A and B be two compact-open local bisections. By part (2) of Lemma 2, both A and B are clopen. Thus $A \cap B$ is clopen. But by part (3) of Lemma 2, $A \cap B$ is also compact, and so it is a compact-open local bisection. The result, that S is a meet-semigroup, follows by Proposition 6. □

The above result was essentially the basis of my paper [24]. This paper won the Mahony–Neumann–Room Prize of the Australian Mathematical Society in 2017.

Recall that an inverse semigroup is *fundamental* if the only elements that commute with all idempotents are themselves idempotents. If we denote the *centralizer of the idempotents* by $\mathsf{Z}(\mathsf{E}(S))$ then an inverse semigroup is fundamental when $\mathsf{E}(S) = \mathsf{Z}(\mathsf{E}(S))$. An étale topological groupoid G is said to be *effective* if the interior of the isometry groupoid consists simply of the space of identities. Fundamental inverse semigroups were introduced for purely algebraic reasons, but they have an important topological counterpart.

Proposition 11 *Let S be a Boolean inverse semigroup. Then S is fundamental if and only if $\mathsf{G}(S)$ is effective.*

Proof We prove first that $a \in \mathsf{Z}(\mathsf{E}(S))$ if and only if $\mathscr{U}_a \subseteq \mathrm{Iso}(\mathsf{G}(S))$. Suppose that a centralizes all the idempotents of S. Then, in particular, a centralizes aa^{-1}. Thus $aaa^{-1} = aa^{-1}a = a$. It follows that $a^{-1}aaa^{-1} = aa^{-1}$. This proves that $aa^{-1} \leq a^{-1}a$. By symmetry, $a^{-1}a \leq aa^{-1}$ and so $a^{-1}a = aa^{-1}$. Let $A \in \mathscr{U}_a$. We need to prove that $\mathbf{d}(A) = \mathbf{r}(A)$. Let $x \in \mathbf{d}(A)$. Then there is an idempotent $e \in \mathbf{d}(A)$ such that $e \leq x$. By assumption, $a \in A$ and so $a^{-1}a \in \mathbf{d}(A)$. It follows that $ea^{-1}a \in A$ and clearly $ea^{-1}a \leq x$. But $ea^{-1}a = eaa^{-1} = aea^{-1}$, where we have again used the fact that a commutes with all idempotents. It follows that $aea^{-1} \leq x$. Now $ae \in A \cdot \mathbf{d}(A) = A$. It follows that $x \in \mathsf{r}(A)$. We have therefore proved that $\mathbf{d}(A) \subseteq \mathbf{r}(A)$. The proof of the reverse inclusion follows by symmetry. We have therefore proved that $\mathbf{d}(A) = \mathbf{r}(A)$. It follows that $a \in \mathsf{Z}(\mathsf{E}(S))$ implies that $\mathscr{U}_a \subseteq \mathrm{Iso}(\mathsf{G}(S))$. Now, suppose that $\mathscr{U}_a \subseteq \mathrm{Iso}(\mathsf{G}(S))$. Let e be any idempotent. We shall prove that $\mathscr{U}_{ae} = \mathscr{U}_{ea}$ and the result will follow by part (2) of Lemma 53. Suppose that $A \in \mathscr{U}_{ae}$. Then $ae \in A$ and so $a \in A$. By assumption, $\mathbf{d}(A) = \mathbf{r}(A)$. Now, $ae \in A$ and so $(ae)^{-1}ae = ea^{-1}a$. It follows that $ea^{-1}a \in \mathbf{r}(A)$. Hence $a^{-1}aea \in A$. It follows that $ea \in A$. We have therefore proved that $A \in \mathscr{U}_{ea}$. By symmetry, $\mathscr{U}_{ae} = \mathscr{U}_{ea}$.

We now prove the claim. Suppose that S is fundamental. Let $\mathscr{U}_a \subseteq \mathrm{Iso}\,(\mathsf{G}\,(S))$. Then a must commute with all idempotents and so is itself an idempotent. It follows that the only open sets in $\mathrm{Iso}\,(\mathsf{G}\,(S))$ are open sets of identities. It follows that $\mathsf{G}\,(S)$ is effective. Conversely, suppose that $\mathsf{G}\,(S)$ is effective. Let a commute with all idempotents. Then $\mathscr{U}_a \subseteq \mathrm{Iso}\,(\mathsf{G}\,(S))$. It follows that every element of $\mathscr{U}_a$ is an idempotent ultrafilter. Thus by part (4) of Lemma 53, we deduce that a is an idempotent and so S is fundamental. □

Fundamental inverse semigroups have the important property that there are no non-trivial *idempotent-separating* homomorphisms.

Recall that an *infinitesimal* in an inverse semigroup S with zero is a non-zero element a such that $a^2 = 0$. We say that a Boolean inverse semigroup is *basic* if every element is a finite join of infinitesimals and an idempotent. You can check that basic inverse semigroups always have meets; see [24, Lemma 4.30].

Lemma 57 *Let S be a Boolean inverse semigroup. Then every ultrafilter A such that $\mathbf{d}\,(A) \neq \mathbf{r}\,(A)$ contains an infinitesimal.*

Proof Let A be an ultrafilter such that $\mathbf{d}\,(A) \neq \mathbf{r}\,(A)$. Then $\mathbf{d}\,(A) = E^{\uparrow}$ and $\mathbf{r}\,(A) = F^{\uparrow}$ where E and F are ultrafilters in the generalized Boolean algebra $\mathsf{E}\,(S)$. By classical Stone duality extended to generalized Boolean algebras, we know that the structure space of $\mathsf{E}\,(S)$ is Hausdorff. Let $e, f \in \mathsf{E}\,(S)$ be such that $E \in \widetilde{U}_e$, $F \in \widetilde{U}_f$, and $\widetilde{U}_e \cap \widetilde{U}_f = \varnothing$. In particular, $ef = 0$. Let $a \in A$. Then $eaf \in A$. But eaf is an infinitesimal. □

The following was stated in [27, Proposition 4.31], but the condition that the Boolean inverse semigroup be a meet-monoid was omitted.

Proposition 12 *Let S be a Boolean inverse meet-semigroup. Then S is basic if and only if $\mathsf{G}\,(S)$ is a principal groupoid.*

Proof The proof that basic implies principal does not require the assumption that the Boolean inverse semigroup have meets. Since ultrafilters are prime in Boolean inverse semigroups, it follows that if A is an element of a local group of the groupoid, then it cannot contain infinitesimals and so must contain an idempotent from which it follows that A is an idempotent ultrafilter. We now prove the converse and use a different approach from the one adopted in [26]. We are given that $\mathsf{G}\,(S)$ is principal and Hausdorff, and we shall prove that S is basic. Let a be any element of S. Then, by assumption, $\phi\,(a) \leq a$ is the largest idempotent less than or equal to a. Observe that $a = \phi\,(a) \vee (a \setminus \phi\,(a))$ is an orthogonal join and that $\phi\,(a \setminus \phi\,(a)) = 0$. With this in mind, let $b \in S$ be an element such that $\phi\,(b) = 0$. We shall be done if we prove that b is a finite join of infinitesimals. Consider the set $\mathscr{U}_b$ of all prime filters that contain b. Let $A \in \mathscr{U}_b$. The groupoid $\mathsf{G}\,(S)$ is principal. If $\mathbf{d}\,(A) = \mathbf{r}\,(A)$ then A must be an identity prime filter and so contains non-zero idempotents. It follows that b must be above a non-zero idempotent, which is a contradiction. Thus $\mathbf{d}\,(A) \neq \mathbf{r}\,(A)$. We now use Lemma 57 to deduce that when the groupoid $\mathsf{G}\,(S)$ is principal each non-identity prime filter A must contain infinitesimals. But any element below an infinitesimal is either zero or an infinitesimal. Thus b must lie above an infinitesimal. It follows that $\mathscr{U}_b = \bigcup_{x \leq b, x^2=0, x \neq 0} \mathscr{U}_x$. We now use compactness of U_b to deduce that $\mathscr{U}_b = \bigcup_{i=1}^{m} \mathscr{U}_{x_i}$ where each x_i is an infinitesimal $x_i \leq b$. It follows by part (2) of Lemma 53 that b is a join of a finite number of infinitesimals. We have therefore proved that S is basic. □

A subset of a groupoid is said to be *invariant* if it is a union of connected components. The following combines results to be found in [26, 35].

Lemma 58 *Let S be a Boolean inverse semigroup. Then there is an order-isomorphism between the set of additive ideals of S and the set of open invariant subsets of* $\mathsf{G}(S)$.

An étale groupoid is said to be *minimal* if it contains exactly two open invariant subsets. The following was essentially proved as [24, Corollary 4.8], but we do not need the assumption that the semigroup is a meet-semigroup.

Proposition 13 *A Boolean inverse semigroup S is* 0*-simplifying if and only if* $\mathsf{G}(S)$ *is minimal.*

A Boolean inverse semigroup which is fundamental and 0-simplifying is said to be *simple*. The terminology is explained by the following result.

Lemma 59 *Let $\theta \colon S \to T$ be a morphism of Boolean inverse semigroups where S is simple. Then θ is injective.*

Proof We prove first that θ is injective on idempotents, in other words that θ is idempotent-separating. Let e and f be idempotents such that $\theta(e) = \theta(f)$. Then $e \wedge f \leq e$ and $\theta(e \setminus (e \wedge f)) = 0$. By assumption, the semigroup S is 0-simplifying and so $e = e \wedge f$. By symmetry $f = e \wedge f$. It follows that $e = f$. We have proved that θ is idempotent-separating. However, we are assuming also that S is fundamental. This means that θ is actually injective. □

An inverse semigroup $S \neq \{0\}$ is said to be 0-*simple* if its only semigroup ideals are $\{0\}$ and S itself. Observe that being 0-simple is a stronger condition than being 0-simplifying. The following was proved as [20, Proposition 3.2.10].

Lemma 60 *Let S be an inverse semigroup with zero. Then it is* 0*-simple if and only if for any two idempotents e and f there exists an idempotent i such that $e \,\mathscr{D}\, i \leq f$.*

A non-zero idempotent e is said to be *properly infinite* if we may find orthogonal idempotents i and j such that $e \,\mathscr{D}\, i$ and $f \,\mathscr{D}\, j$ and $i, j \leq e$. An inverse semigroup with zero is said to be *purely infinite* if every non-zero idempotent is properly infinite. The proof of the following is [26, Lemma 4.11].

Lemma 61 *Let S be a* 0*-simple Tarski monoid. Then S is purely infinite.*

The proof of the following is [26, Theorem 4.16].

Proposition 14 *Let S be a Tarski monoid. Then S is 0-simple if and only if S is 0-simplifying and purely infinite.*

Proof One direction is proved by Lemma 61. The other direction is proved in [26], but since the proof is slightly garbled we give the complete proof here. It is just a translation of [39, Proposition 4.11]. Let e and f be any non-zero idempotents. Under the assumption that S is 0-simplifying, we may find elements $w_1, \ldots, w_n$ such that $e = \bigoplus_{i=1}^{n} \mathbf{d}(w_i)$ and $\mathbf{r}(w_i) \leq f$. Using the fact that the semigroup is purely infinite, we may find elements a and b such that $\mathbf{d}(a) = f = \mathbf{d}(b)$, $\mathbf{r}(a) \perp \mathbf{r}(b)$ and $\mathbf{r}(a), \mathbf{r}(b) \leq f$. Define elements $v_1, \ldots, v_n$ as follows: $v_1 = a$, $v_2 = ba$, $v_3 = b^2 a$, ..., $v_n = b^{n-1}a$. Observe that $\mathbf{d}(v_1) = \mathbf{d}(v_2) = \cdots = \mathbf{d}(v_n) = f$, and that $\mathbf{r}(v_i) \leq f$ for $1 \leq i \leq n$. The elements $\mathbf{r}(v_i)$ are pairwise orthogonal. Consider now the elements $v_1 w_1, \ldots, v_n w_n$. Observe that $\mathbf{d}(v_i) \geq \mathbf{r}(w_i)$. It follows that the domains of these elements are pairwise orthogonal as indeed are their ranges. These elements are compatible and so we may form their (orthogonal) join: $w = \bigoplus_{i=1}^{n} v_i w_i$. Observe that $\mathbf{d}(w) = e$ and that the ranges, being orthogonal and each less than or equal to f, must have a join which is less than or equal to f. □

A Boolean inverse semigroup that is fundamental and 0-simple is *congruence-free*; what we mean by this terminology is that there are no non-trivial congruences on S of any description. See [20].

If S is a Boolean inverse semigroup, then we may always write $S = \bigcup_{e \in \mathsf{E}(S)} eSe$. A Boolean inverse semigroup is said to be *σ-unital*[6] if there is a non-decreasing sequence of idempotents $e_1 \leq e_2 \leq \cdots$ such that $S = \bigcup_{i=1}^{\infty} e_i S e_i$. We call $\{e_i : i \in \mathbb{N} \setminus \{0\}\}$ the *σ-unit*. For example, let S be any Boolean inverse monoid. Then the Boolean semigroup $M_\omega(S)$ is σ-unital.

A topological space is said to be *σ-compact* if it is a union of countably many compact spaces.

Proposition 15 *Let S be a Boolean inverse semigroup. Then S is σ-unital if and only if the identity space of $\mathsf{G}(S)$ is σ-compact.*

Proof Suppose first that S is σ-compact. If A is an ultrafilter that is also idempotent then it contains an idempotent. It is immediate that $\mathsf{G}(S)_o = \bigcup_{i=1}^{\infty} \mathscr{U}_{e_i}$. Thus $\mathsf{G}(S)_o$ is σ-compact. Conversely, suppose that $\mathsf{G}(S)_o = \bigcup_{i=1}^{\infty} U_i$, where each U_i is compact. Observe that $\mathsf{G}(S)_o = \bigcup_{f \in \mathsf{E}(S)} \mathscr{U}_f$. Then we can choose f_i such that $\mathsf{G}(S)_o = \bigcup_{i=1}^{\infty} \mathscr{U}_{f_i}$. Define $e_1 = f_1, e_2 = f_1 \vee f_2, e_3 = f_1 \vee f_2 \vee f_3, \ldots$. Let $s \in S$. Then we can find an idempotent such that $s = isi$; for example, $i = \mathbf{d}(s) \vee \mathbf{r}(s)$ will work. All the ultrafilters in $\mathscr{U}_i$ are idempotent. It follows that $\mathscr{U}_i \subseteq \mathscr{U}_{e_p}$ for some p. Thus by part (1) of Lemma 53, we have that $i \leq e_p$. We have therefore shown that $s \in e_p S e_p$. □

[6] A term taken from ring theory.

The following table summarizes the different aspects of non-commutative Stone duality we have proved:

Boolean inverse semigroup	**Boolean groupoid**
Boolean inverse monoid	Boolean groupoid with a compact identity space
Group of units of monoid	Topological full group
Countable	Second-countable
Tarski algebra of idempotents	Cantor space of identities
Tarski monoid	Tarski groupoid
Semisimple	Discrete
Meet-semigroup	Hausdorff
Fundamental	Effective
Basic	Principal and Hausdorff
0-simplifying	Minimal
Simple	Minimal and effective
0-simple Tarski monoid	Purely infinite and minimal Tarski groupoid
Congruence-free Tarski monoid	Purely infinite, minimal, and effective Tarski groupoid
σ-unital	Identity space is σ-compact

9 Unitization

Every Boolean inverse semigroup can be embedded (in a nice way) into a Boolean inverse monoid [58, Definition 6.6.1]. We shall now obtain this result using our non-commutative Stone duality.

We begin with Lemma 23. Let X be a locally compact Boolean space. Put $X^{\infty} = X \cup \{\infty\}$ and endow X^{∞} with the topology that consists of all the open subsets of X together with the complements in X^{∞} of the compact sets of X together with X^{∞} itself. The following lemma is useful in proving that this really is a topology and will also be needed later.

Lemma 62 *The set $U \cap (X^{\infty} \setminus V)$, where U is open in X and V is compact in X, is open in X.*

Proof Observe that since $U \subseteq X$, the intersection in question is actually $U \cap (X \setminus V)$. But if U is open in X there is a closed set U_1 in X such that $U = X \setminus U_1$. It follows that the intersection is $X \setminus (U_1 \cup V)$. But V is a compact subspace of a Hausdorff space and so is closed by part (2) of Lemma 2. However, $U_1 \cup V$ is closed in X and so $X \setminus (U_1 \cup V)$ is an open subset of X. □

The key result is the following (which, of course, is well-known).

Proposition 16 (One-point compactification) *Let X be a locally compact Hausdorff space. Then X^{∞}, with the above topology, is a compact Hausdorff space that contains X as an open subset. If X is 0-dimensional so too is X^{∞}.*

Proof The proof of the first claim can be found in [52, Sect. 37]. The proof of the second claim is well-known, but we give a proof anyway. We begin by describing the clopen subsets of X^∞. These are of two types:

1. Those that do not contain ∞ are precisely the compact-open subsets of X.
2. Those that do contain ∞ are precisely of the form $X^\infty \setminus U$ where U is a compact-open subset of X.

We now give the proofs of these two claims. (1) Suppose that U is a clopen subset of X^∞ where $\infty \notin U$. Thus, in particular, $U \subseteq X$. Now, U is a closed subset of X^∞, which is compact. It follows that U is compact in X^∞ and so must be compact in X. It is open by definition. Thus U is compact-open in X. We now go in the other direction. Let U be compact-open in X. Then U is open in X^∞ by definition. It remains to prove that U is closed in X^∞. Since U is compact in X, it follows by the definition of the topology that $X^\infty \setminus U$ is open in X^∞. It follows that U is closed in X^∞. Thus U is clopen in X^∞. (2) Let U be a clopen subset of X^∞ that contains ∞. Since U is open and contains ∞, we may write $U = X^\infty \setminus K$ where K is a compact subset of X. Since U is closed in X^∞, there is an open subset $V \subseteq X^\infty$ such that $U = X^\infty \setminus V$. Observe that V cannot contain ∞ and so must be an open subset of X. It follows that U is the complement in X^∞ of a compact-open subset of X. We now go in the opposite direction. Let U be a compact-open subset of X. We prove that $X^\infty \setminus U$ is clopen in X^∞. Since U is compact in X, the set $X^\infty \setminus U$ is open by definition. Since U is open in X, it is open in X^∞ by definition. It follows that $X^\infty \setminus U$ is closed. We have proved that $X^\infty \setminus U$ is clopen.

It remains to prove that the clopen sets form a base for the topology on X^∞ if X is 0-dimensional. We prove that every open subset of X^∞ is a union of clopen subsets. Let U be an open set of X^∞. There are two possibilities. The first is that $U \subseteq X$, then U is the union of the compact-open subsets of X by Lemma 19; thus it is a union of clopen sets of X^*. The second is that $U = X^\infty \setminus K$, where K is a compact subspace of X. Observe first that K is contained in a compact-open subset of X. We prove that K is in fact equal to the intersection of all the compact-open subsets that contain it. Let $x \in X$ but $x \notin K$. Then since X is Hausdorff, there are open sets U and V such that $x \in U$, $K \subseteq V$, and $U \cap V = \varnothing$; we have used [52, Theorem 26.C]. Since V is open, it is a union of compact-open subsets. These cover K which is compact. We can therefore assume that V is compact-open. Thus for every point in the complement of K, we can find a compact-open set that omits that point and contains K. It follows that K is equal to the intersection of all the compact-open sets that contain K, whence $X^\infty \setminus K = \bigcup_i (X^\infty \setminus V_i)$ where each V_i is compact-open in X and contains K. □

The above result can be used to embed generalized Boolean algebras into Boolean algebras using Stone duality; this was first proved in [55] and was touched upon in Lemma 23. Let B be a generalized Boolean algebra. Its Stone space $\mathsf{X}(B)$ is a 0-dimensional locally compact Hausdorff space. Construct the one-point compactification of this space to get a 0-dimensional compact Hausdorff space $\mathsf{X}(B)^\infty$. This gives rise to a Boolean algebra $\mathsf{B}\left(\mathsf{X}(B)^\infty\right)$ into which B can be embedded. This embedding is summarized by the following (well-known) result.

Proposition 17 *Let B be a generalized Boolean algebra without a top element. Then there is a Boolean algebra C which has B as a subalgebra and order-ideal such that the elements of C are either e or e', where $e \in B$.*

The following example is an illustration of Proposition 17. Let B be the generalized Boolean algebra of all finite subsets of the set $\mathbb{N}$. With respect to the discrete topology, $\mathbb{N}$ is a locally compact Boolean space. This space can be embedded in the Boolean space $\mathbb{N}^{\infty} = \mathbb{N} \cup \{\infty\}$ which has as open sets all finite subsets of $\mathbb{N}$ together with all cofinite subsets of $\mathbb{N}$ with ∞ adjoined. The set of all finite subsets of $\mathbb{N}^{\infty}$ which omit ∞ together with all cofinite subsets of $\mathbb{N}^{\infty}$ that contain ∞ form a Boolean algebra. This is isomorphic with the Boolean algebra of all finite subsets of $\mathbb{N}$ together with the cofinite subsets of $\mathbb{N}$.

We shall now extend Proposition 17 to Boolean inverse semigroups.

Let G be a Boolean groupoid where the space G_o is a locally compact Boolean space. Denote by G^{∞} the groupoid $G \cup \{\infty\}$ where ∞ is a new identity such that $(G^{\infty})_o = G_o \cup \{\infty\}$. Endow G^{∞} with the topology generated by the base $\Omega(G) \cup \Omega\left(G_o^{\infty}\right)$ (it remains to show that this really is a base); every open subset of G^{∞} is therefore the union of an open subset of G and an open subset of G_o^{∞}.

Lemma 63 *With the above definitions, G^{∞} is a Boolean groupoid, the identity space of which is compact.*

Proof We show first that $\Omega(G) \cup \Omega\left(G_o^{\infty}\right)$ is a base for a topology. This boils down to checking that if U is an open set of G and V is an open set of G_o^{∞} then $U \cap V$ is an open set. There are two possibilities for V. If V is an open subset of G_o then it is also an open subset of G, since G is an étale groupoid, and so its space of identities is an open subset. It follows that $U \cap V$ is open in G and so belongs to our topology. The other possibility is that $V = G_0^{\infty} \setminus K$ where K is a compact subset of G_0. But U, being in G, does not contain ∞ so $U \cap V = U \cap (G_0 \setminus K)$. Since G is an étale groupoid, G_o is an open subset of G. We have that $U \cap V = (U \cap G_o) \cap (G_0 \setminus K)$. We now apply Lemma 62, to deduce that $U \cap V$ is an open subset of G_o^{∞}.

Next we show that with respect to this topology, G^{∞} is a topological groupoid. It is clear that $g \mapsto g^{-1}$ is a homeomorphism. We prove that multiplication

$$\mathbf{m}_{\infty} \colon G^{\infty} * G^{\infty} \to G^{\infty}$$

is continuous. We denote the multiplication on G by $\mathbf{m}$. The basic open sets of G^{∞} are of two kinds. Those in G and those in G_o^{∞}. The former cause us no problems since they do not contain ∞ and so the result follows from the fact that $\mathbf{m}$ is continuous. The open sets of G_o^{∞} are of two kinds. Those which are simply open subsets of G_o and so open subsets of G (since G is étale G_o is an open subset of G) are dealt with the above since they do not contain ∞. Thus, the only case we have to deal with are those subsets of the form $U = G_o^{\infty} \setminus K$ where K is a compact subset of G_o. We have to prove that $\mathbf{m}_{\infty}^{-1}(U)$ is an open subset of $G^{\infty} * G^{\infty}$. Observe first that

$$\mathbf{m}_{\infty}^{-1}(U) = [\left((G * G) \setminus \mathbf{m}^{-1}(K)\right) \cap \mathbf{m}^{-1}(G_o)] \cup [(U \times U) \cap \left(G^{\infty} * G^{\infty}\right)].$$

It is easy to show that the left-hand side is contained in the right-hand side; just recall that the elements of $\mathbf{m}_{\infty}^{-1}(U)$ are of two types: either ordered pairs $(g, h) \in G \times G$ or (∞, ∞). It is also easy to show that the right-hand side is contained in the left-hand side. It remains to be shown that the right-hand side above is an open subset of $G^{\infty} * G^{\infty}$; to do this we use the product topology on $G^{\infty} \times G^{\infty}$. The set K is compact and so it is a closed subset of G_o. Thus $\mathbf{m}^{-1}(K)$ is a closed subset of $G * G$. Since G is étale, we know that G_o is an open subset of G and so $\mathbf{m}^{-1}(G_o)$ is an open subset of $G * G$. It follows that the first term is an open subset of $G * G$ which is an open subset of $G^{\infty} * G^{\infty}$. The second term is just the intersection of $U \times U$, which is an open subset of $G^{\infty} \times G^{\infty}$, with $G^{\infty} * G^{\infty}$ which gives us an open subset of $G^{\infty} * G^{\infty}$. We have therefore proved that G^{∞} is a topological groupoid. It follows that G^{∞} is an étale groupoid (because G is), and it is Boolean by construction with a compact identity space by Proposition 16. □

By the above result, $\mathsf{KB}(G^{\infty})$ is a Boolean inverse monoid. It is clear that $\mathsf{KB}(G)$ embeds into $\mathsf{KB}(G^{\infty})$. Observe that $\mathsf{KB}(G)$ is closed under binary joins taken in $\mathsf{KB}(G^{\infty})$. The generalized Boolean algebra $\mathsf{B}(G_o)$ is a subalgebra of the Boolean algebra $\mathsf{B}\left(G_o^{\infty}\right)$. We have therefore proved the following.

Lemma 64 *With the above definition,* $\mathsf{KB}(G)$ *is a subalgebra of* $\mathsf{KB}(G^{\infty})$.

We have therefore embedded a Boolean inverse semigroup into a Boolean inverse monoid where the generalized Boolean algebra of the former is embedded into the Boolean algebra of the latter.

We now describe the elements $A \in \mathsf{KB}(G^{\infty})$, the compact-open local bisections of G^{∞}. This will provide the connection with [58, Definition 6.6.1]. There are two cases: either $\infty \notin A$ or $\infty \in A$. In the former case, A is just a compact-open local bisection of G and so an element of $\mathsf{KB}(G)$. We therefore deal with the latter case. Let $A \in \mathsf{KB}(G^{\infty})$ contain ∞. Since it is open, A can be written as a union $A = U \cup V$ where U is an open subset of G and $V = G_o \setminus K$, where K is a compact subset of G_o. We now use the fact that G^{∞} is an étale groupoid and so has a base consisting of compact-open local bisections. We may therefore write $U = \bigcup_{i \in I} B_i$ and $V = \bigcup_{j \in J} C_j$ where the B_i and C_j are compact-open local bisections of G^{∞}. We now use the fact that A is compact to deduce that $A = (B_1 \cup \cdots B_m) \cup (C_1 \cup \cdots \cup C_n)$. Now, $B_1, \ldots, B_m \subseteq U$ and so each subset omits ∞. In addition, $B_i \subseteq A$ so they are pairwise compatible. It follows that $B = B_1 \cup \cdots \cup B_m$ is a compact-open local bisection of G. The union $C_1 \cup \cdots \cup C_n$ is a compact-open local bisection which is an idempotent in $\mathsf{KB}(G^{\infty})$ and contains ∞. It follows that each $C_i \subseteq G_o^{\infty}$ and is clopen. We now use the description of the clopen subsets of G_o^{∞} given in Proposition 16. Those C_i which are compact-open subsets of G_o can be absorbed into B. We may therefore assume that each $C_i = G_o^{\infty} \setminus K_i$, where each K_i is compact-open in G_o^{∞}. It follows that $A = B \cup \left(G_o^{\infty} \setminus D\right)$, where B is an element of $\mathsf{KB}(G)$ and D is a compact-open bisection and idempotent $\mathsf{KB}(G)$. The comparison with Wehrung's construction, [58, Definition 6.6.1], follows from the next lemma.

Lemma 65 *Let S be a Boolean inverse monoid. Let $x = e' \vee a$ where $e \in \mathsf{E}(S)$. Then $x = e' \vee eae$, where clearly $e' \perp eae$.*

Proof By assumption $e' \sim a$. It follows that $e'a$ and $e'a^{-1} = ae'$ are both idempotents. We have that $1 = e \vee e'$. Thus $x = (e \vee e')(e' \vee a) = e' \vee ea \vee e'a$. But $e'a$ is an idempotent less than e. It follows that $x = e' \vee ea$, whence $x = (e' \vee ea)(e \vee e') = e' \vee eae \vee eae'$. But ae' is an idempotent and so $eae' = (ae')e = 0$. We have therefore proved that $x = e' \vee eae$. □

Taking into account Lemmas 64 and 65 and using our non-commutative Stone duality, we have therefore proved the following, a result first established by Wehrung.

Proposition 18 (Unitization) *Let S be a Boolean inverse semigroup which is not a monoid. Then there is a Boolean inverse monoid T containing S as a subalgebra and ideal such that each element of $T \setminus S$ is of the form $e' \vee s$ where $e \in \mathsf{E}(S)$ and $s \in eSe$.*

Our definition of G^{∞} is by means of a base extension as in [37], although our approach is quite different.

We can define the group of units of a Boolean inverse *semigroup* S as follows, independently of what we did above. For each idempotent $e \in \mathsf{E}(S)$, define the group G_e to be the group of units of the Boolean inverse monoid eSe. If $e \leq f$ define a map $\phi^e_f \colon G_e \to G_f$ by $a \mapsto a \vee (f \setminus e)$. It is easy to check that this is a well-defined injective function that maps e to f and is a group homomorphism. Observe that ϕ^e_e is the identity function on G_e, and if $e \leq f \leq g$ then $(g \setminus e) = (g \setminus f) \vee (f \setminus e)$ and so $\phi^e_g = \phi^f_g \phi^e_f$. It follows that we have a (E-unitary) strong semilattice of groups $\{G_e, \phi^e_f : e, f \in \mathsf{E}(S)\}$. Such a system gives rise to an inverse semigroup with central idempotents as follows. Put $T = \bigcup_{e \in \mathsf{E}(S)} G_e$, a disjoint union, with product $\circ$ defined as follows:

$$a \circ b = \phi^e_{e \vee f}(a)\, \phi^f_{e \wedge f}(b)$$

where $a \in G_e$ and $b \in G_f$. See [20, Sect. 5.2, p. 144] for details. Put $\mathsf{C}(S) = (T, \circ)$. Observe that $a \leq b$ in $\mathsf{C}(S)$ precisely when $a \in G_e$ and $b \in G_f$ and $e \leq f$ in S and $\phi^e_{e \vee f}(a) = b$. An inverse semigroup with central idempotents is called a *Clifford semigroup*. An inverse semigroup is said to be *E-unitary* if $e \leq a$, where e is an idempotent, implies that a is an idempotent.

Proposition 19 *With each Boolean inverse semigroup S, we can associate an E-unitary Clifford semigroup $\mathsf{C}(S)$. The meet semilattice of this semigroup is $(\mathsf{E}(S), \vee)$.*

With each inverse semigroup S, we can define the *minimum group congruence* σ which has the property that S/σ is a group. See [20, Sect. 2.4]. In the case that S is E-unitary, it turns out that $\sigma = \sim$. See [20, Theorem 2.4.6].

Definition Let S be a Boolean inverse semigroup. We define the *group of units* of S, denoted by $\mathsf{U}(S)$, to be the group $\mathsf{C}(S)/\sigma$.

It is routine to check that in the case where S is a Boolean inverse *monoid*, the usual definition of the group of units is returned. The group of units we have defined in terms of Boolean inverse semigroups is the same as the group defined in [41, Remark 3.10]. The following lemmas will needed for the proof of the last proposition of this section. The proof of the next lemma uses Lemma 24 and is routine.

Lemma 66 *Let S be a Boolean inverse monoid where e and e_1 are idempotents. Suppose that $a = b \oplus e'$, where $\mathbf{d}(b) = \mathbf{r}(b) = e$, and $a = b_1 \oplus e_1'$, where $\mathbf{d}(b_1) = \mathbf{r}(b_1) = e_1$. Then $b \sim b_1$ in S. Put $x = b \wedge b_1$. Then $\mathbf{d}(x) = \mathbf{r}(x) = ee_1$. We have that $b = x \vee e(ee_1)'$ and $b_1 = x \vee e_1(ee_1)'$.*

The proof of the following is routine.

Lemma 67 *Let S be a Boolean inverse monoid where e and e_1 are idempotents. Suppose that $\mathbf{d}(b) = \mathbf{r}(b) = e$, and $\mathbf{d}(b_1) = \mathbf{r}(b_1) = e_1$ and there is an element x, such that $\mathbf{d}(x) = \mathbf{r}(x) = y \geq e, e_1$, where $b = x \vee ey'$ and $b_1 = x \vee e_1 y'$. Then $b \vee e' = b_1 \vee e_1'$.*

The proof of the next result is by means of a direct verification.

Lemma 68 *Let S be a Boolean inverse monoid where e and f are idempotents. Suppose that $a = a_1 \oplus e'$, where $\mathbf{d}(a_1) = \mathbf{r}(a_1) = e$, and $b = b_1 \oplus f'$, where $\mathbf{d}(b_1) = \mathbf{r}(b_1) = f$. Then*

$$\left(a_1 \oplus (e \vee f) e'\right)\left(b_1 \oplus (e \vee f) f'\right) \oplus e' f' = ab.$$

Proposition 20 *Let G be a Boolean groupoid, the identity space of which is locally compact and put $S = \mathsf{KB}(G)$. Then the group of units of $\mathsf{KB}(G^{\infty})$ (as defined above) is isomorphic to the topological full group of G^{∞}.*

Proof We first of all establish a bijection between the elements of the topological full group and the elements of the group of units. We begin by describing the elements of the topological full group of G^{∞}. A compact-open bisection A of G^{∞} is a compact-open local bisection such that $A^{-1}A = AA^{-1} = G_o^{\infty}$. We know also that A can be written in the form $A = B \cup C$, a disjoint union (by Lemma 63), where B is a compact-open local bisection of G and C is of the form $C = G_o^{\infty} \setminus D$ where D is a compact-open subset of G_o and $B = DBD$. Because of the disjointness, we have that $B^{-1}B = BB^{-1} = E$ and $C = G_o^{\infty} \setminus E$. The above representation for A is, of course, not unique. Suppose that $A = B_1 \cup C_1$, where $B_1^{-1}B_1 = B_1 B_1^{-1} = E_1$ and $C_1 = G_o^{\infty} \setminus E_1$. The fact that $B \sim B_1$ in $\mathsf{C}(S)$ follows by Lemma 66. We may therefore map A to $\sigma(B)$. Now, suppose that B and B_1 are compact-open local bisections of G such that $B^{-1}B = BB^{-1} = E$, $B_1^{-1}B_1 = B_1 B_1^{-1} = E_1$ and $\sigma(B) = \sigma(B_1)$. Then by Lemma 67, we have that

$$A = B \cup \left(G_o^{\infty} \setminus E\right) = B_1 \cup \left(G_o^{\infty} \setminus E_1\right)$$

where A is a compact-open bisection of G^{∞} such that A maps to $\sigma(B) = \sigma(B_1)$. We have therefore established our bijection. The proof that we have a homomorphism and so an isomorphism follows by Lemma 68. □

10 Final Remarks

A *frame* is a complete infinitely distributive lattice [12, p. 39]. A *completely prime filter* in a frame L is a proper filter F such that $\bigvee_{i \in I} a_i \in F$ implies that $a_i \in F$ for some $i \in I$. If X is a topological space then its set of open subsets $\Omega(X)$ is a frame. If $x \in X$, denote by $\mathscr{G}_x$ the set of all open subsets of X that contains x. The set $\mathscr{G}_x$ is a completely prime filter of $\Omega(X)$. We refer to the completely prime filters in $\Omega(X)$ as *points* and denote the set of all points in X by $\mathrm{pt}(\Omega(X))$. We have therefore defined a map $X \to \mathrm{pt}(\Omega(X))$ given by $x \mapsto \mathscr{G}_x$. We say that the space X is *sober* if this map is a bijection. See [12, p. 43]. We say that a frame L is *spatial* if for any elements $a, b \in L$ such that $a \nleq b$ then there is a completely prime filter that contains a but omits b. See [12, p. 43].

We say that a topological space is *spectral* if it is sober and has a base of compact-open subsets with the additional property that the intersection of any two compact-open subsets is itself compact-open. This is different from the definition given in [12] since we do not assume that the space be compact. A *spectral groupoid* is an étale topological groupoid whose space of identities forms a spectral space. If S is a distributive inverse semigroup then $\mathsf{G}(S)$, the set of prime filters of S, is a spectral groupoid; if G is a spectral groupoid then $\mathsf{KB}(G)$, the set of compact-open local bisections of G, is a distributive inverse semigroup. The following is proved as [30, Theorem 3.17].

Theorem 8 (Non-commutative Stone duality for distributive inverse semigroups) *The category of distributive inverse semigroups and their callitic morphisms is dually equivalent to the category coherent continuous covering functors.*

By Lemma 19 and the fact that Hausdorff spaces are sober [12, part (ii) of Lemma II.1.6], a special case of the above theorem is Theorem 7 since the Hausdorff spectral spaces are precisely the locally compact Boolean spaces.

By a *pseudogroup*, we mean an inverse semigroup which has arbitrary compatible joins and multiplication distributes over such joins. Observe that pseudogroups are automatically meet-monoids. In addition, if S is a pseudogroup then $\mathsf{E}(S)$ is a frame. Pseudogroups were studied historically by Boris Schein [49] and more recently by Resende [48]. The connection between pseudogroups and distributive inverse semigroups is provided by the notion of 'coherence'. Let S be a pseudogroup. An element $a \in S$ is said to be *finite* if $a \leq \bigvee_{i \in I} x_i$, where $\{x_i : i \in I\}$ is any compatible subset of S, then there exists a finite subset $x_1, \ldots, x_n$ such that $a \leq \vee_{i=1}^{n} x_i$. Denote the set of finite elements of S by $\mathsf{K}(S)$. We say that the pseudogroup S is *coherent* if $\mathsf{K}(S)$ is a distributive inverse semigroup, and every element of S is a join of a compatible subset of $\mathsf{K}(S)$. It can be shown that every distributive inverse semigroup arises from some pseudogroup as its set of finite elements [30, Proposition 3.5].

Let G be an étale groupoid. Then the set of open local bisections of G, denoted by $\mathsf{B}(G)$, forms a pseudogroup [30, Proposition 2.1]. Let S be a pseudogroup. A *completely prime filter* $A \subseteq S$ is a proper filter in S with the property that $\bigvee_{i \in I} x_i \in A$ implies that $x_i \in A$ for some i. Denote the set of completely prime filters on S by

$\mathsf{G}_{CP}(S)$. Then $\mathsf{G}_{CP}(S)$ is an étale groupoid [30, Proposition 2.8]. A homomorphism $\theta \colon S \to T$ is said to be *hypercallitic*[7] if it preserves arbitrary joins, preserves binary meets, and has the property that for each $t \in T$ we may write $t = \bigvee_{i \in I} t_i$ where $t_i \leq \theta(s_i)$ for each $i \in I$. Denote by **Pseudo** the category of pseudogroups and hypercallitic maps. Denote by **Etale** the category of étale groupoids and continuous covering functors. If $\theta \colon S \to T$ is hypercallitic, define $\mathsf{G}_{CP}(\theta)$ to be θ^{-1}; by [30, Lemmas 2.14 and 2.16], it follows that $\mathsf{G}_{CP}(\theta)$ is a continuous covering functor from $\mathsf{G}_{CP}(T)$ to $\mathsf{G}_{CP}(S)$. On the other hand, if $\phi \colon G \to H$ is a continuous covering functor, then $\mathsf{B}(\phi) = \phi^{-1}$ is a hypercallitic map from $\mathsf{B}(H)$ to $\mathsf{B}(G)$ by [30, Lemma 2.19]. The following is immediate by [30, Corollary 2.18, Theorem 2.22].

Theorem 9 (The adjunction theorem) *The functor* $\mathsf{G}_{CP} \colon \mathbf{Pseudo}^{op} \to \mathbf{Etale}$ *is right adjoint to the functor* $\mathsf{B} \colon \mathbf{Etale} \to \mathbf{Pseudo}^{op}$.

The above theorem is a generalization of [12, Theorem II.1.4] from frames/locales to pseudogroups. It can also be used to prove Theorem 8, which is the approach adopted in [30].

Let S be a pseudogroup. If $s \in S$, denote by $\mathscr{X}_s$ the set of all completely prime filters in S containing the element s. The function $\varepsilon \colon S \to \mathsf{B}(\mathsf{G}_{CP}(S))$ given by $s \mapsto \mathscr{X}_s$ is a surjective callitic morphism [30, Proposition 2.9, part (1) of Proposition 2.12, Corollary 2.18, Lemma 2.21]. We say that S is *spatial* if ε is injective. Denote by $\mathbf{Pseudo}_{sp}$ the category of spatial pseudogroups and hyperelliptic morphisms.

Lemma 69 *The pseudogroup S is spatial if and only if the frame $\mathsf{E}(S)$ is spatial as a frame.*

Proof We denote by $\mathscr{X}_s$ the set of all completely prime filters in S that contain s. We denote by $\widetilde{\mathscr{X}}_e$ the set of all completely prime filters in $\mathsf{E}(S)$ that contain e. Assume first that S is spatial. Suppose that for idempotents e and f we have that $\widetilde{\mathscr{X}}_e = \widetilde{\mathscr{X}}_f$. We use the fact that every idempotent filter in S is determined by the idempotents it contains. It follows that $\mathscr{X}_e = \mathscr{X}_f$ and so, by assumption, $e = f$. We now assume that $\mathsf{E}(S)$ is spatial. Suppose that a and b are elements of S such that $\mathscr{X}_a = \mathscr{X}_b$. Using [30, part (2) and (3) of Lemma 2.6], it follows that every element of $\mathscr{X}_{a^{-1}b}$ contains an idempotent. It is similar for $\mathscr{X}_{ab^{-1}}$. It follows that $\mathscr{X}_{a^{-1}b} = \mathscr{X}_{a^{-1}b \wedge 1}$ and $\mathscr{X}_{ab^{-1}} = \mathscr{X}_{ab^{-1} \wedge 1}$ where $a^{-1}b \wedge 1$ and $ab^{-1} \wedge 1$ are idempotents. We have that $\mathscr{X}_{a^{-1}b} = \mathscr{X}_{a^{-1}a}$. We now use [30, part (2) of Lemma 2.2] and the fact that $\mathsf{E}(S)$ is spatial to deduce that $a^{-1}b \wedge 1 = a^{-1}a$. Similarly, $ab^{-1} \wedge 1 = bb^{-1}$. We have that $a^{-1}a \leq a^{-1}b$ from which we deduce that $a \leq b$. Similarly, $bb^{-1} \leq ab^{-1}$ from which we deduce that $b \leq a$. It follows that $a = b$, as required. □

Let G be an étale groupoid. For each $g \in G$, denote by $\mathscr{C}_g$ the set of all open local bisections of G containing the element g. The function $\eta \colon G \to \mathsf{G}_{CP}(\mathsf{B}(G))$ given by $g \mapsto \mathscr{C}_g$ is a continuous covering functor by [30, Proposition 2.11]. We say that G is *sober* if η is a homeomorphism. Denote by $\mathbf{Etale}_{so}$ the category of sober étale groupoids and continuous covering functors.

[7] The definition we have given here looks different from the one given in [30] but is equivalent by using [47].

Lemma 70 *Let G be an étale groupoid.*

1. *Each open set in a completely prime filter of open subsets F contains as a subset an open local bisection also in F.*
2. *Completely prime filters of open subsets are determined by the open local bisections they contain.*
3. *If F is a completely prime filter of open sets in G then*

$$\mathbf{d}(F) = \{\mathbf{d}(U) : U \in F\}$$

is a completely prime filter of open sets in G_o.

4. *If F is a completely prime filter of open sets and $U \in F$ is an open local bisection then $U\mathbf{d}(F)$ consists of open local bisections and $F = (U\mathbf{d}(F))^{\uparrow}$.*
5. *G is a T_0-space if and only if G_o is a T_0-space.*
6. *G is sober if and only if G_o is sober.*

Proof (1) We use the fact that in an étale topological groupoid, the open local bisections form a basis for the topology. Let F be a completely prime filter of open sets and let $U \in F$ be any element. Then $U = \bigcup_{i \in I} U_i$, where the U_i are open local bisections. But F is completely prime and so $U_i \in F$ for some $i \in I$. Thus $U_i \subseteq U$ is an open local bisection and also belongs to F.

(2) Denote by F' the set of all open local bisections in the completely prime filter of open sets F. Then F' is closed under finite intersections and $(F')^{\uparrow} = F$. It is now easy to check that if A and B are completely prime filters and $A = (A')^{\uparrow}$ and $B = (B')^{\uparrow}$ then $A = B$ if and only if $A' = B'$.

(3) Observe first that $\mathbf{d}(F) = \{\mathbf{d}(U) : U \in F\}$ is a set of open subsets of G_o because G is étale and so $\mathbf{d}$ is a local homeomorphism and so an open map. Let $\mathbf{d}(U) \subseteq X$ where X is an open subset of G_o. Then GX is an open subset of G since G is étale and $\mathbf{d}(GX) = X$. But $U = U\mathbf{d}(U) \subseteq GX$ and so $GX \in F$. It follows that $X \in \mathbf{d}(F)$. Let $\mathbf{d}(U), \mathbf{d}(V) \in \mathbf{d}(F)$. Then $\mathbf{d}(U \cap V) \subseteq \mathbf{d}(U) \cap \mathbf{d}(V)$. It follows that $\mathbf{d}(U) \cap \mathbf{d}(V) \in \mathbf{d}(F)$. Let $\bigcup_{i \in X} X_i \in \mathbf{d}(F)$. Then there exists $U \in F$ such that $\mathbf{d}(U) = \bigcup_{i \in X} X_i \in \mathbf{d}(F)$. But $U = \bigcup_{i \in I} UX_i$ and the result now follows.

(4) Let F be any completely prime filter of open subsets. By part (3), we have proved that $\mathbf{d}(F)$ is a completely prime filter of open subsets of G_o. By part (1), let $U \in F$ be any open local bisection. Observe that $U = UU^{-1}U$ and that, more generally, if $U, V, W \in F$ are open local bisections then $UV^{-1}W \in F$. In addition, when W is an open local bisection we have that $\mathbf{d}(W) = W^{-1}W$. We prove that if $V \in F$ then $U\mathbf{d}(V) \in F$. Since F is completely prime and G is étale, we can find an open local bisection W such that $W \subseteq V$. Thus $V\mathbf{d}(W) = VW^{-1}W \in F$. But $V\mathbf{d}(W) \subseteq U\mathbf{d}(V)$ and so $U\mathbf{d}(V) \in F$. We have shown that $U\mathbf{d}(A) \subseteq A'$. Now let $W \in F$ be any open local bisection. Then $U \cap W \in F$ is an open local bisection. But $U \cap W = U\mathbf{d}(U \cap W)$. Thus W contains as a subset an element of $U\mathbf{d}(A)$. But every element of F contains as a subset an open local bisection in F. It follows that $F = (U\mathbf{d}(F))^{\uparrow} \subseteq F$.

(5) If G is T_0, then it is immediate that G_o is T_0 because in an étale groupoid the space of identities forms an open subspace. Suppose now that G_o is T_0. We shall prove that G is T_0. Let $g, h \in G$ be distinct elements of G. There are two cases. First, suppose that $\mathbf{d}(g) \neq \mathbf{d}(h)$. Since G_o is T_0 we can, without loss of generality, assume that there is an open subset $X \subseteq G_o$ that contains $\mathbf{d}(g)$ but does not contain $\mathbf{d}(h)$. Put $Y = \mathbf{d}^{-1}(X)$. Then Y is an open set that contains g but does not contain h. Second, suppose that $\mathbf{d}(g) = \mathbf{d}(h)$. If the set of all open sets that contains g were the same as the set all open sets that contains h, *then there would be an open local bisection that contained both g and h.* This cannot happen because $\mathbf{d}(g) = \mathbf{d}(h)$ and g and h are distinct. It follows that there must be an open set that contains one of g or h but not the other.

(6) If G is sober, it is easy to check that G_o is sober. Suppose now that G_o is sober. We prove that G is sober. Let F be a completely prime filter of open subsets of G. Then by part (3), $\mathbf{d}(F)$ is a completely prime filter of open subsets of G_o. From the assumption that G_o is sober, there is a unique identity $e \in G_o$ such that $\mathbf{d}(F)$ is precisely the set of all open subsets of G_o that contain e. Choose any $U \in F$ an open local bisection by part (1). Then $e \in \mathbf{d}(U)$. But U is a local bisection and so there is a unique $g \in U$ such that $\mathbf{d}(g) = e$. Let V be any other open local bisection in F. Then there is a unique element $h \in V$ such that $\mathbf{d}(h) = e$. But $U \cap V$ is also an open local bisection in F. Thus, there is a unique element $k \in U \cap V$ such that $\mathbf{d}(k) = e$. But $k \in U$ and $k \in V$ thus $g = k = h$. It follows that all the open local bisections in F contain g and so all elements of F contain g by part (4). We have therefore proved that $F \subseteq F_g$. But $\mathbf{d}(F) = \mathbf{d}\left(F_g\right)$. Thus by part (4), we must have that $F = F_g$. We have proved that every completely prime filter of open sets in G is determined by an element of G. To complete the proof, suppose that $F_g = F_h$. Then $g = h$ because by part (5), the space G is T_0. □

The following was proved as [30, Proposition 2.12].

Lemma 71 *1. For each pseudogroup S, the étale groupoid $\mathsf{G}_{CP}(S)$ is sober.*
2. For each étale groupoid G, the pseudogroup $\mathsf{B}(G)$ is spatial.

From Theorem 9 and what we have said above, we obtain the following [30, Theorem 2.23] which is the basis of all of our duality theorems.

Theorem 10 (Duality theorem between spatial pseudogroups and sober étale groupoids) *The category $\mathbf{Pseudo}_{sp}^{op}$ is equivalent to the category $\mathbf{Etale}_{so}$.*

Our whole approach is predicated on the idea that suitable classes of inverse semigroups can be viewed as generalizations of suitable classes of lattices. The following table summarizes our approach:

Lattices	Non-commutative lattices
Meet semilattices	Inverse semigroups
Frames	Pseudogroups
Distributive lattices	Distributive inverse semigroups
Generalized Boolean algebras	Boolean inverse semigroups

The theory of coverages on inverse semigroups (as a way of constructing pseudogroups) is touched on in [30, Sect. 4]. Weaker axioms for a coverage (which more naturally generalize the meet-semilattice case) are discussed in [3]. Although we discuss Paterson's Universal groupoid in [30, Sect. 5.1] as well as Booleanizations, much better presentations of these results can be found in [28]. Tight completions of inverse semigroups, the subject of [30, Sect. 5.2], are discussed in greater generality in [33].

Acknowledgements Some of the work for this chapter was carried out at LaBRI, Université de Bordeaux, during April 2018 while visiting David Janin. I am also grateful to Phil Scott for alerting me to typos. None of this work would have been possible without my collaboration with Daniel Lenz, and some very timely conversations with Pedro Resende. This chapter is dedicated to the memory of Iain Currie, colleague and friend.

References

1. Burris, S.: The laws of Boole's thought. http://www.math.uwaterloo.ca/~snburris/htdocs/MYWORKS/PREPRINTS/aboole.pdf
2. Burris, S., Sankappanavar, H.P.: A course in universal algebra, The Millennium edn. Freely available from http://math.hawaii.edu/~ralph/Classes/619/
3. de Castro, G.G.: Coverages on inverse semigroups. Semigroup Forum **102**, 375–396 (2021)
4. Doctor, H.P.: The categories of Boolean lattices, Boolean rings and Boolean spaces. Canad. Math. Bull. **7**, 245–252 (1964)
5. Ehresmann, Ch.: Oeuvres complètes et commentées, In: A.C. Ehresmann (ed.) Supplements to Cah. Topologie Géométrie Différentielle Catégoriques, Amiens, 1980–1983
6. Exel. R.: Inverse semigroups and combinatorial C^*-algebras. Bull. Braz. Maths. Soc. (N.S.) **39**, 191–313 (2008)
7. Foster, A.L.: The idempotent elements of a commutative ring form a Boolean algebra; ring duality and transformation theory. Duke Math. J. **12**, 143–152 (1945)
8. Gehrke, M., Grigorieff, S., Pin, J.-E.: Duality and equational theory of regular languages. Lecture Notes in Computer Science 5126, pp. 246–257. Springer (2008)
9. Givant, S., Halmos, P.: Introduction to Boolean Algebras. Springer (2009)
10. Higgins, Ph.J.: Categories and Groupoids. Van Nostrand Reinhold Company, London (1971)
11. Hailperin, T.: Boole's algebra isn't Boolean algebra. Math. Mag. **54**, 172–184 (1981)
12. Johnstone, P.T.: Stone spaces. CUP (1986)
13. Kellendonk, J.: The local structure of tilings and their integer groups of coinvariants. Commun. Math. Phys. **187**, 115–157 (1997)
14. Kellendonk, J.: Topological equivalence of tilings, J. Math. Phys. **38**, 1823–1842 (1997)
15. Koppelberg, S.: Handbook of Boolean Algebra, vol. 1. North-Holland (1989)
16. Kudryavtseva, G., Lawson, M.V., Lenz, D.H., Resende, P.: Invariant means on Boolean inverse monoids. Semigroup Forum **92**, 77–101 (2016)

17. Kudryavtseva, G., Lawson, M.V.: Perspectives on non-commutative frame theory. Adv. Math. **311**, 378–468 (2017)
18. Kumjian. A.: On localizations and simple C^*-algebras. Pacific J. Math. **112**, 141–192 (1984)
19. Lawson, M.V.: Coverings and embeddings of inverse semigroups. Proc. Edinb. Math. Soc. **36**, 399–419 (1996)
20. Lawson, M.V.: Inverse Semigroups: The Theory of Partial Symmetries. World Scientific (1998)
21. Lawson, M.V.: Finite Automata. Chapman and Hall/CRC (2003)
22. Lawson, M.V.: The polycyclic monoids P_n and the Thompson groups $V_{n,1}$. Comm. Algebra **35**, 4068–4087 (2007)
23. Lawson, M.V.: A class of subgroups of Thompson's group V. Semigroup Forum **75**, 241–252 (2007)
24. Lawson, M.V.: A non-commutative generalization of Stone duality. J. Aust. Math. Soc. **88**, 385–404 (2010)
25. Lawson, M.V.: Non-commutative Stone duality: inverse semigroups, topological groupoids and C^*-algebras. Int. J. Algebra Comput. **22**, 1250058 (2012). https://doi.org/10.1142/S0218196712500580
26. Lawson, M.V.: Subgroups of the group of homeomorphisms of the Cantor space and a duality between a class of inverse monoids and a class of Hausdorff étale groupoids. J. Algebra **462**, 77–114 (2016)
27. Lawson, M.V.: Tarski monoids: Matui's spatial realization theorem. Semigroup Forum **95**, 379–404 (2017)
28. Lawson, M.V.: The Booleanization of an inverse semigroup. Semigroup Forum **100**, 283–314 (2020)
29. Lawson, M.V.: Finite and semisimple Boolean inverse monoids. arXiv:2102.12931
30. Lawson, M.V., Lenz, D.H.: Pseudogroups and their étale groupoids. Adv. Math. **244**, 117–170 (2013)
31. Lawson, M.V., Margolis, S.W., Steinberg, B.: The étale groupoid of an inverse semigroup as a groupoid of filters. J. Aust. Math. Soc. **94**, 234–256 (2014)
32. Lawson, M.V., Scott, P.: AF inverse monoids and the structure of countable MV-algebras. J. Pure Appl. Algebra **221**, 45–74 (2017)
33. Lawson, M.V., Vdovina, A.: The universal Boolean inverse semigroup presented by the abstact Cuntz-Krieger relations. J. Noncommut. Geom. **15**, 279–304 (2021)
34. Leech, J.: Inverse monoids with a natural semilattice ordering. Proc. London Math. Soc. (3) **70**, 146–182 (1995)
35. Lenz, D.H.: On an order-based construction of a topological groupoid from an inverse semigroup. Proc. Edinb. Math. Soc. **51**, 387–406 (2008)
36. Malandro, M.E.: Fast Fourier transforms for finite inverse semigroups. J. Algebra **324**, 282–312 (2010)
37. Matsnev, D., Resende, P.: Etale groupoids as germ groupoids and their base extensions. Proc. Edinb. math. Soc. **53**, 765–785 (2010)
38. Matui, H.: Homology and topological full groups of étale groupoids on totally disconnected spaces. Proc. London Math. Soc. (3) **104**, 27–56 (2012)
39. Matui, H.: Topological full groups of one-sided shifts of finite type. J. Reine Angew. Math. **705**, 35–84 (2015)
40. Mac Lane, S.: Categories for the Working Mathematician, 2nd edn. Springer (1998)
41. Nyland, P., Ortega, E.: Topological full groups of ample groupoids with applications to graph algebras. Int. J. Math. **30**, 1950018 (2019)
42. Paterson, A.L.T.: Groupoids, Inverse Semigroups, and Their Operator Algebras. Progress in Mathematics, vol. 170. Birkhäuser, Boston (1998)
43. Pin, J.-E.: Dual space of a lattice as the completion of a Pervin space. In: P. Höfner et al. (ed.) RAMiCS 2017. LNCS 10226, pp. 24–40 (2017)
44. Pippenger, N.: Regular languages and stone duality. Theory Comput. Syst. **30**, 121–134 (1997)
45. Renault, J.: A Groupoid Approach to C^*-Algebras. Lecture Notes in Mathematics, vol. 793. Springer (1980)

46. Resende, P.: Lectures on Étale groupoids, inverse semigroups and quantales. Lecture Notes for the GAMAP IP Meeting, Antwerp, 4–18 September, 2006, 115pp. https://www.math.tecnico.ulisboa.pt/~pmr/poci55958/gncg51gamap-version2.pdf
47. Resende, P.: A note on infinitely distributive inverse semigroups. Semigroup Forum **73**, 156–158 (2006)
48. Resende, P.: Etale groupoids and their quantales. Adv. Math. **208**, 147–209 (2007)
49. Schein, B.: Completions, translational hulls, and ideal extensions of inverse semigroups. Czechoslovak Math. J. **23**, 575–610 (1973)
50. Sikorski, R.: Boolean Algebras, 3rd edn. Springer (1969)
51. Solomon, L.: Representations of rook monoids. J. Algebra **256**, 309–342 (2002)
52. Simmons, G.F.: Introduction to Topology and Analysis. McGraw-Hill Kogakusha, Ltd, (1963)
53. Sims, A.: Hausdorff étale groupoids and their C^*-algebra. arXiv:1710.10897
54. Stone, M.H.: The theory of representations for Boolean algebras. Trans. Amer. Math. Soc. **40**, 37–111 (1936)
55. Stone, M.H.: Applications of the theory of Boolean rings to general topology. Trans. Amer. Math. Soc. **41**, 375–481 (1937)
56. Stone, M.H.: Topological representations of distributive lattices and Brouwerian logics. Časopis Pro Pestovani Matematiky a Fysiky **67**, 1–25 (1937)
57. Vickers, S.: Topology via Logic. CUP (1989)
58. Wehrung. F.: Refinement Monoids, Equidecomposability Types, and Boolean Inverse Monoids. Lecture Notes in Mathematics, vol. 2188. Springer (2017)
59. Willard, S.: General Topology. Dover (2004)

Group Lattices Over Division Rings

Alanka Thomas and P. G. Romeo

A tribute to the renowned semigroup theorist K.S.S.Nambooripad.

Abstract K.S.S. Nambooripad introduced group lattice as a lattice together with a group acting on it (Nambooripad in Monash conference on semigroup theory in Honor of G. B. Preston. World scientific Publishing Co. Pte. Ltd., Singapore (1991), [1]), (Nambooripad in Group Lattices (1992), [2]). He studied group actions on the lattice of submodules of a module over a division ring, which he called the group lattices over division rings, and explored the correspondence between the theory of group lattices, group representation theory, and module theory. Nambooripad established a one-to-one correspondence between group lattices over division rings and semilinear projective representations of a group over a division ring. Also, there is a one-to-one correspondence between group lattices over division rings and modules over twisted group rings. In this article, we review the mathematical work of Nambooripad on group lattices and give supporting examples to illustrate the theory.

Keywords Group · Lattice · Group lattice · Schreier extension

---

A. Thomas (✉) · P. G. Romeo
Department of Mathematics, Cochin University of Science and Technology, 682022 Kochi, Kerala, India
e-mail: alankathomas1@gmail.com

P. G. Romeo
e-mail: romeo_parackal@yahoo.com

A. A. Ambily and V. B. Kiran Kumar (eds.), *Semigroups, Algebras and Operator Theory*, Springer Proceedings in Mathematics & Statistics 436,
https://doi.org/10.1007/978-981-99-6349-2_3

1 Introduction

Prof. K.S.S. Nambooripad (1935–2020) is one of the most celebrated Indian mathematicians who has significantly contributed to the structure theory of regular semigroups. Nambooripad introduced Group lattice as a lattice with a group acting on it. He was interested in G-lattices over the arguesian geomodular lattice, the lattice of submodules of module V over a division ring K with $rank(V) \geq 3$. If G acts on an arguesian geomodular lattice coordinatized by a module V over K, it is called a $G-$lattice over the division ring K. Nambooripad established a one-one correspondence between $G-$lattices over K and representations of G over K with the rank at least 3. Nambooripad used the Schreier extension of K by G while studying G-lattices over K and observed correspondence between Schreier extensions of K by G and $G-$lattices over K via twisted group rings. The results regarding this appeared in the proceedings of the Monash conference on semigroup theory organized in honor of G.B. Preston in 1990 [1].

This paper is motivated by Nambooripad's results on group lattices [1, 2]. We dedicate this to the memory of K.S.S. Nambooripad, the thesis adviser of the second author.

Apart from introduction, we have three sections in this paper. The second section gives the definition and examples of $G-$lattice. The third section is devoted to the discussion of G-lattices over K. For the fourth section, one should have the notion of the Schreier extension. If G and N are two groups, the Schreier extension of N by G is a group H, having N as a normal subgroup and $H/N \cong G$ [3]. The fourth section studies Schreier extensions, twisted group rings, and their correspondence with group lattices over a division ring.

2 Group Lattices

An action of a group on an algebraic structure is crucial as it gives a representation of the group by automorphisms on the structure. These representations are helpful as they transform a group into a known one. The same thing happens for group lattices also. This section defines group lattices and provides enough examples to support the theory.

Definition 1 Let G be a group and $(L, \leq, \wedge, \vee)$ be a lattice. An action of G on L is a map from $G \times L$ to L, which assigns $gm \in L$ for each $(g, m) \in G \times L$ and satisfies the following, for each $g, h \in G$ and $m, n \in L$

1. $g(hm) = (gh)m$
2. $em = m$ where e is the identity in G
3. $m \leq n$ if and only if $gm \leq gn$

If a group G acts on L, L is called a $G - lattice$. A $group\, lattice$ is a G-lattice for some group G.

Proposition 1 *Let G be a group and $(L, \leq, \wedge, \vee)$ be a G-lattice. Then*

1. $g(m \wedge n) = gm \wedge gn$
2. $g(m \vee n) = gm \vee gn$

for each $g \in G$ and $m, n \in L$.

Proof For each $g \in G$ and $m, n \in L$, we have $g(m \wedge n) \leq gm \wedge gn$. So,

$$m \wedge n = g^{-1}(g(m \wedge n)) \leq g^{-1}(gm \wedge gn) \leq g^{-1}gm \wedge g^{-1}gn = m \wedge n$$

So we can replace every inequality with equality, and so, $g(m \wedge n) = gm \wedge gn$. The proof for the second part is similar. □

Example

Let G be a group and $L(G)$ be the lattice of all subgroups of G. Then $L(G)$ is a $G - lattice$ under conjugation, $gH = gHg^{-1} = \{ghg^{-1} \mid h \in H\}$ for all $g \in G$ and $H \leqslant G$.

A $G - sublattice$ L' of a G-lattice L is a sublattice of L which is itself a G-lattice under the same product as that of L. For a $G-$sublattice L', $gm \in L'$ for all $g \in G$ and $m \in L'$ and the properties (1)–(3) of Definition 1 inherit from L. In the above example, the sublattice $LN(G)$ of all normal subgroups of G is a G-sublattice of $L(G)$.

Example

If X is a $G - set$, the power set $\mathscr{P}(X)$ is a $G-$lattice with respect to the product $gA = \{ga \mid a \in A\}$ for all $g \in G$ and $A \subseteq X$.

Before moving to more examples, recall some preliminary results required.

Definition 2 Let V be a module over a division ring K. A semilinear transformation on V is the pair (f, θ_f), where f is a map on V and θ_f is an automorphism on K such that,

1. $f(u + v) = f(u) + f(v)$
2. $f(\alpha v) = \theta_f(\alpha) f(v)$

for all $u, v \in V$ and $\alpha \in K$.

A bijective semilinear transformation is called a semilinear automorphism. The set of all semilinear automorphisms on V form a group under composition, denoted by $SGL(V)$. If θ_f is the identity map on K, the semilinear transformation is a linear transformation on V. Linear automorphisms are bijective linear transformations, and the set of all linear automorphisms $GL(V)$ is a subgroup of $SGL(V)$.

Definition 3 A semilinear projective representation of a group G over a division ring K is a map $\rho : G \to SGL(V)$ where V is a module over K such that, for each $g, h \in G$ there exist $\alpha(g, h) \in K^* = K\backslash\{0\}$ such that $\rho(g)\rho(h) = \alpha(g, h)\rho(gh)$.

The map ρ in Definition 3 is not a group homomorphism in general, that is $\rho(g)\rho(h)$ is not equal to $\rho(gh)$, but a scalar multiple of $\rho(gh)$. If $\alpha(g, h) = 1$ for all $g, h \in G$, ρ is a group homomorphism and the representation is called a semilinear representation of G over K.

A projective linear representation of G over K is a semilinear projective representation for which, each $\rho(g)$ is a linear automorphism. Here the codomain of ρ is in $GL(V)$. A linear representation of a group G over K is a map $\rho : G \to GL(V)$ such that $\rho(g)\rho(h) = \rho(gh)$ for all $g, h \in G$. A linear representation is a projective and semilinear representation at the same time.

Two semilinear projective representations ρ and $\tilde{\rho}$ of G over K are equivalent, if there exists a map $\eta : G \to K^*$ such that $\tilde{\rho}(g) = \eta(g)\rho(g)$ for all $g \in G$.

Example

Let V be module over a division ring K and $\rho : G \to SGL(V)$ be a semilinear projective representation of G over K. Then, the lattice of submodules $L(V)$ of V, is a $G-$lattice with respect to the product defined as, $gW = \rho(g)W = \{\rho(g)w \mid w \in W\}$ for all $g \in G$ and $W \in L(V)$.

The G-lattice $L(V)$ defined above is called a semilinear projective G-lattice. Similarly we have linear, projective, and semilinear G-lattices on $L(V)$, according to the choice of representations.

It is well known that every group action provides a representation of the group by automorphisms of the respective structure and vice versa. We can observe the same for group lattices also. For that, consider a representation ρ of G by automorphisms on L, which is a homomorphism from G to the group of all lattice automorphisms on L, $AutL$. Define the product $gm = \rho(g)m$ for all $g \in G$ and $m \in L$. Since ρ is a group homomorphism, the product satisfies properties (1) and (2) in Definition 1 and the remaining follows as each $\rho(g)$ is a lattice automorphism. Hence L is a $G-$lattice with respect to this action.

For the converse part, consider a $G-$lattice L. For each $g \in G$ define $\rho(g) : L \to L$ such that $\rho(g)m = gm$. Clearly $\rho(g)$ is a lattice homomorphism. $\rho(g)m = \rho(g)n$ implies $gm = gn$ and so $m = (g^{-1}g)m = g^{-1}(gm) = g^{-1}(gn) = (g^{-1}g)n = n$. So $\rho(g)$ is injective. $\rho(g)$ is surjective, since $g^{-1}m$ is the preimage of m under $\rho(g)$. Hence $\rho(g)$ is a lattice automorphism for each $g \in G$. From property (1) of Definition 1, $\rho(gh) = \rho(g)\rho(h)$ and so the map $\rho : G \to AutL$ which takes each $g \in G$ to $\rho(g)$ is a group homomorphism. That is, ρ is a representation of G by automorphisms on L. Hence there is a one-one correspondence between the $G-$lattices and the representations of G by lattice automorphisms. Thus we can summarize this observation as the following theorem.

Theorem 1 *Let G be a group and L be a lattice. If $\rho : G \to Aut L$ is a representation of G by automorphisms on L, then L is a G-lattice with respect to the product $gm = \rho(g)m$ for each $g \in G$ and $m \in L$. Conversely if L is a G-lattice, the map $\rho : G \to Aut L$ such that $\rho(g)m = gm$ is a representation of G by automorphisms on L.*

3 G-Lattices Over a Division Ring

By a G-lattice over a division ring K, we mean a G-lattice $L(V)$ of submodules of a K-module V. Let us first recall the terms and notations used in the sequel. A lattice L is said to be coordinatized by a module V over K if $L \cong L(V)$, the lattice of all submodules of V over K. If $rank(V) \geq 3$, $L(V)$ is called an arguesian geomodular lattice. A lattice morphism $f : L(V) \to L(V)$ is coordinatizable if there exists a semilinear transformation $(f', \theta_{f'})$ on V such that $fW = f'W = \{f'w \mid w \in W\}$. We assume, all lattice morphisms between arguesian geomodular lattices considered here are coordinatizable. If a group G acts on an arguesian geomodular lattice, it is called a G-lattice over K, more generally a group lattice over the division ring K.

The following theorem establishes a one-one correspondence between G-lattices over K and representations of G over K.

Theorem 2 *Let G be a group and K be a division ring. Every representation of G over K determines a G-lattice over K. Conversely, let L be a G-lattice over K, then there is a representation ρ of G over K that determines L.*

Proof Let $\rho : G \to SGL(V)$ be a representation of G over K. Then $L(V)$ is a $G-$lattice with respect to the product $gW = \rho(g)W$ for all $g \in G$ and $W \in L(V)$.

Conversely, L be a G-lattice over K. Then $L \cong L(V)$ where V is a module over K having $rank(V) \geq 3$. By Theorem 1 there exists a representation $\rho : G \to Aut L$ that determines the $G-$lattice L. $(\tilde{\rho}(g), \theta_{\rho(g)})$ be semilinear automorphisms coordinatizing $\rho(g)$ for each $g \in G$. Hence the action of G on L can be rewritten as, $gW = \rho(g)W = \tilde{\rho}(g)W$ for all $g \in G$ and $W \in L$. Let $\langle v \rangle$ be the submodule generated by $v \in V$, then

$$\tilde{\rho}(g)\tilde{\rho}(h)\langle v \rangle = \rho(g)\rho(h)\langle v \rangle = \rho(gh)\langle v \rangle = \tilde{\rho}(gh)\langle v \rangle \tag{1}$$

and so $\langle \tilde{\rho}(g)\tilde{\rho}(h)v \rangle = \langle \tilde{\rho}(gh)v \rangle$. Then there exists an $\alpha_v \in K^*$ such that $\tilde{\rho}(g)\tilde{\rho}(h)v = \alpha_v \tilde{\rho}(gh)v$. Let $w \in V$ be linearly independant with v. Since $\tilde{\rho}(gh)$ is an automorphism, $\tilde{\rho}(gh)v$ and $\tilde{\rho}(gh)w$ are also linearly independent. Consider

$$\begin{aligned} \alpha_{v+w}\tilde{\rho}(gh)v + \alpha_{v+w}\tilde{\rho}(gh)w &= \alpha_{v+w}\tilde{\rho}(gh)(v + w) = \tilde{\rho}(g)\tilde{\rho}(h)(v + w) \\ &= \tilde{\rho}(g)\tilde{\rho}(h)v + \tilde{\rho}(g)\tilde{\rho}(h)w \\ &= \alpha_v\tilde{\rho}(gh)v + \alpha_w\tilde{\rho}(gh)w \end{aligned} \tag{2}$$

then $\alpha_v = \alpha_{v+w} = \alpha_w$. Similarly for any scalar β and w linearly independant with βv we get $\alpha_v = \alpha_{\beta v} = \alpha_w$. So the scalar $\alpha(g, h) = \alpha_v$ is independent of the choice of v. Hence $\tilde{\rho}(g)\tilde{\rho}(h) = \alpha(g, h)\tilde{\rho}(gh)$ and so $\tilde{\rho} : G \to SGL(V)$ is a semilinear projective representation of G over K that determines the G lattice L. □

Theorem 3 *Semilinear projective representations ρ and $\tilde{\rho}$ of G over K determine the same $G-$lattice if and only if they are equivalent.*

Proof If ρ and $\tilde{\rho}$ are equivalent, there exist $\eta : G \to K^*$ such that $\tilde{\rho}(g) = \eta(g)\rho(g)$. Then $\tilde{\rho}(g)W = \eta(g)\rho(g)W = \rho(g)W$ for all $g \in G$ and $W \in L(V)$ and so ρ and $\tilde{\rho}$ induce the same $G-$lattice.

Conversely if ρ and $\tilde{\rho}$ induce the same $G-$lattice, then for any $v \in V$, $\langle \rho(g)v \rangle = \rho(g)\langle v \rangle = \tilde{\rho}(g)\langle v \rangle = \langle \tilde{\rho}(g)v \rangle$. Similarly as in the proof of Theorem 2, there exist $\alpha_g \in K^*$ independent of v such that $\tilde{\rho}(g)v = \alpha_g \rho(g)v$ for all $v \in V$. Define $\eta : G \to K^*$ such that $\eta(g) = \alpha_g$. Then $\tilde{\rho}(g) = \eta(g)\rho(g)$ and ρ and $\tilde{\rho}$ are equivalent. □

4 Schreier Extension and Associated Group Lattices

Let N and G be groups. A Schreier extension of N by G is a group H having N as a normal subgroup and $H/N \cong G$ [3]. Let H be a group having N as a normal subgroup and $H/N \cong G$. Then $H = \bigcup_{g \in G} N\overline{g} = \{a\overline{g} \mid a \in N \text{ and } g \in G\}$, where $\{\overline{g} \mid g \in G\}$ is the set of all representatives from each coset of N. For each $g \in G$ the map $b \to \overline{g}b\overline{g}^{-1}$ is a group automorphism in N and let it be denoted by $\chi(g)$. Also since $\overline{gh}$ is the representative of the coset containing $\overline{g}\overline{h}$ there exist $[g, h] \in N$ such that $\overline{g}\overline{h} = [g, h]\overline{gh}$. Then $(a\overline{g})(b\overline{h}) = a\overline{g}b\overline{g}^{-1}\overline{g}\overline{h} = a\chi(g)(b)[g, h]\overline{gh}$ and H is a group with respect to this binary operation. Hence we have functions,

1. $\chi : G \to AutN$ defined by $\chi(g)(a) = \overline{g}a\overline{g}^{-1}$
2. $[\_,\_] : G \times G \to N$ such that $\overline{g}\overline{h} = [g, h]\overline{gh}$.

and we can see that χ and $[\_,\_]$ has the following properties

$(E1)$ $\chi(g)\chi(h) = [g, h]\chi(gh)[g, h]^{-1}$
$(E2)$ $[g, h][gh, k] = \chi(g)([h, k])[g, hk]$
$(E3)$ $[1, 1] = 1$.

In general, if $\chi : G \to AutN$ and $[\_,\_] : G \times G \to N$ are function satisfying above properties, $H = \{a\overline{g} \mid a \in N \text{ and } g \in G\}$ is a group with respect to the binary operation $(a\overline{g})(b\overline{h}) = a\chi(g)(b)[g, h]\overline{gh}$. This is the construction of Schreier extension of N by G. The pair of functions $(\chi, [\_,\_])$ called a factor system for (N, G) and the extension is denoted by $H(\chi, [\_,\_])$. Factor systems $(\chi, [\_,\_])$ and $(\chi', [\_,\_]')$ are equivalent, if there exists a map $\mu : G \to N$ such that

$(E4)$ $\chi'(g) = \mu(g)^{-1}\chi(g)\mu(g)$
$(E5)$ $[g, h]\mu(gh) = \mu(g)\chi'(g)(\mu(h))[g, h]'$
$(E6)$ $\mu(1) = 1$.

If $(\chi, [\_,\_])$ and $(\chi', [\_,\_]')$ are equivalent, the map $a\overline{g} \to a\mu(g)\tilde{g}$ from $H(\chi, [\_,\_])$ to $H(\chi', [\_,\_]')$ is an isomorphism. Two-factor systems are equivalent if and only if corresponding extensions are isomorphic.

A central extension of N by G is an extension $H(\chi, [\_,\_])$ for which, $[g, h]$ are in the center of N, for each g, h in G. Using $(E1)$, it is seen that the map $\chi : G \to Aut N$ is a homomorphism for a central extension. If N is an abelian group, every Schreier extension of N is a central extension.

If $\chi(g) = I_N$ for all $g \in G$, the binary operation in H reduces to $(a\overline{g})(b\overline{h}) = ab[g, h]\overline{gh}$ and H is called a projective extension of N by G. If $[g, h] = 1$ for all $g, h \in G$, then H is the semidirect product of N and G, this type of extensions are called split extensions or semidirect products. It is easy to see that projective extensions and semidirect products are central extensions. The direct product of N and G is the Schreier extension of N by G which is both projective and split.

Here we deal with Schreier extension of K^* by G, where K^* is the group of nonzero elements of a division ring under multiplication. We shall say that $H(\chi, [\_,\_])$ is an extension of K by G and that $(\chi, [\_,\_])$ is a factor system for (K, G), if it is a factor system for (K^*, G) with the property that $\chi(g) \in Aut K$ for each $g \in G$. For a semilinear projective representation ρ, each $\rho(g)$ is a semilinear transformation. So, there exists automorphism $\theta_{\rho(g)}$ on K such that $\rho(g)(\alpha v) = \theta_{\rho(g)}(\alpha)\rho(g)(v)$. For brevity, let it be denoted by $\chi(g)$. Since the representation is projective, for each $g, h \in G$ there exist $[g, h] \in K^*$ such that $\rho(g)\rho(h) = [g, h]\rho(gh)$. We will prove in Proposition 2 that this χ and $[\_,\_]$ together form a factor system for an extension of K by G.

Proposition 2 *Let ρ be a semilinear projective representation of G over K. Then $(\chi, [\_,\_])$ is a factor system for (K, G).*

Proof Given ρ is a semilinear projective representation of G over K. For each $g \in G$ there exists an automorphism $\chi(g)$ of K such that $\rho(g)(\alpha v) = \chi(g)(\alpha)\rho(g)(v)$ and $\rho(g)\rho(h) = [g, h]\rho(gh)$ for all $g, h \in G$. $(\chi, [\_,\_])$ is a factor system if it satisfies properties $(E1) - (E3)$.

$$\begin{aligned}\chi(g)(\chi(h)(\alpha))\rho(g)\rho(h)(v) &= \rho(g)\rho(h)(\alpha v) = [g, h]\chi(gh)(\alpha)\rho(gh)(v)\\ &= [g, h]\chi(gh)(\alpha)[g, h]^{-1}[g, h]\rho(gh)(v)\\ &= [g, h]\chi(gh)(\alpha)[g, h]^{-1}\rho(g)\rho(h)(v)\end{aligned} \tag{3}$$

Hence $\chi(g)\chi(h) = [g, h]\chi(gh)[g, h]^{-1}$ for all $g, h \in G$.

$$\begin{aligned}[g, h][gh, k]\rho(ghk)(v) &= [g, h]\rho(gh)(\rho(k)(v) = \big(\rho(g)\rho(h)\big)\rho(k)(v)\\ &= \rho(g)\big(\rho(h)\rho(k)\big)(v) = \rho(g)\big([h, k]\rho(hk)\big)(v)\\ &= \chi(g)([h, k])\rho(g)\rho(hk)(v)\\ &= \chi(g)([h, k])[g, hk]\rho(ghk)(v).\end{aligned} \tag{4}$$

Hence $[g, h][gh, k] = \chi(g)([h, k])[g, hk]$. It is clear that $[1, 1] = 1$. So $(\chi, [\_,\_])$ is a factor system for (K, G). □

A semilinear projective representation ρ of G over K is said to be associated with a factor system $(\chi, [\_,\_])$, if $\chi(g) = \theta_{\rho(g)}$ and $\rho(g)\rho(h) = [g, h]\rho(gh)$ for each $g, h \in G$. A representation is projective if and only if the Schreier extension associated with it is a projective extension. Similarly, a representation is semilinear if and only if the Schreier extension associated with it is a semidirect product. So, direct product $K \times G$ is the Schreier extension of K by G associated with a linear representation of G over K. Equivalent representations always give equivalent factor systems. Proposition 3 deals with this observation.

Proposition 3 *Let* $(\chi, [\_,\_])$ *and* $(\chi', [\_,\_]')$ *be factor systems for* (K, G) *and* ρ *be the representation of* G *over* K *associated with* $(\chi, [\_,\_])$. *Then* $(\chi, [\_,\_])$ *is equivalent to* $(\chi', [\_,\_]')$ *if and only if there is a representation of* $\tilde{\rho}$ *associated with* $(\chi', [\_,\_]')$ *which is equivalent to* ρ.

Proof Let ρ and $\tilde{\rho}$ be representation of G over K associated with factor systems $(\chi, [\_,\_])$ and $(\chi', [\_,\_]')$ for (K, G) respectively. If ρ is equivalent to $\tilde{\rho}$, there exists a map $\mu : G \to K^*$ such that $\rho(g) = \mu(g)\tilde{\rho}(g)$. Then μ is an equivalence of the factor system $(\chi, [\_,\_])$ with $(\chi', [\_,\_]')$; for, consider,

$$\mu(g)\chi'(g)(\alpha)\tilde{\rho}(g)(v) = \mu(g)\tilde{\rho}(g)(\alpha v) = \rho(g)(\alpha v) = \chi(g)(\alpha)\mu(g)\tilde{\rho}(g)(v). \quad (5)$$

So $\mu(g)\chi'(g) = \chi(g)\mu(g)$ for all $g \in G$ and this proves $(E4)$. Now consider

$$\begin{aligned} [g, h]\mu(gh)\tilde{\rho}(gh)(v) &= \rho(g)\rho(h)(v) = \mu(g)\tilde{\rho}(g)\mu(h)\tilde{\rho}(h)(v) \\ &= \mu(g)\chi'(g)(\mu(h))[g, h]'\tilde{\rho}(gh)(v) \end{aligned} \quad (6)$$

Hence $[g, h]\mu(gh) = \mu(g)\chi'(g)(\mu(h))[g, h]'$ and this proves $(E5)$. Thus μ satisfies $(E4) - (E6)$ and so μ is an equivalence between the factor systems $(\chi, [\_,\_])$ and $(\chi', [\_,\_]')$.

Conversely let $(\chi, [\_,\_])$ and $(\chi', [\_,\_]')$ be equivalent and ρ be a representation of G over K associated with $(\chi, [\_,\_])$. There exists an equivalence $\mu : G \to K^*$ from $(\chi', [\_,\_]')$ and $(\chi, [\_,\_])$ satisfying $(E4) - (E6)$. For each $g \in G$ define $\tilde{\rho}(g) = \mu(g)\rho(g)$, then $\tilde{\rho}$ is a representation of G over K. Since each $g \in G$, $\tilde{\rho}(g)$ preserves addition as in the case of $\rho(g)$. Now using $(E4)$ we get

$$\tilde{\rho}(g)(\alpha v) = \mu(g)\chi(g)(\alpha)\rho(g)(v) = \chi'(g)(\alpha)\mu(g)\rho(g)(v) = \chi'(g)(\alpha)\tilde{\rho}(g)(v) \quad (7)$$

and so $\tilde{\rho}(g)$ is a semilinear transformation for each $g \in G$. Using the bijectivity of $\rho(g)$, $\tilde{\rho}(g)$ is also bijective and so $\tilde{\rho}(g)$ is a semilinear automorphism for each $g \in G$.

$$\begin{aligned} \tilde{\rho}(g)\tilde{\rho}(h)(v) &= \mu(g)\rho(g)\mu(h)\rho(h)(v) = \mu(g)\chi(g)(\mu(h))[g, h]\rho(gh)(v) \\ &= [g, h]'\mu(gh)\rho(gh)(v) = [g, h]'\tilde{\rho}(gh)(v) \end{aligned} \quad (8)$$

Hence $\tilde{\rho}(g)\tilde{\rho}(h) = [g, h]'\tilde{\rho}(gh)$ and $\tilde{\rho}$ is a representation of G over K. It is clear that $\tilde{\rho}$ is associated with the factor system $(\chi', [\_,\_])$. Since $\mu : G \to K^*$ is a map such that $\tilde{\rho}(g) = \mu(g)\rho(g)$ for all $g \in G$, ρ and $\tilde{\rho}$ are equivalent representations of G over K. Hence the proof. □

Theorem 2 and Proposition 2 together tells us that every G-lattice over K gives us a Schreier extension of K by G. The converse is also true. We need to have the following definition of a twisted group ring to prove that.

Definition 4 Let $H = H(\chi, [\_,\_])$ be a Schreier extension of a division ring K by a group G. The twisted group ring $K(G; H)$ is the free module over K, generated by the set $\{\bar{g} \mid g \in G\}$ of all coset representatives of K^* in H.

For $u = \sum_{g\in G} a_g\bar{g}$ and $v = \sum_{h\in G} b_h\bar{h}$ in $K(G; H)$ and $c \in K$ we have,

$$u + v = \sum_{g\in G}(a_g + b_g)\bar{g} \qquad cu = \sum_{g\in G} ca_g\bar{g} \tag{9}$$

$$uv = \sum_{k\in G} c_k\bar{k} \text{ where } c_k = \sum_{gh=k} a_g\chi(g)(b_h)[g, h] \tag{10}$$

The above product is obtained by distributively extending the product in H. $K(G; H)$ is a ring and need not be an algebra in general. If factor systems $(\chi, [\_,\_])$ and $(\chi', [\_,\_]')$ are equivalent, $H(\chi, [\_,\_])$ and $H(\chi', [\_,\_]')$ are isomorphic and so $K(G; H)$ and $K(G; H')$ are also isomorphic. Hence $K(G; H)$ is uniquely determined by an extension H.

Theorem 4 *Let K be a division ring and G a group. Every $G-$lattice L over K determines a unique extension of K by G. Conversely, if H is an extension of K by G, there is a $G-$lattice L over K such that H is determined by L.*

Proof Corresponding to each $G-$lattice over K, we have a representation of G over K and vice versa. So the first part is clear. Let $H = H(\chi, [\_,\_])$ be an extension of K by G. For each $g \in G$ define $\rho(g) : K(G; H) \to K(G; H)$ such that for each $v \in K(G; H)$, $\rho(g)(v) = (1\bar{g})v$ where the product on the right is the product in $K(G; H)$. For $v \in K(G; H)$ and $\alpha \in K$,

$$\rho(g)(\alpha v) = 1\bar{g}(\alpha v) = \chi(g)(\alpha)\bar{g}v = \chi(g)(\alpha)\rho(g)(v). \tag{11}$$

Hence $\rho(g)$ is a semilinear automorphism having $\theta_{\rho(g)} = \chi(g)$. Also

$$\rho(g)\rho(h)(v) = \bar{g}(\bar{h}(v)) = \overline{g}\overline{h}(v) = [g, h]\overline{gh}(v) = [g, h]\rho(gh)(v) \tag{12}$$

and so $\rho : G \to SGL(K(G; H))$ is a representation associated with the factor system $(\chi, [\_,\_])$. The lattice $L(K(G; H))$ of all K-submodules of $K(G; H)$ is a $G-$lattice with the action induced by ρ. It is clear that the extension determined by the G-lattice $L(K(G; H))$ is the given H. □

The following example illustrate Theorem 4 for the C_3-lattice $L(\mathbb{R}^3)$.

Example

Consider $\mathbb{R}^3$ and the lattice $L(\mathbb{R}^3)$ of its subspaces. Let $C_3 = \{1, a, a^2\}$ be the cyclic group of order 3 generated by a. $L(\mathbb{R}^3)$ is a C_3-lattice with respect to the action defined as follows.

1. $1W = W$ for all $W \in L(\mathbb{R}^3)$.
2. aW is the subspace of $\mathbb{R}^3$ obtained by shifting the components of each vector in W one position to the right. For example, $W = \{(x, y, z) \in \mathbb{R}^3 \mid x + y = z\}$ is a subspace of $\mathbb{R}^3$ and $aW = \{(x, y, z) \in \mathbb{R}^3 \mid x + z = y\}$.
3. a^2W is the subspace of $\mathbb{R}^3$ obtained by shifting the components of each vector in W two positions to the right. We have $a^2W = \{(x, y, z) \in \mathbb{R}^3 \mid y + z = x\}$ for the above subspace.

Let $v = (x, y, z) \in \mathbb{R}^3$ and $\langle v\rangle = \{(kx, ky, kz) \mid k \in \mathbb{R}\}$ be the subspace spanned by v. Then $1\langle v\rangle = \langle v\rangle = \langle (x, y, z)\rangle$,$a\langle v\rangle = \{(kz, kx, ky) \mid k \in \mathbb{R}\} = \langle (z, x, y)\rangle$ and $a^2\langle v\rangle = \{(ky, kz, kx) \mid k \in \mathbb{R}\} = \langle (y, z, x)\rangle$. Define $\rho(1)(x, y, z) = (x, y, z)$, $\rho(a)$ $(x, y, z) = (z, x, y)$ and $\rho(a^2)(x, y, z) =(y, z, x)$. It is easy to see that $\rho(g)$ is an automorphism in $\mathbb{R}^3$ for each $g \in C_3$ and $\rho : C_3 \to Aut\mathbb{R}^3$ which takes each $g \in C_3$ to $\rho(g)$ defined above is a group homomorphism. Hence ρ is a linear representation of C_3 over $\mathbb{R}$ and is the representation corresponding to the C_3-lattice $L(\mathbb{R}^3)$.

Now we discuss the Schreier extension associated with ρ. Since ρ is a linear representation, $\chi(g) = \theta_{\rho(g)} = I_{\mathbb{R}}$ and $[g, h] = 1$ for all $g, h \in C_3$. So Schreier extension of $\mathbb{R}^*$ by C_3 is $H = \{k\bar{1} \mid k \in \mathbb{R}^*\} \cup \{k\bar{a} \mid k \in \mathbb{R}^*\} \cup \{k\bar{a^2} \mid k \in \mathbb{R}^*\}$ and the binary operation in H is given by $k\bar{g}k'\bar{h} = k\chi(g)(k')[g, h]\overline{gh} = kk'\overline{gh}$. Define a map $\phi : \mathbb{R}^* \times C_3 \to H$ such that $\phi(k, g) = k\bar{g}$. ϕ is a group homomorphism since $\phi(k, g)\phi(k', h) = k\bar{g}k'\bar{h} = kk'\overline{gh} = \phi(kk', gh)$. $\phi(k, g) = \phi(k', h)$ means that $k\bar{g} = k'\bar{h}$, so $k = k'$ and $g = h$ which implies $(k, g) = (k', h)$ and hence ϕ is injective. ϕ is surjective since for $k\bar{g} \in H$ there is always $(k, g) \in \mathbb{R}^* \times C_3$ such that $\phi(k, g) = k\bar{g}$. Hence H is isomorphic to $\mathbb{R}^* \times C_3$.

The twisted group ring $\mathbb{R}(C_3, \mathbb{R}^* \times C_3) = \{x\bar{1} + y\bar{a} + z\bar{a^2}\}$ is a vector space with respect to the addition and scalar multiplication defined by

$$\left(x_1\bar{1}+y_1\bar{a} + z_1\bar{a^2}\right)+\left(x_2\bar{1} + y_2\bar{a} + z_2\bar{a^2}\right) = (x_1 + x_2)\bar{1} + (y_1 + y_2)\bar{a} + (z_1 + z_2)\bar{a^2}$$

$$k\left(x\bar{1} + y\bar{a} + z\bar{a^2}\right) = kx\bar{1} + ky\bar{a} + kz\bar{a^2}$$

Then $\theta : \mathbb{R}^3 \to \mathbb{R}(C_3, \mathbb{R}^* \times C_3)$ given by $\theta(x, y, z) = x\bar{1} + y\bar{a} + z\bar{a^2}$ is an isomorphism. So $L(\mathbb{R}^3) \cong L(\mathbb{R}(C_3, \mathbb{R}^* \times C_3))$. Note that $L(\mathbb{R}(C_3, \mathbb{R}^* \times C_3))$ is a C_3-lattice with respect to the product defined as,

1. $1W = W$
2. $aW = \{\bar{a}(x\bar{1} + y\bar{a} + z\bar{a^2}) \mid x\bar{1} + y\bar{a} + z\bar{a^2} \in W\} = \{z\bar{1} + x\bar{a} + y\bar{a^2} \mid x\bar{1}+ y\bar{a} + z\bar{a^2} \in W\}$

3. $a^2W = \{\overline{a^2}(x\bar{1} + y\bar{a} + z\overline{a^2}) \mid x\bar{1} + y\bar{a} + z\overline{a^2} \in W\} = \{y\bar{1} + z\bar{a} + x\overline{a^2} \mid x\bar{1} + y\bar{a} + z\overline{a^2} \in W\}$

Hence C_3-lattices $L(\mathbb{R}(C_3, \mathbb{R}^* \times C_3))$ and $L(\mathbb{R}^3)$ are isomorphic.

Remark 1 V need not be isomorphic with $K(G; H)$ in general. If we replace C_3 by S_3 in the above example and define an action of S_3 on $L(\mathbb{R}^3)$ as above, the twisted group ring $\mathbb{R}(S_3; \mathbb{R}^* \times S_3)$ is isomorphic with $\mathbb{R}^6$ not with $\mathbb{R}^3$.

If K is a field, the twisted group ring is a vector space over K. By distributively extending the binary operation in H, there is a product in $K(G; H)$. But $K(G; H)$ is not an algebra in general. The following theorem gives a necessary and sufficient condition so that the twisted group ring is an algebra.

Theorem 5 *Let L be a $G-$lattice over the field K and H be the extension determined by L. Then, the twisted group ring $K(G; H)$ is an algebra over K if and only if L is a projective $G-$lattice over K.*

Proof Assume that L is a projective $G-$lattice over the field K. To show that $K(G; H)$ an algebra it remains to prove that $(au)v = u(av) = a(uv)$ for all $a \in K$ and $u, v \in K(G; H)$. Since the product in $K(G; H)$ is obtained by distributively extending the product in H, it is enough to check the above property for the elements in the basis $\{\bar{g} \mid g \in G\}$ of $K(G; H)$. $\bar{g}(a\bar{h}) = a(\overline{gh}) = (a\bar{g})\bar{h}$ since H is a projective extension, this proves that $K(G; H)$ is a $K-$algebra.

Conversely let $K(G; H)$ be an algebra. Then $ba\bar{1} = (b\bar{1})(a\bar{1}) = a(b\bar{1}) = ab\bar{1}$ which implies that $ab = ba$ for all $a, b \in K^*$ and so K is a field. $\chi(g)(a)\bar{g} = \bar{g}(a\bar{1}) = a\bar{g}$ and so $\chi(g)(a) = a$ for all $a \in K^*$ and $g \in G$. Hence $\chi(g) = I_K$ for each $g \in G$ and so H is a projective extension of K by G. □

Every semilinear projective representation $\rho : G \to SGL(V)$ defines an action of G on $L(V)$. But the product $gv = \rho(g)(v)$ is not an action of G on V in general. However, it may turn out to be an action of $K(G; H)$ by distributively extending the multiplication.

Proposition 4 *Let G be a group, K be a division ring, and V be a K-module. If $\rho : G \to SGL(V)$ is the representation associated with the extension H of K by G, then V is a $K(G; H)$-module.*

Proof Assume that $\rho : G \to SGL(V)$ is a representation of G over K.

For $s = \sum\limits_{g \in G} a_g\bar{g}$ and $v \in V$ define $sv = \sum\limits_{g \in G} a_g\rho(g)(v)$. Then V is a $K(G; H)$-module with respect to this scalar multiplication. Consider $s = \sum\limits_{g \in G} a_g\bar{g}, t = \sum\limits_{g \in G} b_g\bar{g}$ in $K(G; H)$ and $u, v \in V$. Using the semilinearity of $\rho(g)$ and the $K-$module structure on V we get,

1. $s(u + v) = \sum\limits_{g \in G} a_g\big(\rho(g)(u) + \rho(g)(v)\big) = su + sv.$

2. $(s+t)v = \sum_{g\in G} a_g\rho(g)(v) + b_g\rho(g)(v) = sv + tv.$
3. $s(tv) = \sum_{g\in G} a_g\rho(g)\left(\sum_{h\in G} b_h\rho(h)(v)\right) = \sum_{g\in G}\sum_{h\in G} a_g\chi(g)(b_h)\rho(g)\rho(h)(v)$

   $=\sum_{k\in G}\left(\sum_{gh=k} a_g\chi(g)(b_h)[g,h]\right)\rho(k)(v) = (st)v.$
4. $\bar{1}v = \rho(1)(v) = v.$
5. $(bs)v = \sum_{g\in G} ba_g\rho(g)(v) = b\sum_{g\in G} a_g\rho(g)(v) = b(sv).$

Hence V is a $K(G;H)$-module. □

Converse of Proposition 4 is also true. See Proposition 5.

Proposition 5 *Let G be a group, K a division ring, and H be a Schreier extension of K by G. Let V be a $K(G;H)$-module. Then V is K-module, and $\rho : G \to SGL(V)$ defined by $\rho(g)(v) = (1\bar{g})v$ for each $g \in G$ and $v \in V$, is a semilinear projective representation associated with the extension H.*

Proof We can find a copy of K in $K(G;H)$ by identifying $k \in K$ with $k\bar{1}$ in $K(G;H)$. Thus by restricting the scalar multiplication of $K(G;H)$ on V to K, we get V as a K-module.

For each $g \in G$, define $\rho(g) : V \to V$ such that $\rho(g)(v) = (1\bar{g})v$ for each $v \in V$. Then for $v \in K(G;H)$ and $\alpha \in K$,

$$\rho(g)(\alpha v) = 1\bar{g}(\alpha v) = \chi(g)(\alpha)\bar{g}v = \chi(g)(\alpha)\rho(g)(v). \tag{13}$$

Hence $\rho(g)$ is a semilinear automorphism having $\theta_{\rho(g)} = \chi(g)$. Also

$$\rho(g)\rho(h)(v) = \bar{g}(\bar{h}(v)) = \bar{g}\bar{h}(v) = [g,h]\overline{gh}(v) = [g,h]\rho(gh)(v) \tag{14}$$

and so $\rho : G \to SGL(V)$ is a representation associated with the extension H. □

Thus we summarize that given an extension H of K by G, every $K(G;H)$-module V induce a $G-$lattice on $L(V)$ and conversely, every $G-$lattice over K induces an extension H of K by G and a $K(G;H)$-module. Moreover, when H and H' are isomorphic extensions, $K(G;H)$ and $K(G;H')$ are isomorphic, and they induce the same action on the module V corresponding to the $G-$lattice $L(V)$.

References

1. Nambooripad, K.S.S.: Group lattices and semigroups. In: Hall, T.E., Jones, P.R., Meakin, J.C. (eds.) Monash Conference on Semigroup Theory in Honour of G. B. Preston, pp. 224–245. World scientific Publishing Co. Pte. Ltd., Singapore (1991)
2. Nambooripad, K.S.S.: Group-Lattices (1992). (Preprint)
3. Hall, M.: The Theory of Groups. The Macmillan Company, New York (1963)

Identities in Twisted Brauer Monoids

Nikita V. Kitov and Mikhail V. Volkov

Abstract We show that it is co-NP-hard to check whether a given semigroup identity holds in the twisted Brauer monoid $\mathcal{B}_n^\tau$ with $n \geq 5$.

Keywords Twisted Brauer monoid · Identity checking problem

1 Introduction

A *semigroup word* is merely a finite sequence of symbols, called *letters*. An *identity* is a pair of semigroup words, traditionally written as a formal equality. We write identities using the sign $\simeq$, so that the pair (w, w') is written as $w \simeq w'$, and reserve the usual equality sign $=$ for 'genuine' equalities. For a semigroup word w, the set of all letters that occur in w is denoted by $\text{alph}(w)$. If $\mathcal{S}$ is a semigroup, any map $\varphi\colon \text{alph}(w) \to \mathcal{S}$ is called a *substitution*; the element of $\mathcal{S}$ that one gets by substituting $\varphi(x)$ for each letter $x \in \text{alph}(w)$ and computing the product in $\mathcal{S}$ is denoted by $\varphi(w)$ and called the *value* of w under φ.

Let $w \simeq w'$ be an identity, and let $X = \text{alph}(ww')$. We say that a semigroup $\mathcal{S}$ *satisfies* $w \simeq w'$ (or $w \simeq w'$ *holds* in $\mathcal{S}$) if $\varphi(w) = \varphi(w')$ for every substitution $\varphi\colon X \to \mathcal{S}$, that is, each substitution of elements in $\mathcal{S}$ for letters in X yields equal values to w and w'.

Given a semigroup $\mathcal{S}$, its *identity checking problem*, denoted CHECK- ID($\mathcal{S}$), is a combinatorial decision problem whose instance is an identity $w \simeq w'$; the answer to the instance $w \simeq w'$ is 'YES' if $\mathcal{S}$ satisfies $w \simeq w'$ and 'NO' otherwise. An alternative

The authors were supported by the Ministry of Science and Higher Education of the Russian Federation, project FEUZ-2023-2022.

N. V. Kitov · M. V. Volkov (✉)
Institute of Natural Sciences and Mathematics, Ural Federal University,
620000 Ekaterinburg, Russia
e-mail: m.v.volkov@urfu.ru

N. V. Kitov
e-mail: n.v.kitov@urfu.ru

A. A. Ambily and V. B. Kiran Kumar (eds.), *Semigroups, Algebras and Operator Theory*,
Springer Proceedings in Mathematics & Statistics 436,
https://doi.org/10.1007/978-981-99-6349-2_4

name for this problem that sometimes appears in the literature is the '*term equivalence problem*'.

The identity checking problem is obviously decidable for finite semigroups. An active research direction aims at classifying finite semigroups $\mathcal{S}$ according to the computational complexity of CHECK- ID($\mathcal{S}$); see [27, Sect. 1] for a brief overview and references. For an infinite semigroup, the identity checking problem can be undecidable; for an example of such a semigroup, see [37]. On the other hand, many infinite semigroups that naturally arise in mathematics such as semigroups of matrices over an infinite field, or semigroups of relations on an infinite domain, or semigroups of transformations of an infinite set satisfy only identities of the form $w \approx w$, and hence, the identity checking problem for such 'big' semigroups is trivially decidable in linear time. Another family of natural infinite semigroups with linear time identity checking comes from various additive and multiplicative structures in arithmetics and commutative algebra, typical representatives being the semigroups of positive integers under addition or multiplication. It is folklore that these commutative semigroups satisfy exactly so-called balanced identities. (An identity $w \approx w'$ is *balanced* if every letter occurs in w and w' the same number of times. Clearly, the balancedness of $w \approx w'$ can be verified in linear in $|ww'|$ time.)

The latter example shows in a nutshell a common approach to identity checking in semigroups. Given a semigroup $\mathcal{S}$, one looks for a *combinatorial characterization* of the identities holding in $\mathcal{S}$ that could be effectively verified. Recently such characterizations have been found for some infinite semigroups of interest, including, e.g., the free 2-generated semiband $\mathcal{J}_\infty = \langle e, f \mid e^2 = e,\ f^2 = f \rangle$ [41], the bicyclic monoid $\mathcal{B} = \langle p, q \mid qp = 1 \rangle$ [11], the Kauffman monoids $\mathcal{K}_3$ and $\mathcal{K}_4$ [10, 27], and several monoids originated in combinatorics of tableaux such as hypoplactic, stalactic, taiga, sylvester, and Baxter monoids [7–9, 18]. Therefore the identity checking problem in each of these semigroups is solvable in polynomial time. On the other hand, no natural examples of infinite semigroups with decidable but computationally hard identity checking seem to have been published so far. The aim of the present paper is to exhibit a series of such examples. Namely, we show that for the twisted Brauer monoid $\mathcal{B}_n^\tau$ with $n \geq 5$, the problem CHECK- ID($\mathcal{B}_n^\tau$) is co-NP-hard.

The paper is structured as follows. In Sect. 2 we first recall the definition of the twisted Brauer monoids $\mathcal{B}_n^\tau$. Then we show that for each n, the monoid $\mathcal{B}_n^\tau$ embeds into a regular monoid that has much better structure properties albeit it satisfies exactly the same identities as $\mathcal{B}_n^\tau$. In Sect. 3 we modify an approach devised in [1] to deal with the identity checking problem for finite semigroups so that the modified version applies to infinite semigroups subject to some finiteness conditions. In Sect. 4 we prove our main result (Theorem 2), and Sect. 5 collects some additional remarks and discusses future work.

We assume the reader's acquaintance with a few basic concepts of semigroup theory, including Green's relations and presentations of semigroups via generators and relations. The first chapters of Howie's classic textbook [21] contain everything we need. For computational complexity notions, we refer the reader to Papadimitriou's textbook [39].

2 Twisted Brauer Monoids

2.1 Definition

Twisted Brauer monoids can be defined in various ways. Here we give their geometric definition, following [2] (where the name 'wire monoids' was used).

Let $[n] = \{1, \dots, n\}$ and let $[n]' = \{1', \dots, n'\}$ be a disjoint copy of $[n]$. Consider the set $\mathcal{B}_n^\tau$ of all pairs $(\pi; s)$ where π is a partition of the $2n$-element set $[n] \cup [n]'$ into 2-element blocks and s is a non-negative integer. Such a pair is represented by a *diagram* as shown in Fig. 1 (borrowed from [2]). We represent the elements of $[n]$ by points on the left-hand side of the diagram (*left points*) while the elements of $[n]'$ are represented by points on the right-hand side of the diagram (*right points*). For $(\pi; s) \in \mathcal{B}_n^\tau$, we represent the number s by s closed curves (called *circles* or *floating components*) and each block of the partition π is represented by a line referred to as a *wire*. Thus, each wire connects two points; it is called an *ℓ-wire* if it connects two left points, an *r-wire* if it connects two right points, and a *t-wire* if it connects a left point with a right point. The diagram in Fig. 1 has three wires of each type and three circles; it corresponds to the pair

$$\Big(\big\{\{1, 5'\}, \{2, 4\}, \{3, 5\}, \{6, 9'\}, \{7, 9\}, \{8, 8'\}, \{1', 2'\}, \{3', 4'\}, \{6', 7'\}\big\};\ 3\Big).$$

In what follows we use 'vertical' diagrams as the one in Fig. 1 but in the literature (see, e.g., [12]) the reader can also meet representations of pairs from $\mathcal{B}_n^\tau$ by 'horizontal' diagrams like the one in Fig. 2. Of course, the 'vertical' and 'horizontal' viewpoints are fully equivalent. We also stress that only two things matter in any diagrammatic representation of the elements of $\mathcal{B}_n^\tau$, namely, (1) which points are connected and (2) the number of circles; neither the shape nor the relative position

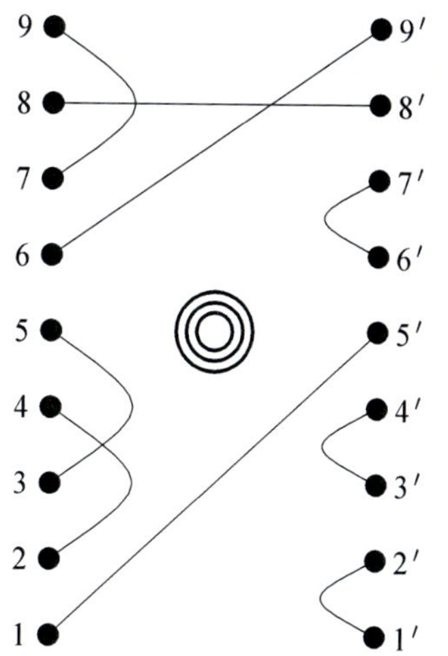

Fig. 1 Diagram representing an element of $\mathcal{B}_9^\tau$

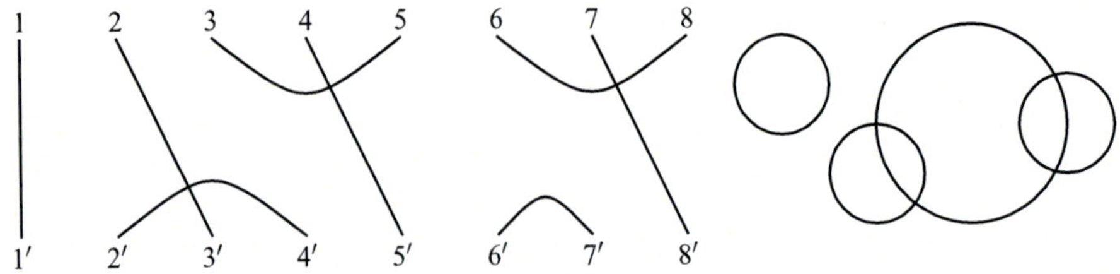

Fig. 2 Diagram representing an element of $\mathcal{B}_8^\tau$

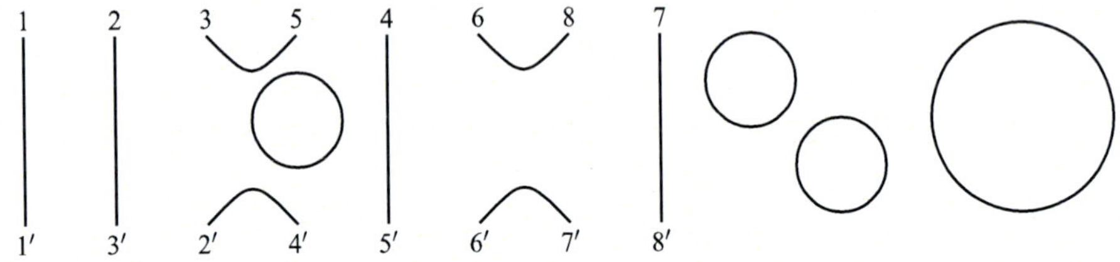

Fig. 3 Diagram of Fig. 2 redrawn

of the wires and circles matters. For instance, the diagram in Fig. 3 represents the same element of $\mathcal{B}_8^\tau$ as the diagram in Fig. 2.

Now we define a multiplication in $\mathcal{B}_n^\tau$. Pictorially, in order to multiply two diagrams, we glue their wires together by identifying each right point u' of the first diagram with the corresponding left point u of the second diagram. This way we obtain a new diagram whose left (respectively, right) points are the left (respectively, right) points of the first (respectively, second) diagram. Two points of this new diagram are connected in it if one can reach one of them from the other by walking along a sequence of consecutive wires of the factors, see Fig. 4 (where the labels $1, 2, \dots, 9, 1', 2', \dots, 9'$ are omitted but they are assumed to go up in the consecutive order). All circles of the factors are inherited by the product; in addition, some extra circles may arise from r-wires of the first diagram combined with ℓ-wires of the second diagram.

In more precise terms, if $\xi = (\pi_1; s_1)$, $\eta = (\pi_2; s_2)$, then a left point p and a right point q' of the product $\xi\eta$ are connected by a t-wire if and only if one of the following holds:

- $p — u'$ is a t-wire in ξ and $u — q'$ is a t-wire in η for some $u \in [n]$;
- for some $s > 1$ and some $u_1, v_1, u_2, \dots, v_{s-1}, u_s \in [n]$ (all pairwise distinct), $p — u_1'$ is a t-wire in ξ and $u_s — q'$ is a t-wire in η, while $u_i — v_i$ is an ℓ-wire in η and $v_i' — u_{i+1}'$ is an r-wire in ξ for each $i = 1, \dots, s-1$.
(The reader may trace an application of the second rule in Fig. 4, in which such a 'composite' t-wire connects 1 and 3′ in the product diagram.)

Analogous characterizations hold for the ℓ-wires and r-wires of $\xi\eta$. Here we include only the rules for forming the ℓ-wires as the r-wires of the product are obtained in a perfectly symmetric way.

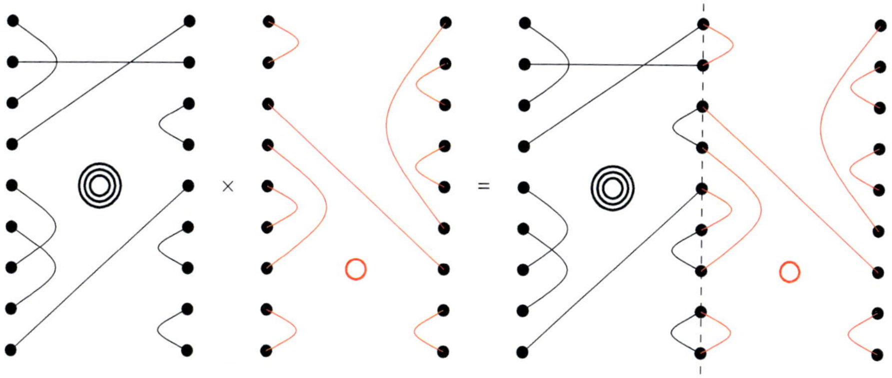

Fig. 4 Multiplication of diagrams

Two left points p and q of $\xi\eta$ are connected by an ℓ-wire if and only if one of the following holds:

- $p — q$ is an ℓ-wire in ξ;
- for some $s \geq 1$ and some $u_1, v_1, u_2, \ldots, v_s \in [n]$ (all pairwise distinct), $p — u'_1$ and $q — v'_s$ are t-wires in ξ, while $u_i — v_i$ is an ℓ-wire in η for each $i = 1, \ldots, s$ and if $s > 1$, then $v'_i — u'_{i+1}$ is an r-wire in ξ for each $i = 1, \ldots, s-1$.
(Again, Fig. 4 provides an instance of the second rule: look at the ℓ-wire that connects 6 and 8 in the product diagram.)

Finally, each circle of the product $\xi\eta$ corresponds to either a circle in ξ or η or a sequence $u_1, v_1, \ldots, u_s, v_s \in [n]$ with $s \geq 1$ and pairwise distinct $u_1, v_1, \ldots, u_s, v_s$ such that all $u_i — v_i$ are ℓ-wires in η, while all $v'_i — u'_{i+1}$ and $v'_s — u'_1$ are r-wires in ξ.
(In Fig. 4, one sees such a 'new' circle formed by the ℓ-wire $1 — 2$ of the second factor glued to the r-wire $2' — 1'$ of the first factor.)

The above-defined multiplication in $\mathcal{B}_n^\tau$ is easily seen to be associative and the diagram with 0 circles and the n horizontal t-wires $1 — 1'$, ..., $n — n'$ is the identity element with respect to the multiplication. Thus, $\mathcal{B}_n^\tau$ is a monoid called the *twisted Brauer monoid*.

2.2 *Background*

We refer to [12] for a throughout analysis of the semigroup-theoretic properties of twisted Brauer monoids. Here we explain the terminology and relate the monoids $\mathcal{B}_n^\tau$ to the representation theory of classical groups. Some parts of our discussion involve concepts from beyond semigroup theory. These parts are not used in subsequent proofs so that the reader who is only interested in our main result can safely skip them.

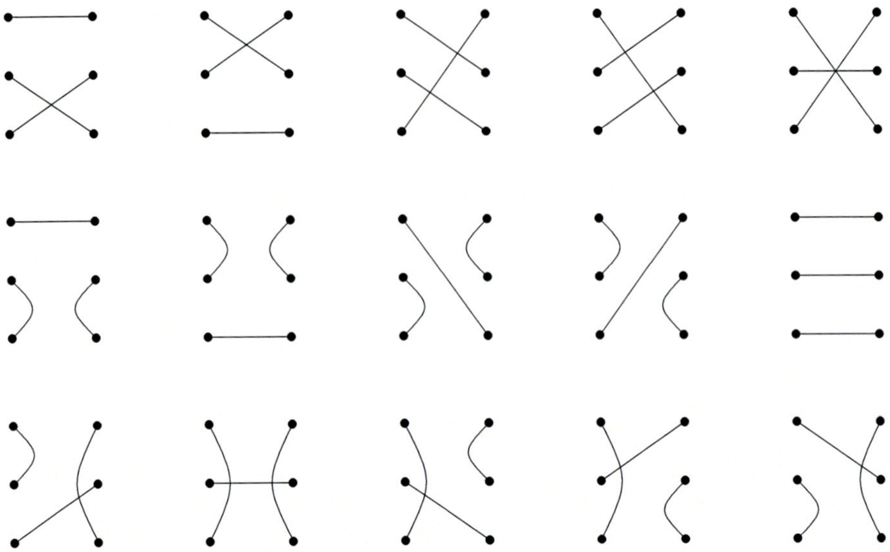

Fig. 5 Diagrams representing the elements of $\mathcal{B}_3$

Let R be a commutative ring with 1 and $\mathcal{S}$ a semigroup. Following [44], we say that a map $\tau\colon \mathcal{S}\times\mathcal{S}\to R$ is a *twisting* from $\mathcal{S}$ to R if

$$\tau(s,t)\tau(st,u)=\tau(s,tu)\tau(t,u) \text{ for all } s,t,u\in\mathcal{S}. \tag{1}$$

The *twisted semigroup algebra* of $\mathcal{S}$ over R, with twisting τ, denoted by $R^\tau[\mathcal{S}]$, is the free R-module spanned by $\mathcal{S}$ as a basis with multiplication $\circ$ defined by

$$s\circ t=\tau(s,t)st \quad \text{for all } s,t\in\mathcal{S}, \tag{2}$$

and extended by linearity. Condition (1) readily implies that this multiplication is associative. If $\tau(s,t)=1$ for all $s,t\in\mathcal{S}$, then the twisted semigroup algebra $R^\tau[\mathcal{S}]$ is nothing but the usual semigroup algebra $R[\mathcal{S}]$. Thus, twisted semigroup algebras provide a vast generalization of semigroup algebras while retaining many useful properties of the latter, in particular, those important for representation theory.

Having clarified the meaning of 'twisted', let us explain what the Brauer monoid is. Denote by $\mathcal{B}_n$ the set of all partitions of the $2n$-element set $[n]\cup[n]'$ into 2-element blocks; we visualize them as diagrams similar to the one in Fig. 1 but without floating components. For instance, Fig. 5 shows the 15 diagrams representing the partitions from $\mathcal{B}_3$.

Identifying each partition $\pi\in\mathcal{B}_n$ with the pair $(\pi;0)\in\mathcal{B}_n^\tau$, we may treat $\mathcal{B}_n$ as a subset of $\mathcal{B}_n^\tau$. The multiplication in $\mathcal{B}_n^\tau$ induces a multiplication in $\mathcal{B}_n$ as follows: given two partitions $\pi_1,\pi_2\in\mathcal{B}_n$, one computes the product of the pairs $(\pi_1;0)$ and $(\pi_2;0)$ in $\mathcal{B}_n^\tau$ and if $(\pi_1;0)(\pi_2;0)=(\pi;s)$, one lets π be the product of π_1 and π_2 in

$\mathcal{B}_n$. In other words, one multiplies the diagrams of π_1 and π_2, using the multiplication rules in Sect. 2.1, and then discards all 'new' circles if they arise.

Under the above-defined multiplication, $\mathcal{B}_n$ constitutes a monoid called the *Brauer monoid*. This family of monoids was invented by Brauer [5] back in 1937, hence the name. Brauer used the monoid $\mathcal{B}_n$ to study the linear representations of orthogonal and symplectic groups.

Observe that $\mathcal{B}_n$ is **not** a submonoid of $\mathcal{B}_n^\tau$; at the same time, $\mathcal{B}_n$ is easily seen to be the homomorphic image of $\mathcal{B}_n^\tau$ under the 'forgetting' homomorphism $(\pi; s) \mapsto \pi$ that removes the circles from the diagrams in $\mathcal{B}_n^\tau$.

Given two partitions $\pi_1, \pi_2 \in \mathcal{B}_n$, we denote by $\langle \pi_1, \pi_2 \rangle$ the number of cycles that arise when the pairs $(\pi_1; 0)$ and $(\pi_2; 0)$ are multiplied in $\mathcal{B}_n^\tau$. Using this notation, we have the following useful formula expressing the multiplication in $\mathcal{B}_n^\tau$ via that in $\mathcal{B}_n$:

$$(\pi_1; s_1)(\pi_2; s_2) = (\pi_1\pi_2; s_1 + s_2 + \langle \pi_1, \pi_2 \rangle), \tag{3}$$

where the product $\pi_1\pi_2$ in the right-hand side is computed in $\mathcal{B}_n$.

Now let F be a field of characteristic 0. Fix an element $\theta \in F \setminus \{0\}$ and consider the map $\tau \colon \mathcal{B}_n \times \mathcal{B}_n \to F$ defined by

$$\tau(\pi_1, \pi_2) = \theta^{\langle \pi_1, \pi_2 \rangle}.$$

It is known (and easy to verify) that τ satisfies (1) so the map is a twisting from $\mathcal{B}_n$ to F. Hence, one can construct the twisted semigroup algebra $F^\tau[\mathcal{B}_n]$. In the literature, the notation and the name for this algebra vary; we denote it by $B_n(\theta)$ as in [26] and call it *Brauer's centralizer algebra* as, e.g., in [19]. In [5], Brauer's centralizer algebra $B_n(m)$, where m is a positive integer, was used to study the natural representation of the orthogonal group O_m on the n-th tensor power $(F^m)^{\otimes n}$ of the space F^m. (The algebra $B_n(m)$ is exactly the centralizer of the diagonal action of O_m on $(F^m)^{\otimes n}$, hence the name.) When m is an even positive integer, Brauer's centralizer algebra $B_n(-m)$ allowed for a similar study of the representation of the symplectic group Sp_m on $(F^m)^{\otimes n}$. It was also present in [5], albeit implicitly; see [19] for a detailed analysis. For the case where the parameter θ is arbitrary, the ring-theoretic structure of the algebra $B_n(\theta)$ has been determined by Wenzl [43] (in particular, he has proved that if θ is not an integer, then the algebra $B_n(\theta)$ is semisimple).

In a simplified form, the approach used in [43] to give a uniform treatment of the algebras $B_n(\theta)$ for various θ can be stated as follows. Define a twisting from $\mathcal{B}_n$ to the polynomial ring $F[X]$ by[1]

$$\tau(\pi_1, \pi_2) = X^{\langle \pi_1, \pi_2 \rangle}. \tag{4}$$

Then one gets the twisted semigroup algebra $(F[X])^\tau[\mathcal{B}_n]$ that can be denoted by $B_n(X)$. For each $\theta \in F \setminus \{0\}$, evaluating X at θ gives rise to a homomorphism from

[1] Wenzl [43] employed the same twisting but to the field of rational functions rather than the polynomial ring.

$B_n(X)$ onto $B_n(\theta)$ so that the algebra $B_n(X)$ is a kind of mother of all Brauer's centralizer algebras.

On the other hand, $B_n(X)$ is nothing but the usual (non-twisted) semigroup algebra $F[\mathcal{B}_n^\tau]$. Indeed, it is easy to show that the bijection $\beta\colon (\pi; s) \mapsto X^s\pi$ is a semigroup isomorphism between the F-basis $\mathcal{B}_n^\tau$ of $F[\mathcal{B}_n^\tau]$ and the F-basis $\{X^s\pi \mid s \in \mathbb{Z}_{\geq 0},\ \pi \in \mathcal{B}_n\}$ of $B_n(X)$ where the latter basis is equipped with multiplication $\circ$ as in (2):

$$\begin{aligned}
\beta\big((\pi_1; s_1)(\pi_2; s_2)\big) &= \beta\big((\pi_1\pi_2; s_1 + s_2 + \langle \pi_1, \pi_2\rangle)\big) && \text{by (3)}\\
&= X^{s_1+s_2+\langle \pi_1,\pi_2\rangle}\pi_1\pi_2 && \text{by the definition of } \beta\\
&= X^{s_1}\pi_1 \circ X^{s_2}\pi_2 && \text{by the definition of } \circ\text{; see (2)}\\
&= \beta(\pi_1; s_1) \circ \beta(\pi_2; s_2) && \text{by the definition of } \beta.
\end{aligned}$$

The isomorphism $(F[X])^\tau[\mathcal{B}_n] \cong F[\mathcal{B}_n^\tau]$, which moves the twist from the outer ring to the inner semigroup, stands behind our terminology: instead of twisting the semigroup algebra of the Brauer monoid, we twist the monoid itself, thus getting the twisted Brauer monoid.

2.3 Presentation for $\mathcal{B}_n^\tau$

It is known that the twisted Brauer monoid $\mathcal{B}_n^\tau$ can be generated by the following $2n - 1$ pairs:

- *transpositions* $t_i = \Big(\big\{i — (i+1)',\ i' — i+1,\ j — j' \mid \text{for } j \neq i, i+1\big\};\ 0\Big)$, $i = 1, \dots, n-1$,
- *hooks* $h_i = \Big(\big\{i — i+1,\ i' — (i+1)',\ j — j' \mid \text{for } j \neq i, i+1\big\};\ 0\Big)$, $i = 1, \dots, n-1$,
- and the *circle* $c = \Big(\big\{j — j' \mid \text{for } j = 1, \dots, n\big\};\ 1\Big)$.

For an illustration, see Fig. 5: the first two diagrams in the top row represent the transpositions t_1 and t_2, and the first two diagrams in the middle row represent the hooks h_1 and h_2. (The omitted labels $1, 2, 3, 1', 2', 3'$ are assumed to go up in the consecutive order.)

For all $i, j = 1, \dots, n-1$, the generators $t_1, \dots, t_{n-1}, h_1, \dots, h_{n-1}, c$ satisfy the following:

$$t_i^2 = 1, \tag{5}$$

$$t_i t_j = t_j t_i \quad \text{if } |i-j| \geq 2, \tag{6}$$

$$t_i t_j t_i = t_j t_i t_j \quad \text{if } |i-j| = 1, \tag{7}$$

$$c t_i = t_i c, \tag{8}$$

$$h_i h_j = h_j h_i \quad \text{if } |i-j| \geq 2, \tag{9}$$

$$h_i h_j h_i = h_i \quad \text{if } |i-j| = 1, \tag{10}$$

$$h_i^2 = ch_i = h_i c, \tag{11}$$
$$h_i t_i = t_i h_i = h_i, \tag{12}$$
$$h_i t_j = t_j h_i \qquad \text{if } |i-j| \geq 2, \tag{13}$$
$$t_i h_j h_i = t_j h_i, \ h_i h_j t_i = h_i t_j \qquad \text{if } |i-j| = 1. \tag{14}$$

Proposition 1 *The relations (5)–(14) form a monoid presentation of the twisted Brauer monoid $\mathcal{B}_n^\tau$ with respect to the generators $t_1, \ldots, t_{n-1}, h_1, \ldots, h_{n-1}, c$.*

Even though Proposition 1 does not seem to have been registered in the literature, its result is not essentially new as it is an immediate combination of two known ingredients: (1) a presentation for the Brauer monoid $\mathcal{B}_n$ [28, Section 3], and (2) a method for 'twisting' presentations, that is, obtaining a presentation for a twisted semigroup algebra from a given presentation of the underlying semigroup of the algebra [13, Section 6 and Remark 45].[2]

The presentation of Proposition 1 is not the most economical one in terms of the number of generators (in fact, for each n, the monoid $\mathcal{B}_n^\tau$ can be generated by just four elements, see [12, Proposition 3.11]). It is, however, quite transparent and conveniently reveals some structural components of $\mathcal{B}_n^\tau$.

For instance, the relations (5)–(7) are nothing but Moore's classical relations [35, Theorem A] for the symmetric group $\mathbb{S}_n$. Since both sides of any other relation involve some generator beside the transpositions $t_1, \ldots, t_{n-1}$, no other relation can be applied to a word composed of transpositions only. Therefore, the transpositions generate in $\mathcal{B}_n^\tau$ a subgroup isomorphic to $\mathbb{S}_n$; this subgroup is actually the group of units of $\mathcal{B}_n^\tau$. The reader may use Fig. 5 as an illustration for $n = 3$: the five diagrams on the top row, together with the last diagram of the middle row, represent the elements of $\mathbb{S}_3$.

On the other hand, the relations (9)–(11) that do not involve $t_1, \ldots, t_{n-1}$ are the so-called Temperley–Lieb relations constituting a presentation for the *Kauffman monoid* $\mathcal{K}_n$. (We mentioned the monoids $\mathcal{K}_3$ and $\mathcal{K}_4$ from this family in the introduction.) Again, Fig. 5 provides an illustration for $n = 3$: the five diagrams of the middle row represent the partitions π such that $(\pi; s) \in \mathcal{K}_3$ for any $s \in \mathbb{Z}_{\geq 0}$. One sees that these five diagrams are exactly those whose wires do not cross. It is this property that Kauffman [25] used to introduce the monoids $\mathcal{K}_n$; namely, he defined $\mathcal{K}_n$ as the submonoid of $\mathcal{B}_n^\tau$ consisting of all elements of $\mathcal{B}_n^\tau$ that have a representation as a diagram in which the labels $1, 2, \ldots, n, 1', 2', \ldots, n'$ go up in the consecutive order and wires do not cross. The fact that the submonoid can be identified with the monoid generated by $h_1, \ldots, h_{n-1}, c$ subject to the relations (9)–(11) was stated in [25] with a proof sketch; for a detailed proof, see [4] or [14].

---

[2] The authors are grateful to Dr. James East who drew their attention to this combination.

2.4 The Monoid $\mathcal{B}_n^{\pm\tau}$ and its Identities

Following an idea by Karl Auinger (personal communication), we embed the monoid $\mathcal{B}_n^\tau$ into a larger monoid $\mathcal{B}_n^{\pm\tau}$ that shares the identities with $\mathcal{B}_n^\tau$ but has much better structure properties. In [27], we applied the same trick to Kauffman monoids.

In terms of generators and relations, the *±-twisted Brauer monoid* $\mathcal{B}_n^{\pm\tau}$ can be defined as the monoid with $2n$ generators $t_1, \dots, t_{n-1}, h_1, \dots, h_{n-1}, c, d$ subject to the relations (5)–(14) and the additional relations

$$cd = dc = 1. \tag{15}$$

Observe that (8) and (15) imply that $dt_i = t_i d$ for each $i = 1, \dots, n-1$. Indeed,

$$\begin{aligned} dt_i &= d^2ct_i && \text{since } dc = 1\\ &= d^2t_ic && \text{since } ct_i = t_ic\\ &= d^2t_ic^2d && \text{since } cd = 1\\ &= d^2c^2t_id && \text{since } c^2t_i = t_ic^2\\ &= t_id && \text{since } d^2c^2 = 1. \end{aligned}$$

Similarly, the relations (11) and (15) imply that $dh_i = h_id$ for each $i = 1, \dots, n-1$.

It is easy to see that the submonoid of $\mathcal{B}_n^{\pm\tau}$ generated by $t_1, \dots, t_{n-1}, h_1, \dots, h_{n-1}, c$ is isomorphic to $\mathcal{B}_n^\tau$. The generator d commutes with every element of this submonoid since it commutes with each of its generators.

To interpret the ±-twisted Brauer monoid in terms of diagrams, we introduce two sorts of circles: positive and negative. Each diagram may contain only circles of one sort. When two diagrams are multiplied, the following two rules are obeyed: all 'new' circles (the ones that arise when the diagrams are glued together) are positive; in addition, if the product diagram inherits some negative circles from its factors, then pairs of 'opposite' circles are consecutively removed until only circles of a single sort (or no circles at all) remain. The twisted Brauer monoid $\mathcal{B}_n^\tau$ is then nothing but the submonoid of all diagrams having only positive circles or no circles at all. Thus, if elements of $\mathcal{B}_n^{\pm\tau}$ are presented as pairs $(\pi; s)$ with $\pi \in \mathcal{B}_n$ and $s \in \mathbb{Z}$, then the multiplication formula (3) persists. Also, the 'forgetting' homomorphism $(\pi; s) \mapsto \pi$ of $\mathcal{B}_n^\tau$ onto the Brauer monoid $\mathcal{B}_n$ extends to the monoid $\mathcal{B}_n^{\pm\tau}$.

A further interpretation of the ±-twisted Brauer monoid comes from the considerations in Sect. 2.2. Recall that the twisted Brauer monoid $\mathcal{B}_n^\tau$ has been identified with the F-basis of the twisted semigroup algebra $(F[X])^\tau[\mathcal{B}_n]$ where the twisting τ from $\mathcal{B}_n$ to the polynomial ring $F[X]$ is defined by (4). If one substitutes $F[X]$ by the ring $F[X, X^{-1}]$ of Laurent polynomials and uses the same twisting τ, the F-basis of the twisted semigroup algebra $(F[X, X^{-1}])^\tau[\mathcal{B}_n]$ can be identified with the monoid $\mathcal{B}_n^{\pm\tau}$.

Finally, in terms of 'classical' semigroup theory, $\mathcal{B}_n^{\pm\tau}$ is the semigroup of quotients of $\mathcal{B}_n^\tau$ in the sense of Murata [36]. In ring theory, it is known that ring identities are preserved by passing to *central localizations*, that is, rings of quotients over central subsemigroups, see, e.g., [40, Theorem 3.1]. A similar general result holds for semigroups, but we state only a special case that is sufficient for our purposes.

Recall that an identity $w \simeq w'$ is *balanced* if for every letter in $\mathrm{alph}(ww')$, the number of its occurrences in w is equal to the number of its occurrences in w'.

Lemma 1 ([42, Lemma 1]) *Suppose that a monoid $\mathcal{S}$ has a submonoid $\mathcal{T}$ such that $\mathcal{S}$ is generated by $\mathcal{T} \cup \{d\}$ for some element d that commutes with every element of $\mathcal{T}$. Then all balanced identities satisfied by $\mathcal{T}$ hold in $\mathcal{S}$ as well.*

Corollary 1 *The monoids $\mathcal{B}_n^{\pm\tau}$ and $\mathcal{B}_n^{\tau}$ satisfy the same identities.*

Proof As observed above, the monoid $\mathcal{B}_n^{\pm\tau}$ has a submonoid isomorphic to $\mathcal{B}_n^{\tau}$, and the element d commutes with every element of this submonoid and generates $\mathcal{B}_n^{\pm\tau}$ together with this submonoid. By Lemma 1, $\mathcal{B}_n^{\pm\tau}$ satisfies all balanced identities that hold in $\mathcal{B}_n^{\tau}$. Obviously, the set $\{c^r \mid r \in \mathbb{Z}_{\geq 0}\}$ forms a submonoid in $\mathcal{B}_n^{\tau}$ and this submonoid is isomorphic to the additive monoid of non-negative integers. It is well known that every identity satisfied by the latter monoid is balanced. Hence, so is every identity that holds in $\mathcal{B}_n^{\tau}$, and we conclude that $\mathcal{B}_n^{\pm\tau}$ satisfies all identities of the monoid $\mathcal{B}_n^{\tau}$.

The converse statement is obvious as identities are inherited by submonoids. □

2.5 Structure Properties of the Monoid $\mathcal{B}_n^{\pm\tau}$

We have already mentioned that the ±-twisted Brauer monoid $\mathcal{B}_n^{\pm\tau}$ has a 'prettier' structure in comparison with $\mathcal{B}_n^{\tau}$. Here we present some of semigroup-theoretic features of $\mathcal{B}_n^{\pm\tau}$, restricting ourselves to those that are employed in the proof of our main result.

Recall that an element a of a semigroup $\mathcal{S}$ is said to be *regular* if there exists an element $b \in \mathcal{S}$ satisfying $aba = a$. A semigroup is called *regular* if every its element is regular. As the name suggests, regularity is a sort of 'positive' property. Our first structure observation about ±-twisted Brauer monoids is that they are regular (unlike twisted Brauer monoids that are known to miss this property). We employ the map $\pi \mapsto \pi^*$ on the Brauer monoid $\mathcal{B}_n$ defined as follows. Consider the permutation * on $[n] \cup [n]'$ that swaps primed with unprimed elements, that is, set $k^* = k'$, $(k')^* = k$ for all $k \in [n]$. Then define, for $\pi \in \mathcal{B}_n$,

$$p\ \pi^*\ q \Leftrightarrow p^*\ \pi\ q^* \quad \text{for all} \quad p, q \in [n] \cup [n]'.$$

Thus, π^* is obtained from π by interchanging the primed with the unprimed elements. In the geometrical representation of partitions in $\mathcal{B}_n$ via diagrams, the application of * can be visualized as the reflection along the axis between $[n]$ and $[n]'$.

Proposition 2 *The monoid $\mathcal{B}_n^{\pm\tau}$ is regular.*

Proof Take an arbitrary element $\xi = (\pi; s) \in \mathcal{B}_n^{\pm\tau}$. Denote by k the number of t-wires of the partition π. Then $n - k$ is even and the number of ℓ-wires [r-wires] in π is $m = \frac{n-k}{2}$. If p—q' is a t-wire in π, then q—p' is a t-wire in π^* whence p—p' is a t-wire in $\pi\pi^*$ and p—q' is a t-wire in $\pi\pi^*\pi$. This implies that π and $\pi\pi^*\pi$ have the same t-wires. Hence the number of ℓ-wires [r-wires] in $\pi\pi^*\pi$ is m. Since all ℓ-wires and r-wires of π are inherited by $\pi\pi^*\pi$, we conclude that π and $\pi\pi^*\pi$ have the same ℓ-wires and the same r-wires. Therefore, $\pi\pi^*\pi = \pi$ in $\mathcal{B}_n$.

If u'—v' is an r-wire in π, then u—v is an ℓ-wire in π^*, and gluing these two wires together yields a circle. We see that multiplying π by π^* on the right produces m circles, that is, $\langle\pi, \pi^*\rangle = m$. As observed in the previous paragraph, $\pi\pi^*$ has t-wires of the form p—p' where p—q' is a t-wire in π, whence the number of t-wires of $\pi\pi^*$ is at least k, and therefore, the number of r-wires of $\pi\pi^*$ is at most m. Since all r-wires of π^* are inherited by $\pi\pi^*$, we conclude that the latter partition has no other r-wires. If x'—y' is an r-wire in π^* (and hence in $\pi\pi^*$), then x—y is an ℓ-wire in π. Gluing these two wires together yields a circle, whence multiplying $\pi\pi^*$ by π on the right produces m circles. Thus, $\langle\pi\pi^*, \pi\rangle = m$. Now let $\eta = (\pi^*; -2m - s)$. Then

$$\begin{aligned}
\xi\eta\xi = (\pi; s)(\pi^*; -2m - s)(\pi; s) &= (\pi\pi^*; -2m + \langle\pi, \pi^*\rangle)(\pi; s) && \text{by (3)}\\
&= (\pi\pi^*; -m)(\pi; s) && \text{since } \langle\pi, \pi^*\rangle = m\\
&= (\pi\pi^*\pi; -m + s + \langle\pi\pi^*, \pi\rangle) && \text{by (3)}\\
&= (\pi\pi^*\pi; s) && \text{since } \langle\pi\pi^*, \pi\rangle = m\\
&= (\pi; s) = \xi && \text{since } \pi\pi^*\pi = \pi.
\end{aligned}$$

We see that the element ξ is regular. □

Remark 1 It is known and easy to see that the map $\pi \mapsto \pi^*$ is an *involution* of $\mathcal{B}_n$, that is,

$$\pi^{**} = \pi \text{ and } (\pi_1\pi_2)^* = \pi_2^*\pi_1^* \text{ for all } \pi, \pi_1, \pi_2 \in \mathcal{B}_n.$$

In the proof of Proposition 2, we have verified that $\pi\pi^*\pi = \pi$ for all $\pi \in \mathcal{B}_n$. (This fact is also known, but we have included its proof as the argument helps us to calculate the numbers $\langle\pi, \pi^*\rangle$ and $\langle\pi\pi^*, \pi\rangle$.) The three properties of the map $\pi \mapsto \pi^*$ mean that the Brauer monoid $\mathcal{B}_n$ is a *regular $*$-semigroup* as defined in [38]. This stronger form of regularity does not extend to the $\pm$-twisted Brauer monoid $\mathcal{B}_n^{\pm\tau}$. If, as the proof of Proposition 2 suggests, one defines the map $\xi = (\pi; s) \mapsto \xi^* := (\pi^*; t(\pi) - n - s)$ where $t(\pi)$ is the number of t-wires of the partition π, then the equalities $\xi\xi^*\xi = \xi$, $\xi^*\xi\xi^* = \xi^*$, and $(\xi^*)^* = \xi$ hold, but the equality $(\xi_1\xi_2)^* = \xi_2^*\xi_1^*$ fails in general. In fact, if $\xi_1 = (\pi_1; s_1)$ and $\xi_2 = (\pi_2; s_2)$, then the necessary and sufficient condition for $(\xi_1\xi_2)^* = \xi_2^*\xi_1^*$ to hold is that the number of r-wires of π_1 equals the number of ℓ-wires of π_2 and every r-wire of π_1 is merged with exactly one ℓ-wire of π_2 when the product $\pi_1\pi_2$ is formed.

We proceed with determining the Green structure of the $\pm$-twisted Brauer monoid. Recall the necessary definitions.

Let $\mathcal{S}$ be a semigroup. As usual, $\mathcal{S}^1$ stands for the least monoid containing $\mathcal{S}$ (that is, $\mathcal{S}^1 = \mathcal{S}$ if $\mathcal{S}$ is a monoid and otherwise $\mathcal{S}^1 = \mathcal{S} \cup \{1\}$ where the new symbol 1 behaves as a multiplicative identity element). Define three natural preorders $\leq_{\mathscr{L}}$, $\leq_{\mathscr{R}}$ and $\leq_{\mathscr{J}}$ which are the relations of left, right and bilateral divisibility respectively:

$$a \leq_{\mathscr{L}} b \Leftrightarrow a = sb \text{ for some } s \in \mathcal{S}^1;$$
$$a \leq_{\mathscr{R}} b \Leftrightarrow a = bs \text{ for some } s \in \mathcal{S}^1;$$
$$a \leq_{\mathscr{J}} b \Leftrightarrow a = sbt \text{ for some } s, t \in \mathcal{S}^1.$$

Green's equivalences $\mathscr{L}$, $\mathscr{R}$, and $\mathscr{J}$ are the equivalence relations corresponding to the preorders $\leq_{\mathscr{L}}$, $\leq_{\mathscr{R}}$ and $\leq_{\mathscr{J}}$ (that is, $a \mathrel{\mathscr{L}} b$ if and only if $a \leq_{\mathscr{L}} b \leq_{\mathscr{L}} a$, etc.). In addition, let $\mathscr{H} = \mathscr{L} \cap \mathscr{R}$ and $\mathscr{D} = \mathscr{L}\mathscr{R} = \{(a, b) \in \mathcal{S} \times \mathcal{S} \mid \exists c \in \mathcal{S} : (a, c) \in \mathscr{L} \wedge (c, b) \in \mathscr{R}\}$.

Our description of Green's relations on $\mathcal{B}_n^{\pm\tau}$ has the same form as (and easily follows from) the description of Green's relations on the Brauer monoid $\mathcal{B}_n$ in [34, Sect. 7]. At the same time, it essentially differs from the description of Green's relations on the twisted Brauer monoid that can be found in [12, Sect. 3.1] or [16, Sect. 4].

For a partition $\pi \in \mathcal{B}_n$, let $L(\pi)$ and $R(\pi)$ denote the sets of its ℓ-and, respectively, r-wires, and let $t(\pi)$ denote the number of its t-wires.

Proposition 3 *Elements* $(\pi_1; s_1), (\pi_2; s_2) \in \mathcal{B}_n^{\pm\tau}$ *are*

(L) $\mathscr{L}$*-related if and only if* $R(\pi_1) = R(\pi_2)$*;*
(R) $\mathscr{R}$*-related if and only if* $L(\pi_1) = L(\pi_2)$*;*
(H) $\mathscr{H}$*-related if and only if* $L(\pi_1) = L(\pi_2)$ *and* $R(\pi_1) = R(\pi_2)$*;*
(J) $\mathscr{J}$*-related if and only if they are* $\mathscr{D}$*-related if and only if* $t(\pi_1) = t(\pi_2)$*.*

Proof (L) If $(\pi_1; s_1)$ and $(\pi_2; s_2)$ are $\mathscr{L}$-related in $\mathcal{B}_n^{\pm\tau}$, then their images π_1 and π_2 under the 'forgetting' homomorphism $(\pi; s) \mapsto \pi$ are $\mathscr{L}$-related in $\mathcal{B}_n$. By [34, Theorem 7(1)], this implies $R(\pi_1) = R(\pi_2)$.

Conversely, suppose that $R(\pi_1) = R(\pi_2)$. Then π_1 and π_2 are $\mathscr{L}$-related in $\mathcal{B}_n$ by [34, Theorem 7(1)]. This means that $\sigma_1\pi_1 = \pi_2$ and $\sigma_2\pi_2 = \pi_1$ for some partitions $\sigma_1, \sigma_2 \in \mathcal{B}_n$. Let $r_1 = \langle\sigma_1, \pi_1\rangle$ and $r_2 = \langle\sigma_2, \pi_2\rangle$. Then, using (3), we get

$$(\sigma_1; s_2 - s_1 - r_1)(\pi_1; s_1) = (\sigma_1\pi_1; s_2 - r_1 + \langle\sigma_1, \pi_1\rangle) = (\pi_2; s_2),$$
$$(\sigma_2; s_1 - s_2 - r_2)(\pi_2; s_2) = (\sigma_2\pi_2; s_1 - r_2 + \langle\sigma_2, \pi_2\rangle) = (\pi_1; s_1).$$

Hence, $(\pi_1; s_1)$ and $(\pi_2; s_2)$ are $\mathscr{L}$-related in $\mathcal{B}_n^{\pm\tau}$.

(R) follows by a symmetric argument.

(H) is clear.

(J) The 'only if' part follows from the fact that $t(\pi\sigma) \leq \min\{t(\pi), t(\sigma)\}$ for all partitions $\pi, \sigma \in \mathcal{B}_n$.

For the 'if' part, suppose that $t(\pi_1) = t(\pi_2) = k$. Then $n - k$ is even and the number of ℓ-wires [r-wires] in π_1 and π_2 is $m = \frac{n-k}{2}$. Construct a partition σ as follows. Take the same r-wires as in π_1 and the same ℓ-wires as in π_2. After that, there remain $k = n - 2m$ 'non-engaged' left points and the same number of 'non-engaged' right points; to complete the construction, we couple them into t-wires in any of $k!$ possible ways. Then $(\pi_1; s)$ and $(\sigma; 0)$ are $\mathscr{L}$-related in $\mathcal{B}_n^{\pm\tau}$ by (L), while $(\sigma, 0)$ and $(\pi_2; s_2)$ are $\mathscr{R}$-related in $\mathcal{B}_n^{\pm\tau}$ by (R). Therefore, $(\pi_1; s_1)$ and $(\pi_2; s_2)$ are $\mathscr{D}$-related in $\mathcal{B}_n^{\pm\tau}$. □

Comparing Proposition 3 and the description of Green's relations on the Brauer monoid in [34, Sect. 7], we state the following.

Corollary 2 *For any Green relation $\mathscr{K} \in \{\mathscr{J}, \mathscr{D}, \mathscr{L}, \mathscr{R}, \mathscr{H}\}$, one has $(\pi_1; s_1)\ \mathscr{K}\ (\pi_2; s_2)$ in $\mathcal{B}_n^{\pm\tau}$ if and only if $\pi_1\ \mathscr{K}\ \pi_2$ in $\mathcal{B}_n$.*

By Proposition 3(J), the monoid $\mathcal{B}_n^{\pm\tau}$ has the following $\mathscr{J}$-classes:

$$J_k = \{(\pi; s) \in \mathcal{B}_n \times \mathbb{Z} \mid t(\pi) = k\}.$$

Here $k = n, n-2, \ldots, 0$ if n is even and $k = n, n-2, \ldots, 1$ if n is odd so the number of $\mathscr{J}$-classes is $\lceil \frac{n+1}{2} \rceil$. For comparison, the twisted Brauer monoid $\mathcal{B}_n^{\tau}$ contains infinite descending chains of $\mathscr{J}$-classes; see [12, Sect. 3.1].

A semigroup $\mathcal{S}$ is called *stable* if for all $a, b \in \mathcal{S}$, the implications $a\ \mathscr{J}\ ab \Rightarrow a\ \mathscr{R}\ ab$ and $a\ \mathscr{J}\ ba \Rightarrow a\ \mathscr{L}\ ba$ hold. It is well known that every finite semigroup is stable, see, e.g., [15].

Corollary 3 *The monoid $\mathcal{B}_n^{\pm\tau}$ is stable.*

Proof Let $a = (\pi_1; s_1)$ and $b = (\pi_2; s_2)$ be such that $a\ \mathscr{J}\ ab$ in $\mathcal{B}_n^{\pm\tau}$. Then $\pi_1\ \mathscr{J}\ \pi_1\pi_2$ in $\mathcal{B}_n$ by Corollary 2 whence $\pi_1\ \mathscr{R}\ \pi_1\pi_2$ in $\mathcal{B}_n$ since the finite monoid $\mathcal{B}_n$ is stable. Now Corollary 2 gives $a\ \mathscr{R}\ ab$ in $\mathcal{B}_n^{\pm\tau}$. The other implication from the definition of stability is verified in the same way. □

The final ingredient that we need is the structure of the maximal subgroups of $\mathcal{B}_n^{\pm\tau}$.

Proposition 4 *For $k > 0$, the maximal subgroups in the $\mathscr{J}$-class J_k of the monoid $\mathcal{B}_n^{\pm\tau}$ are isomorphic to the group $\mathbb{S}_k \times \mathbb{Z}$.*

Proof By Green's Theorem [21, Theorem 2.2.5], the maximal subgroups in J_k are exactly the $\mathscr{H}$-classes in J_k that contain idempotents. If $\epsilon = (\pi; s)$ is an idempotent in J_k, then $(\pi; s) = (\pi; s)^2 = (\pi^2, 2s + \langle \pi, \pi \rangle)$ by (3). Hence $\pi = \pi^2$ is an idempotent in the monoid $\mathcal{B}_n$, and $s = 2s + \langle \pi, \pi \rangle$, whence $\langle \pi, \pi \rangle = -s$. By Corollary 2, the $\mathscr{H}$-class of the idempotent ϵ is the set $\mathcal{H}_\epsilon = \{(\sigma; i) \mid \sigma \in \mathcal{H}_\pi;\ i \in \mathbb{Z}\}$, where $\mathcal{H}_\pi$ is the $\mathscr{H}$-class of the idempotent π in $\mathcal{B}_n$. By Proposition 3(H) all partitions $\sigma_1, \sigma_2 \in \mathcal{H}_\pi$ have the same ℓ- and r-wires as π. Hence $\langle \sigma_1, \sigma_2 \rangle = \langle \pi, \pi \rangle = -s$, and therefore,

$$(\sigma_1; i_1)(\sigma_2; i_2) = (\sigma_1\sigma_2; i_1 + i_2 + \langle \sigma_1, \sigma_2 \rangle) = (\sigma_1\sigma_2; i_1 + i_2 - s).$$

This readily implies that the bijection $\mathcal{H}_\epsilon \to \mathcal{H}_\pi \times \mathbb{Z}$ defined by $(\sigma; i) \mapsto (\sigma; i - s)$ is a group isomorphism. Since the number of t-wires in each partition in $\mathcal{H}_\pi$ is equal to k, [34, Theorem 1] implies that the subgroup $\mathcal{H}_\pi$ is isomorphic to the symmetric group $\mathbb{S}_k$. □

Remark 2 If n is even, the least $\mathscr{J}$-class of the monoid $\mathcal{B}_n^{\pm\tau}$ (with respect to the ordering of $\mathscr{J}$-classes induced by the preorder $\leq_{\mathscr{J}}$) is J_0. Its maximal subgroups are isomorphic to $\mathbb{Z}$. For the sake of uniformity, let $\mathbb{S}_0$ be the trivial group (this complies with the usual convention that $0! = 1$). This way Proposition 4 extends to the case $k = 0$.

3 Reduction Theorem for Identity Checking

We need the following reduction:

Theorem 1 *Let $\mathcal{S}$ be a stable semigroup with finitely many $\mathscr{J}$-classes and $\mathcal{G}$ the direct product of all maximal subgroups of $\mathcal{S}$. Then there exists a polynomial reduction from the problem* CHECK- ID($\mathcal{G}$) *to the problem* CHECK- ID($\mathcal{S}$).

In [1, Theorem 1], the same reduction was proved for **finite** semigroups. In fact, the proof in [1] needs only minor adjustments to work under the premises of Theorem 1. Still, for the reader's convenience, we provide a self-contained argument so that it should be possible to understand the proof of Theorem 1 without any acquaintance with [1].

Proof The existence of a polynomial reduction from CHECK- ID($\mathcal{G}$) to CHECK- ID($\mathcal{S}$) means the following. Given an arbitrary instance of CHECK- ID($\mathcal{G}$), i.e., an arbitrary identity $u \doteq v$, one can construct an identity $U \doteq V$ such that:

(Size) the lengths of the words U and V are bounded by the values of a fixed polynomial in the lengths of the words u and v;
(Equi) the identity $U \doteq V$ holds in $\mathcal{S}$ if and only if the identity $u \doteq v$ holds in $\mathcal{G}$.

Toward the construction, assume that $\Sigma = \text{alph}(uv)$ consists of the letters $x_1, \ldots, x_m$. Let Σ^+ denote the *free semigroup over* Σ, that is, the set of all words built from the letters in Σ and equipped with concatenation as multiplication. It is known (and easy to verify) that Σ^+ has the following *universal property*: every map $\Sigma \to \Sigma^+$ uniquely extends to an endomorphism of the semigroup Σ^+.

Define the following m words:

$$\begin{aligned} w_1 &= x_1^2 x_2 \cdots x_m x_1, \\ w_2 &= x_1 x_2^2 \cdots x_m x_1, \\ &\cdots\cdots\cdots\cdots\cdots\cdots \\ w_{m-1} &= x_1 x_2 \cdots x_{m-1}^2 x_m x_1, \\ w_m &= x_1 x_2 \cdots x_m x_1. \end{aligned} \tag{16}$$

(The reader might suspect a typo in the last line of (16) as the word w_m involves no squared letter, unlike all previous words. No, the expression for w_m is correct, and its distinct role will be revealed shortly.) We denote by φ the endomorphism of Σ^+ that extends the map $x_i \mapsto w_i$, $i = 1, \dots, m$. For each $k = 1, 2, \dots$, let $w_{i,k} = \varphi^k(x_i)$ and let N be the number of $\mathscr{J}$-classes of $\mathcal{S}$. We claim that the identity

$$U = u(w_{1,2N}, \dots, w_{m,2N}) \bumpeq v(w_{1,2N}, \dots, w_{m,2N}) = V$$

possesses the desired properties (Size) and (Equi).

For (Size), observe that the length of each of the words (16) does not exceed $m + 2$, and therefore, the length of each of the words $w_{1,2N}, \dots, w_{m,2N}$ does not exceed $(m + 2)^{2N}$. Here, the number N is defined by the semigroup $\mathcal{S}$ only and does not depend on the words u and v, and the number m does not exceed the maximum of the lengths of u and v. Since the length of the word $U = u(w_{1,2N}, \dots, w_{m,2N})$ (respectively, $V = v(w_{1,2N}, \dots, w_{m,2N})$) does not exceed the product of the maximum length of the words $w_{i,2N}$ and the length of the word u (respectively, v), the polynomial X^{2N+1} witnesses the property (Size).

The verification of (Equi) is more involved. We start with the following observation.

Lemma 2 *If $\mathcal{S}$ a stable semigroup with a finite number N of $\mathscr{J}$-classes, then for every substitution $\Sigma \to \mathcal{S}$, there is a subgroup $\mathcal{H}$ in $\mathcal{S}$ such that the values of all words $w_{1,2N}, \dots, w_{m,2N}$ under this substitution belong to $\mathcal{H}$.* □

Proof Notice that for each $k = 1, 2, \dots$,

$$\begin{aligned} w_{i,k+1} = \varphi^{k+1}(x_i) &= \varphi^k(\varphi(x_i)) = \varphi^k(w_i(x_1, \dots, x_m)) \\ &= w_i(\varphi^k(x_1), \dots, \varphi^k(x_m)) = w_i(w_{1,k}, \dots, w_{m,k}). \end{aligned} \tag{17}$$

Inspecting the definition (16), we see that every letter x_i occurs in each of the words $w_1, \dots, w_m$. Therefore, the equalities (17) imply that the word $w_{i,k}$ appears as a factor in the word $w_{j,k+1}$ for every $k = 1, 2, \dots$ and every $i, j = 1, \dots, m$.

Fix a substitution $\Sigma \to \mathcal{S}$ and denote the value of a word $w \in \Sigma^+$ under this substitution by $\overline{w}$. Since $w_{1,k}$ appears as a factor in $w_{1,k+1}$, the following inequalities hold in $\mathcal{S}$:

$$\overline{w_{1,1}} \geq_{\mathscr{J}} \overline{w_{1,2}} \geq_{\mathscr{J}} \cdots \geq_{\mathscr{J}} \overline{w_{1,2N+1}}.$$

Among these inequalities, at most $N - 1$ can be strict, whence by the pigeonhole principle, the sequence $\overline{w_{1,1}}, \overline{w_{1,2}}, \dots, \overline{w_{1,2N+1}}$ contains three adjacent $\mathscr{J}$-

related elements. Let $k < 2N$ be such that $\overline{w_{1,k}} \mathrel{\mathscr{J}} \overline{w_{1,k+1}} \mathrel{\mathscr{J}} \overline{w_{1,k+2}}$. Again inspecting (16), we see that the word x_1^2 appears as a factor in the word w_1. Hence, by the equalities (17), the word $w_{1,k}^2$ appears as a factor in the word $w_{1,k+1}$, and therefore, we have $\overline{w_{1,k}}^2 \geq_{\mathscr{J}} \overline{w_{1,k+1}} \mathrel{\mathscr{J}} \overline{w_{1,k}}$ in $\mathcal{S}$. Obviously, $\overline{w_{1,k}}^2 \leq_{\mathscr{J}} \overline{w_{1,k}}$ whence $\overline{w_{1,k}}^2 \mathrel{\mathscr{J}} \overline{w_{1,k}}$. Since $\mathcal{S}$ is stable, $\overline{w_{1,k}}^2 \mathrel{\mathscr{J}} \overline{w_{1,k}}$ implies $\overline{w_{1,k}}^2 \mathrel{\mathscr{L}} \overline{w_{1,k}}$ and $\overline{w_{1,k}}^2 \mathrel{\mathscr{R}} \overline{w_{1,k}}$, that is, $\overline{w_{1,k}}^2 \mathrel{\mathscr{H}} \overline{w_{1,k}}$. By Green's Theorem [21, Theorem 2.2.5], the $\mathscr{H}$-class $\mathcal{H}$ of the element $\overline{w_{1,k}}$ is a maximal subgroup of the semigroup $\mathcal{S}$.

Yet another look at (16) reveals that each of the words $w_1, \dots, w_m$ starts and ends with the letter x_1. In view of the equalities (17), the word $w_{1,k}$ appears as a prefix as well as a suffix of each of the words $w_{i,k+1}$, which, in turn, appear as factors in the word $w_{1,k+2}$. Hence $\overline{w_{i,k+1}} = \overline{w_{1,k}}b = a\overline{w_{1,k}}$ for some $a, b \in \mathcal{S}$ and all elements $\overline{w_{i,k+1}}$ lie in the $\mathscr{J}$-class of $\overline{w_{1,k}}$. By stability of the semigroup $\mathcal{S}$, all these elements lie in both the $\mathscr{L}$-class and the $\mathscr{R}$-class of the element $\overline{w_{1,k}}$. Thus, all elements $\overline{w_{i,k+1}}$ lie in the subgroup $\mathcal{H}$, whence the subgroup contains all elements $\overline{w_{i,\ell}}$ for all $\ell > k$. We see that the subgroup $\mathcal{H}$ indeed contains the values of all words $w_{1,2N}, \dots, w_{m,2N}$ under the substitution we consider. □

Now we are in a position to prove that if the identity $u \simeq v$ holds in $\mathcal{G}$, then the identity $U \simeq V$ holds in $\mathcal{S}$. Consider an arbitrary substitution $\zeta\colon \Sigma \to \mathcal{S}$. By Lemma 2, the values of the words $w_{1,2N}, \dots, w_{m,2N}$ under ζ lie in a subgroup $\mathcal{H}$ of the semigroup $\mathcal{S}$. Since $\mathcal{H}$ is a subgroup of $\mathcal{G}$, the identity $u \simeq v$ holds in $\mathcal{H}$, and hence, substituting for $x_1, \dots, x_m$ the values of the words $w_{1,2N}, \dots, w_{m,2N}$ yields the equality

$$u(\zeta(w_{1,2N}), \dots, \zeta(w_{m,2N})) = v(\zeta(w_{1,2N}), \dots, \zeta(w_{m,2N})).$$

in $\mathcal{H}$. However,

$$\begin{aligned} u(\zeta(w_{1,2N}), \dots, \zeta(w_{m,2N})) &= \zeta(u(w_{1,2N}, \dots, w_{m,2N})) = \zeta(U), \\ v(\zeta(w_{1,2N}), \dots, \zeta(w_{m,2N})) &= \zeta(v(w_{1,2N}, \dots, w_{m,2N})) = \zeta(V), \end{aligned}$$

and hence U and V take the same value under ζ. Since the substitution was arbitrary, the identity $U \simeq V$ holds in $\mathcal{S}$.

It remains to verify the converse: if the identity $U \simeq V$ holds in $\mathcal{S}$, then the identity $u \simeq v$ holds in $\mathcal{G}$. As identities are inherited by direct products, it suffices to show that $u \simeq v$ holds in every maximal subgroup $\mathcal{H}$ of $\mathcal{S}$. This amounts to verifying that $u(h_1, \dots, h_m) = v(h_1, \dots, h_m)$ for an arbitrary m-tuple of elements $h_1, \dots, h_m \in \mathcal{H}$.

The free semigroup Σ^+ can be considered as a subsemigroup in the free group $\mathcal{FG}(\Sigma)$ over Σ. The endomorphism $\varphi\colon x_i \mapsto w_i$ of Σ^+ extends to an endomorphism of $\mathcal{FG}(\Sigma)$, still denoted by φ. The words $w_1, \dots, w_m$ defined by (16) generate $\mathcal{FG}(\Sigma)$ since in $\mathcal{FG}(\Sigma)$, one can express $x_1, \dots, x_m$ via $w_1, \dots, w_m$ as follows:

$$
\begin{aligned}
x_1 &= w_1 w_m^{-1},\\
x_2 &= x_1^{-1} w_2 w_m^{-1} x_1,\\
x_3 &= (x_1x_2)^{-1} w_3 w_m^{-1} x_1x_2,\\
&\dots\dots\dots\dots\dots\dots\dots\dots\dots\dots\\
x_{m-1} &= (x_1x_2\cdots x_{m-2})^{-1} w_{m-1} w_m^{-1} x_1x_2\cdots x_{m-2},\\
x_m &= (x_1x_2\cdots x_{m-1})^{-1} w_m x_1^{-1}.
\end{aligned}
$$

(This is where the distinct expression for w_m comes into play!) Hence φ treated as an endomorphism of $\mathcal{FG}(\Sigma)$ is surjective, and so is any power of φ. It is well known (cf. [29, Proposition I.3.5]) that every surjective endomorphism of a finitely generated free group is an automorphism. Denote by φ^{-2N} the inverse of the automorphism φ^{2N} of $\mathcal{FG}(\Sigma)$ and let $g_i = \varphi^{-2N}(x_i)$, $i = 1, \dots, m$. Then

$$
\begin{aligned}
w_{i,2N}(g_1, \dots, g_m) &= w_{i,2N}(\varphi^{-2N}(x_1), \dots, \varphi^{-2N}(x_m))\\
&= \varphi^{-2N}(w_{i,2N}(x_1, \dots, x_m)) = \varphi^{-2N}(\varphi^{2N}(x_i)) = x_i \qquad (18)
\end{aligned}
$$

for all $i = 1, \dots, m$. Since the equalities (18) hold in the free m-generated group, they remain valid under any interpretation of the letters $x_1, \dots, x_m$ by arbitrary m elements of an arbitrary group. Now we define a substitution $\zeta : \Sigma \to \mathcal{H}$ letting

$$
\zeta(x_i) = g_i(h_1, \dots, h_m), \quad i = 1, \dots, m.
$$

Then in view of (18) we have

$$
\begin{aligned}
\zeta(w_{i,2N}(x_1, \dots, x_m)) &= w_{i,2N}(\zeta(x_1), \dots, \zeta(x_m))\\
&= w_{i,2N}\big(g_1(h_1, \dots, h_m), \dots, g_m(h_1, \dots, h_m)\big) = h_i
\end{aligned}
$$

for all $i = 1, \dots, m$. Hence we have

$$
\begin{aligned}
u(h_1, \dots, h_m) &= u(\zeta(w_{1,2N}(x_1, \dots, x_m)), \dots, \zeta(w_{m,2N}(x_1, \dots, x_m)))\\
&= \zeta(u(w_{1,2N}(x_1, \dots, x_m), \dots, w_{m,2N}(x_1, \dots, x_m)))\\
&= \zeta(U(x_1, \dots, x_m)),
\end{aligned}
$$

and, similarly, $v(h_1, \dots, h_m) = \zeta(V(x_1, \dots, x_m))$. Since the identity $U \simeq V$ holds in $\mathcal{S}$, the values of the words U and V under ζ are equal, whence $u(h_1, \dots, h_m) = v(h_1, \dots, h_m)$, as required. This completes the proof of (Equi), and hence, the proof of Theorem 1. □

4 Co-NP-Completeness of Identity Checking in $\mathcal{B}_n^\tau$ with $n \geq 5$

We are ready to prove our main result.

Theorem 2 *For each $n \geq 5$, the problem* CHECK- ID($\mathcal{B}_n^\tau$) *is co-NP-complete.*

Proof Proving that a decision problem P is co-NP-complete amounts to showing that the problem P belongs to the complexity class co-NP and is co-NP-hard, the latter meaning that there exists a polynomial reduction from a co-NP-complete problem to P.

The fact that CHECK- ID($\mathcal{B}_n^\tau$) lies in the class co-NP is easy. The following nondeterministic algorithm has a chance to return the answer 'NO' if and only if it is given an identity $w \doteq w'$ that does not hold in $\mathcal{B}_n^\tau$.

1. If $|\operatorname{alph}(ww')| = k$, guess a k-tuple of elements in $\mathcal{B}_n^\tau$.

2. Substitute the elements from the guessed k-tuple for the letters in $\operatorname{alph}(ww')$ and compute the values of the words w and w'.

3. Return 'NO' if the values are different.

The multiplication in $\mathcal{B}_n^\tau$ is constructive so that the computation in Step 2 takes polynomial (in fact, linear) time in the lengths of w and w'.

In order to prove the co-NP-hardness of CHECK- ID($\mathcal{B}_n^\tau$), we use the reduction of Theorem 1 and the powerful result by Horváth, Lawrence, Merai, and Szabó [20] who discovered that for every nonsolvable finite group $\mathcal{G}$, the problem CHECK- ID($\mathcal{G}$) is co-NP-complete. Already Galois knew that for $n \geq 5$ the group $\mathbb{S}_n$ is nonsolvable so the result of [20] applies to $\mathbb{S}_n$.

Fix an $n \geq 5$. An identity $w \doteq w'$ holds in the group $\mathbb{S}_n$ if and only if so does the identity $w^{n!-1}w' \doteq 1$. The length of the word $w^{n!-1}w'$ is bounded by the value of the polynomial $X^{n!}$ in the maximum length of the words w and w'. Thus, we have a mutual polynomial reduction between the problem CHECK- ID($\mathbb{S}_n$) and the problem of determining whether or not all values of a given semigroup word v in $\mathbb{S}_n$ are equal to the identity of the group. It is well known that the center of $\mathbb{S}_n$ is trivial whence the latter property is equivalent to saying that all values of v in $\mathbb{S}_n$ lie in the center. This, in turn, is equivalent to the fact that the identity $vx \doteq xv$ where $x \notin \operatorname{alph}(v)$ holds in $\mathbb{S}_n$. Clearly, for any word v, the identity $vx \doteq xv$ is balanced. We conclude that the problem CHECK- ID($\mathbb{S}_n$) remains co-NP-complete when restricted to balanced identities.

An identity holds in the direct product of semigroups if and only if it holds in each factor of the product. Applying this to the product $\mathbb{S}_n \times \mathbb{Z}$ and taking into account that $\mathbb{Z}$ satisfies exactly balanced identities, we see that the identities holding in $\mathbb{S}_n \times \mathbb{Z}$ are precisely the balanced identities holding in $\mathbb{S}_n$. Hence the problem CHECK- ID($\mathbb{S}_n \times \mathbb{Z}$) is co-NP-complete.

By Proposition 4 (and Remark 2), the maximal subgroups of the $\pm$-twisted Brauer monoid $\mathcal{B}_n^{\pm\tau}$ are of the form $\mathbb{S}_k \times \mathbb{Z}$, where $k = n, n-2, \dots, 0$ if n is even and $k = n, n-2, \dots, 1$ if n is odd. Any group of this form embeds into $\mathbb{S}_n \times \mathbb{Z}$ whence the identities that hold in each maximal subgroup of $\mathcal{B}_n^{\pm\tau}$ are exactly the identities of $\mathbb{S}_n \times \mathbb{Z}$. We conclude that the identities of the direct product $\mathcal{G}$ of all maximal subgroups of $\mathcal{B}_n^{\pm\tau}$ coincide with the identities of $\mathbb{S}_n \times \mathbb{Z}$. Hence, the problem CHECK-ID($\mathcal{G}$) is co-NP-complete.

By Corollary 3 and Proposition 3, the ±-twisted Brauer monoid is stable and has finitely many $\mathscr{J}$-classes. Thus, Theorem 1 applies to $\mathcal{B}_n^{\pm\tau}$, providing a polynomial reduction from the co-NP-complete problem CHECK- ID($\mathcal{G}$) to the problem CHECK-ID($\mathcal{B}_n^{\pm\tau}$). Hence, the latter problem is co-NP-hard. It remains to refer to Corollary 1 stating that the ±-twisted Brauer monoid $\mathcal{B}_n^{\pm\tau}$ and the twisted Brauer monoid $\mathcal{B}_n^{\tau}$ satisfy the same identities, and therefore, the problem CHECK- ID($\mathcal{B}_n^{\tau}$) is co-NP-hard as well. □

The restriction $n \geq 5$ in Theorem 2 is essential for the above proof. This does not mean, however, that it is necessary for co-NP-completeness of the problem CHECK-ID($\mathcal{B}_n^{\tau}$). In fact, we have proved that identity checking in $\mathcal{B}_4^{\tau}$ remains co-NP-complete. The proof uses a completely different technique, and therefore, it will be published separately.

The case $n = 3$ remains open. As for $n = 1, 2$, the monoid $\mathcal{B}_1^{\tau}$ is trivial, and hence, it satisfies every identity, and the monoid $\mathcal{B}_2^{\tau}$ is commutative and can be easily shown to satisfy exactly balanced identities. Thus, for $n = 1, 2$, the problem CHECK- ID($\mathcal{B}_n^{\tau}$) is polynomial (actually, linear) time decidable.

Remark 3 Up to now, the only available information about the identities of twisted Brauer monoids was [2, Theorem 4.1] showing that no finite set of identities of $\mathcal{B}_n^{\tau}$ with $n \geq 3$ can infer all such identities, in other words, $\mathcal{B}_n^{\tau}$ with $n \geq 3$ has no finite identity basis. This fact was obtained via a 'high-level' argument that allows one to prove, under certain conditions, that a semigroup $\mathcal{S}$ admits no finite identity basis, without writing down any concrete identity holding in $\mathcal{S}$. In contrast, the above proof of Theorem 2 via Theorem 1 is constructive. Following the recipe of Theorem 1, one can use the words (16) to convert any concrete balanced semigroup identity $u \simeq v$ of the group $\mathbb{S}_n$ into an identity $U \simeq V$ of the twisted Brauer monoid $\mathcal{B}_n^{\tau}$. We refer the reader to [6, 24] for recent information about short balanced semigroup identities in symmetric groups.

5 Related Results and Further Work

5.1 Checking Identities in Twisted Partition Monoids

The approach of the present paper can be applied to studying the identities of other interesting families of infinite monoids, in particular, twisted partition monoids. The latter constitute a natural generalization of twisted Brauer monoids and also serve as bases of certain semigroup algebras relevant in statistical mechanics and representation theory, the so-called *partition algebras*. Partition algebras were discovered and studied in depth by Martin [30–33] and, independently, by Jones [22] in the context of statistical mechanics; their remarkable role in representation theory is nicely presented in the introduction of [17].

We define twisted partition monoids, 'twisting' the definition of partition monoids as given in [44]. As in Sect. 2.1, let $[n] = \{1, \ldots, n\}$ and $[n]' = \{1', \ldots, n'\}$. Consider the set $\mathcal{P}_n^\tau$ of all pairs $(\pi; s)$ where π is an arbitrary partition of the $2n$-element set $[n] \cup [n]'$ and s is a non-negative integer. (The difference with $\mathcal{B}_n^\tau$ is that one drops the restriction that all blocks of π consist of two elements.) The product $(\pi; s)$ of two pairs $(\pi_1; s_1), (\pi_2; s_2) \in \mathcal{P}_n^\tau$ is computed in the following six steps.

1. Let $[n]'' = \{1'', \ldots, n''\}$ and define the partition π_2' on $[n]' \cup [n]''$ by
$$x' \; \pi_2' \; y' \Leftrightarrow x \; \pi_2 \; y \text{ for all } x, y \in [n] \cup [n]'.$$
2. Let π'' be the equivalence relation on $[n] \cup [n]' \cup [n]''$ generated by $\pi_1 \cup \pi_2'$, that is, π'' is the transitive closure of $\pi_1 \cup \pi_2'$.
3. Count the number of blocks of π'' that involve only elements from $[n]'$ and denote this number by $\langle \pi_1, \pi_2 \rangle$.
4. Convert π'' into a partition π' on the set $[n] \cup [n]''$ by removing all elements having a single prime $'$ from all blocks; all blocks having only such elements are removed as a whole.
5. Replace double primes with single primes to obtain a partition π, that is, set
$$x \; \pi \; y \Leftrightarrow f(x) \; \pi' \; f(y) \text{ for all } x, y \in [n] \cup [n]'$$
where $f\colon [n] \cup [n]' \to [n] \cup [n]''$ is the bijection $x \mapsto x, x' \mapsto x''$ for all $x \in [n]$.
6. Set $(\pi_1; s_1)(\pi_2; s_2) = (\pi; s_1 + s_2 + \langle \pi_1, \pi_2 \rangle)$.

For an illustration, let $n = 5$ and consider $(\pi_1; s_1), (\pi_2; s_2) \in \mathcal{P}_5^\tau$ with

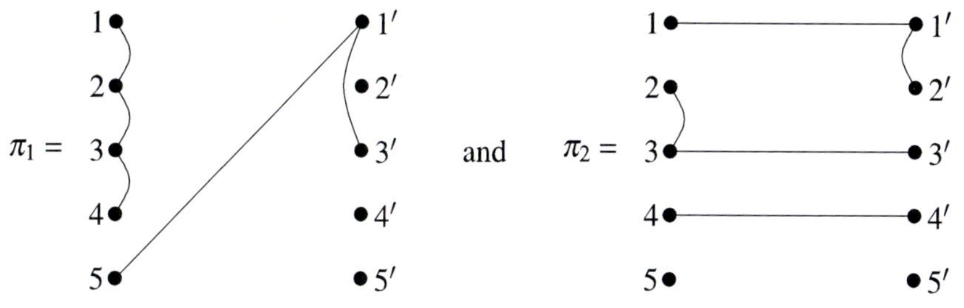

where the partitions are shown as graphs on $[5] \cup [5]'$ whose connected components represent blocks. Then

$\pi'' =$ 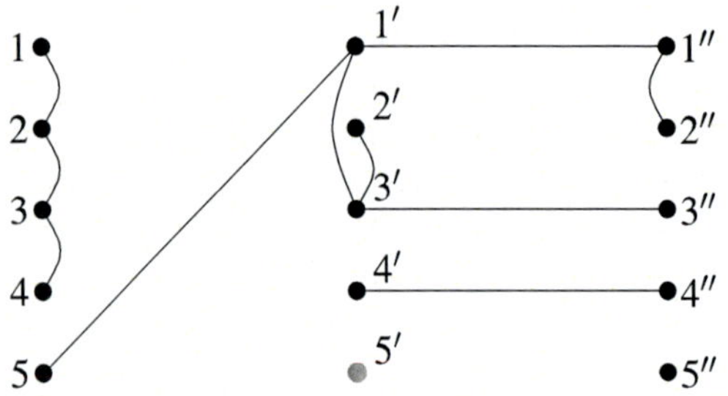

and we see that $\langle \pi_1, \pi_2 \rangle = 1$ as only the singleton light-gray block consists of elements with single prime. Hence

$\pi' =$ 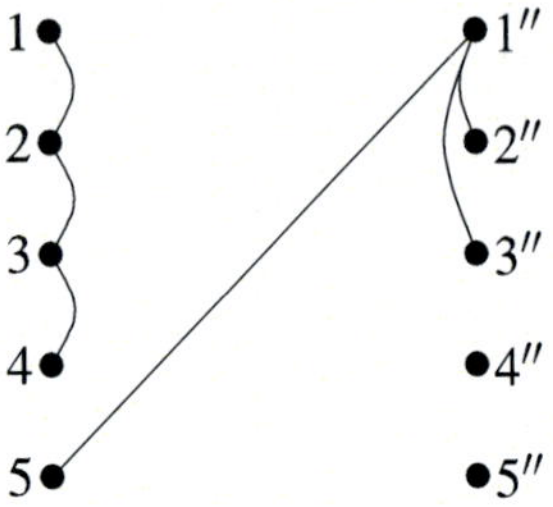

and, finally, $\pi =$ 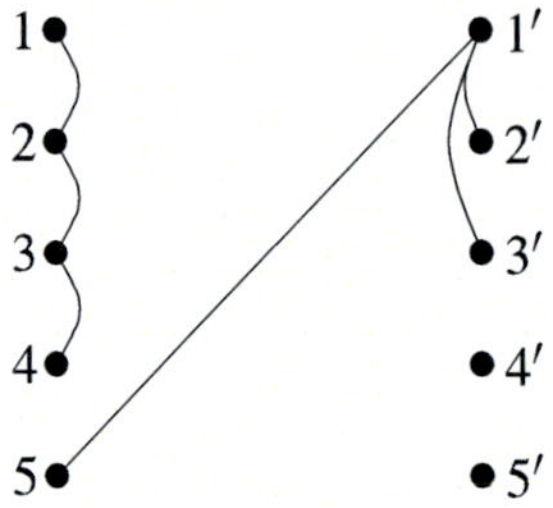

We conclude that $(\pi_1; s_1)(\pi_2; s_2) = (\pi; s_1 + s_2 + 1)$.

The above-defined multiplication in $\mathcal{P}_n^\tau$ is associative [31] and its restriction to $\mathcal{B}_n^\tau$ coincides with the multiplication in the twisted Brauer monoid defined in Sect. 2.1. Hence, $\mathcal{B}_n^\tau$ is a submonoid in $\mathcal{P}_n^\tau$, and it is easy to see that the identity element of $\mathcal{B}_n^\tau$ serves as the identity element for $\mathcal{P}_n^\tau$ as well. Thus, $\mathcal{P}_n^\tau$ is a monoid called the *twisted partition monoid*.

The machinery developed in the present paper works, with minor adjustments, for twisted partition monoids and yields the following analog of Theorem 2.

Theorem 3 *For each $n \geq 5$, the problem* CHECK- ID($\mathcal{P}_n^\tau$) *is co-NP-complete.*

The detailed proof of Theorem 3 will be published elsewhere.

Similar results can be obtained for various submonoids of $\mathcal{P}_n^\tau$, provided that they share the structure of their maximal subgroups with $\mathcal{P}_n^\tau$ and $\mathcal{B}_n^\tau$. For instance, an analog of Theorems 2 and 3 holds for the *twisted partial Brauer monoid*, that is, the submonoid of $\mathcal{P}_n^\tau$ formed by all pairs $(\pi; s)$ such that each block of the partition π consists of *at most* two elements (as opposed to *exactly* two elements for the case of $\mathcal{B}_n^\tau$).

5.2 *Open Questions*

We have already mentioned that the question of the complexity of identity checking in the monoid $\mathcal{B}_3^\tau$ is left open. Another interesting question concerns the complex-

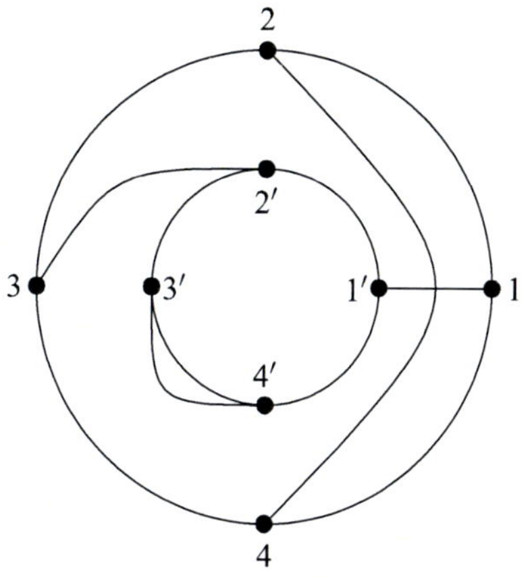

Fig. 6 Annular diagram of a partition of $[4] \cup [4]'$

ity of identity checking in the Kauffman monoids $\mathcal{K}_n$ with $n \geq 5$ (For $n \leq 4$, the problem CHECK- ID($\mathcal{K}_n$) is known to be polynomial time decidable; see [10, 27].) The approach of the present paper does not apply to Kauffman monoids since their subgroups are trivial. So, some fresh ideas are needed to handle this case.

Another natural family of submonoids in twisted Brauer monoids is the twisted version of Jones's *annular monoids* [23]. This version comes from the representation of partitions of $[n] \cup [n]'$ by annular rather than rectangular diagrams. Map the elements of $[n]$ to the n-th roots of unity doubled and the elements of $[n]'$ to the n-th roots of unity:

$$k \mapsto 2e^{\frac{2\pi i(k-1)}{n}} \text{ and } k' \mapsto e^{\frac{2\pi i(k-1)}{n}} \text{ for all } k \in [n].$$

Then the wires of any partition of $[n] \cup [n]'$ can be drawn in the complex plane as lines within the annulus $\{z \mid 1 < |z| < 2\}$ (except for their endpoints). For example, the annular diagram in Fig. 6 (taken from [3]) represents the partition $\{\{1, 1'\}, \{2, 4\}, \{3, 2'\}, \{3', 4'\}\}$.

The *twisted annular monoid* $\mathcal{A}_n^{\tau}$ is the submonoid of $\mathcal{B}_n^{\tau}$ consisting of all elements whose partitions have a representation as an annular diagram whose wires do not cross. The subgroups of $\mathcal{A}_n^{\tau}$ are known to be finite and cyclic, and checking identities in finite cyclic groups is easy: an identity $w \bumpeq w'$ holds in the cyclic group of order m if and only if for every letter in alph(ww'), the number of its occurrences in w is congruent modulo m to the number of its occurrences in w'. Thus, the problem CHECK- ID($\mathcal{A}_n^{\tau}$) also cannot be handled with the approach of the present paper and needs new tools.

Acknowledgements The authors are extremely grateful to the anonymous referee who spotted several inaccuracies and suggested shorter alternative proofs of Corollary 3 and Proposition 4 incorporated in the present version.

References

1. Almeida, J., Volkov, M.V., Goldberg, S.V.: Complexity of the identity checking problem for finite semigroups. Zap. Nauchn. Sem. POMI **358**, 5–22 (2008). (Russian; Engl. translation J. Math. Sci. **158**(5), 605–614 (2009))
2. Auinger, K., Chen, Y., Hu, X., Luo, Y., Volkov, M.V.: The finite basis problem for Kauffman monoids. Algebr. Univers. **74**(3–4), 333–350 (2015)
3. Auinger, K., Dolinka, I., Volkov, M.V.: Equational theories of semigroups with involution. J. Algebra **369**, 203–225 (2012)
4. Borisavljević, M., Došen, K., Petrić, Z.: Kauffman monoids. J. Knot Theory Ramif. **11**, 127–143 (2002)
5. Brauer, R.: On algebras which are connected with the semisimple continuous groups. Ann. Math. **38**(4), 857–872 (1937)
6. Bulatov, A.A., Karpova, O., Shur, A.M., Startsev, K.: Lower bounds on words separation: Are there short identities in transformation semigroups? Electron. J. Comb. **24**(3), article no. 3.35 (2017)
7. Cain, A.J., Johnson, M., Kambites, M., Malheiro, A.: Representations and identities of plactic-like monoids. J. Algebra **606**, 819–850 (2022)
8. Cain, A.J., Malheiro, A., Ribeiro, D.: Identities and bases in the hypoplactic monoid. Commun. Algebra **50**(1), 146–162 (2022)
9. Cain, A.J., Malheiro, A., Ribeiro, D.: Identities and bases in the sylvester and Baxter monoids. J. Algebr. Comb. (2023). https://doi.org/10.1007/s10801-022-01202-6
10. Chen, Y., Hu, X., Kitov, N.V., Luo, Y., Volkov, M.V.: Identities of the Kauffman monoid $\mathcal{K}_3$. Commun. Algebra **48**(5), 1956–1968 (2020)
11. Daviaud, L., Johnson, M., Kambites, M.: Identities in upper triangular tropical matrix semigroups and the bicyclic monoid. J. Algebra **501**, 503–525 (2018)
12. Dolinka, I., East, J.: Twisted Brauer monoids. Proc. Royal Soc. Edinburgh, Ser. A **148A**, 731–750 (2018)
13. East, J.: Generators and relations for partition monoids and algebras. J. Algebra **339**, 1–26 (2011)
14. East, J.: Presentations for Temperley-Lieb algebras. Q. J. Math. **72**(4), 1253–1269 (2021)
15. East, J., Higgins, P.M.: Green's relations and stability for subsemigroups. Semigroup Forum **101**, 77–86 (2020)
16. FitzGerald, D.G., Lau, K.W.: On the partition monoid and some related semigroups. Bull. Aust. Math. Soc. **83**(2), 273–288 (2011)
17. Halverson, T., Ram, A.: Partition algebras. Eur. J. Comb. **26**(6), 869–921 (2005)
18. Han, B.B., Zhang, W.T.: Finite basis problems for stalactic, taiga, sylvester and Baxter monoids. J. Algebra Appl. **22**(10), article no. 2350204 (2023)
19. Hanlon, Ph., Wales, D.: On the decomposition of Brauer's centralizer algebras. J. Algebra **121**(2), 409–445 (1989)
20. Horváth, G., Lawrence, J., Mérai, L., Szabó, C.S.: The complexity of the equivalence problem for nonsolvable groups. Bull. Lond. Math. Soc. **39**(3), 433–438 (2007)
21. Howie, J.M.: Fundamentals of Semigroup Theory. Clarendon Press, Oxford (1995)
22. Jones, V.F.R.: The Potts model and the symmetric group. In: Subfactors: Proceedings of the Taniguchi Symposium on Operator Algebras (Kyuzeso, 1993), World Scientific, River Edge, NJ, pp. 259–267 (1994)
23. Jones, V.F.R.: A quotient of the affine Hecke algebra in the Brauer algebra. Enseign. Math. II. Sér. **40**, 313–344 (1994)
24. Karpova, O., Shur, A.M.: Words separation and positive identities in symmetric groups. J. Automata, Lang. Combi. **26**(1–2), 67–89 (2021)
25. Kauffman, L.: An invariant of regular isotopy. Trans. Am. Math. Soc. **318**, 417–471 (1990)
26. Kerov, S.V.: Realizations of representations of the Brauer semigroup. Zap. Nauchn. Sem. LOMI **164**, 188–193 (1987). (Russian; Engl. translation J. Soviet Math. **47**, 2503–2507 (1989))

27. Kitov, N.V., Volkov, M.V.: Identities of the Kauffman monoid $\mathcal{K}_4$ and of the Jones monoid $\mathcal{J}_4$. In: Blass, A., Cégielski, P., Dershowitz, N., Droste, M., Finkbeiner B. (eds.) Fields of Logic and Computation III. Essays Dedicated to Yuri Gurevich on the Occasion of His 80th Birthday. Lecture Notes in Computer Science, vol. 12180, pp. 156–178. Springer, Cham (2020)
28. Kudryavtseva, G., Mazorchuk, V.: On presentations of Brauer-type monoids. Cent. Eur. J. Math. **4**(3), 413–434 (2006)
29. Lyndon, R.C., Schupp, P.E.: Combinatorial Group Theory. Springer, Berlin (1977)
30. Martin, P.: Potts Models and Related Problems in Statistical Mechanics. Series on Advances in Statistical Mechanics, vol. 5, World Scientific, Teaneck, NJ (1991)
31. Martin, P.: Temperley-Lieb algebras for nonplanar statistical mechanics–the partition algebra construction. J. Knot Theory Ramif. **3**, 51–82 (1994)
32. Martin, P.: The structure of the partition algebras. J. Algebra **183**, 319–358 (1996)
33. Martin, P.: The partition algebra and the Potts model transfer matrix spectrum in high dimensions. J. Phys. A **33**, 3669–3695 (2000)
34. Mazorchuk, V.: On the structure of Brauer semigroup and its partial analogue. Voprosy Algebry (Gomel) **13**, 29–45 (1998)
35. Moore, E.H.: Concerning the abstract groups of order $k!$ and $\frac{1}{2}k!$ holohedrically isomorphic with the symmetric and alternating substitution-groups on k letters. Proc. Lond. Math. Soc. **28**, 357–366 (1897)
36. Murata, K.: On the quotient semi-group of a non-commutative semi-group. Osaka Math. J. **2**, 1–5 (1950)
37. Murskiĭ, V.L.: Several examples of varieties of semigroups. Mat. Zametki **3**(6), 663–670 (1968). (Russian; Engl. translation (entitled *Examples of varieties of semigroups*) Math. Notes **3**(6), 423–427 (1968))
38. Nordahl, T.E., Scheiblich, H.E.: Regular $*$-semigroups. Semigroup Forum **16**(3), 369–377 (1978)
39. Papadimitriou, C.H.: Computational Complexity. Addison-Wesley, Reading, MA (1994)
40. Rowen, L.H.: On rings with central polynomials. J. Algebra **31**(3), 393–426 (1974)
41. Shneerson, L., Volkov, M.V.: The identities of the free product of two trivial semigroups. Semigroup Forum **95**(1), 245–250 (2017)
42. Volkov, M.V.: Remark on the identities of the grammic monoid with three generators. Semigroup Forum **106**(1), 332–337 (2023)
43. Wenzl, H.: On the structure of Brauer's centralizer algebras. Ann. Math. **128**(1), 173–193 (1988)
44. Wilcox, S.: Cellularity of diagram algebras as twisted semigroup algebras. J. Algebra **309**(1), 10–31 (2007)

Structure of the Semi-group of Regular Probability Measures on Locally Compact Hausdorff Topological Semiroups

M. N. N. Namboodiri

Abstract Let G be a locally compact Hausdorff semi-group, and $P(G)$ denotes the class of all regular probability measures on G. It is well known that $P(G)$ is a semi-group under the convolution of measures. This semi-group has been studied elaborately and intensely by many researchers. There are many deep results on the structure of regular probability measures $P(G)$ on compact/locally compact Hausdorff topological groups G. See, for instance, the classic monographs by Parthasarathy [1], Grenander [2] Mukherjea and Tserpes [3], Wendel [4] to quote selected references. In his remarkable paper, Wendel proved many deep theorems in this context. He proved that $P(G)$ is a semi-group which is not a group by proving that the only invertible elements are point mass-supported measures (Dirac delta measures). In a more general setting, such as locally compact Hausdorff semi-groups Mukherjea and Nicolas Terpse investigated the same during 1975-76. The algebraic regularity of $P(G)$ has recently been studied in [5] when G is a compact Hausdorff group. This article attempts for similar study when G is a locally compact Hausdorff semi-group.

Keywords Probability · Measures · Semi-groups · Regular

1 Introduction

As mentioned in the abstract, it is well known that the set $P(G)$ of regular probability measures on a locally compact, Hausdorff topological group G is a semi-group under convolution, which is abelian if and only if the group G is abelian. It is also known that $P(G)$ is a compact convex set under the weak* topology of measures. Wendel [4], in his remarkable paper, established many significant results regarding the algebraic,

M. N. N. Namboodiri (✉)
IMRT, Thiruvananthapuram, India
e-mail: mnnadri@gmail.com

Formerly with Mathematics Department, CUSAT, Kochi, India

A. A. Ambily and V. B. Kiran Kumar (eds.), *Semigroups, Algebras and Operator Theory*, Springer Proceedings in Mathematics & Statistics 436,
https://doi.org/10.1007/978-981-99-6349-2_5

topological as well as geometric structure of $P(G)$. He showed that $P(G)$ is a closed convex semi-group which is not a group except for trivial group $\{e\}$, by showing that the only invertible elements are point mass measures supported on single elements.

The problem we consider is **the algebraic regularity** of $P(G)$ as a semi-group under convolution of measures when G is a locally compact Hausdorff topological semi-group. The main tool used in [5] is the well-known Wendel's product theorem [4] when G is a compact Hausdorff group. Its semi-group versions are known during 1970–76 [3], which is the main tool in the following analysis.

2 Preliminaries

A semi-group is called **algebraically regular** if each of its elements has a generalised inverse.

The main result proved in this article is Theorem 4.11, which states that $P(G)$ is not a regular semi-group unless, of course, for the trivial case $G = \{e\}$. Let G be a locally compact Hausdorff topological semi-group, and $\mathcal{B}$ denotes the σ-algebra of all Borel sets in G. A probability measure μ is a non-negative countably additive function on $\mathcal{B}$ such that the total mass $\mu(G) = 1$. A point mass measure concentrated at a point $x \in G$ or Dirac delta measure is a measure δ_x such that $\delta_x(A) = 1$ if $x \in A$ and zero otherwise, $A \in \mathcal{B}$. One of the interesting results of Wendel is that the only invertible elements in $P(G)$ are Dirac delta measures. The convolution $\star$ in $P(G)$ is defined as follows;

Definition 1 (*Convolution* [3]) Let $\mu, \nu \in P(G)$. Then $\mu \star \nu$ is the probability measure defined as $\mu \star \nu(A) = \int \mu\left(Ax^{-1}\right) d\nu(x)$ where $Ax^{-1} = \{y \in G : yx \in A\}$ for every $A \in \mathcal{B}$.

Definition 2 (*Generalised inverse*) Let $\mathcal{S}$ be a semi-group and let $s \in \mathcal{S}$. An element $s^{\dagger} \in \mathcal{S}$ is called a generalised inverse of s if $ss^{\dagger}s = s$..

A generalised inverse is called a semi inverse if in addition $s^{\dagger}ss^{\dagger} = s^{\dagger}$.

For example, it is well known that the set $M_N(\mathbb{C})$ of all complex matrices of finite order N is a regular semi-group. The property of algebraic regularity is almost essential in the characterisation theorems of Nambooripad [6].

3 Regularity Question

For a compact topological group G, Wendel [4] proved that the set $P(G)$ is a semi-group which is not a group under convolution by proving that the only invertible elements in $P(G)$ are Dirac delta measures. One crucial property needed for measures under consideration is the regularity which is not guaranteed for compact topological groups. Next, we quote a fundamental theorem due to Wendel.

Definition 3 (*Support*) Support $S(\mu)$ of $\mu \in P(G)$ is defined as
$S(\mu) = \{g \in G : \mu(E_g) > 0$ for every neighbourhood E_g of $g \in G\}$.

Theorem 1 (Wendel) *Let $A = S(\mu)$ and $B = S(\nu)$ be supports of two measures μ and ν in $P(G)$. Then the support $S(\mu \star \nu) = AB = \{xy | x \in A, y \in B\}$.*

The following version can be found in [7] for topological semi-groups.

Theorem 2 *Let G be a locally compact Hausdorff topological semi-group and $P(G)$ denotes the class of all regular probability measures on the class $\mathcal{B}$ Borel sets in G. Then for any $\mu, \nu \in P(G)$*

$$S(\mu \star \nu) = closure\{S(\mu).S(\nu)\}, \tag{1}$$

where the closure is taken with respect to the weak topology of measures.*

The following theorem [3] is the semi-group version of the celebrated theorem for compact Hausdorff groups [4].

Theorem 3 *Let G be a locally compact Hausdorff semi-group. If $\mu \in P(G)$ is an idempotent, then $S(\mu)$ is completely simple.*

A locally compact Hausdorff topological semi-group G is said to be completely simple pp. 4 [3], if it is simple and has a primitive idempotent (minimal with respect to the partial order $\leqq$), where two idempotents e, f satisfy the inequality $e \leqq f$ if $ef = fe = e$.

Proposition 1 *Let G be an idempotent semi-group. Then $\{\delta_g : g \in G\}$ is an idempotent sub semi-group(band) of $P(G)$.*

Now we prove the main theorem of this short research article.

Theorem 4 *Let G be a nontrivial locally compact topological semi-group with an identity e. If G is not an idempotent semi-group (band) then $P(G)$ is not regular if there is an element $a \in G$ such that $a^2 \neq e$ and $a^2 \neq a$.*

Proof First of all observe that if $S(\mu) \& S(\nu)$ are finite sets then $S(\mu \star \nu) = S(\mu).S(\nu)$. Let $a \in G$ be such that $a^2 \neq e$ and $a^2 \neq a$. Consider the probability measure $\mu = \frac{1}{2}(\delta_e + \delta_a)$ where δ_x is the Dirac delta measure at x for each $x \in G$. We show that μ does not have a generalised inverse. If possible, a generalised inverse $\mu^\dagger$ of μ exist. Therefore we have

$$\mu \star \mu^\dagger \star \mu = \mu. \tag{2}$$

Now it follows that $\mu \star \mu^\dagger$ is an idempotent. By Theorem 3 above $H = S(\mu \star \mu)$ is a completely simple semi-group. Let $h \in H$. From Eq. 2 above, we get

$$h.\{e, a\} \subset \{e, a\}. \tag{3}$$

Therefore we have the following implications: (i) :$h = e$ or $h = a$;(ii): $ha = e$ or $ha = a$. Now $(i)\&(ii)$ together imply that $a = e$ or $a = a$ or $a^2 = e$ or $a^2 = a$. But the last two implications are not possible by hypothesis. Hence we have $h = e$. Thus we get $H = \{e\}$. Now let $\omega \in S\left(\mu^{\dagger}\right)$. We will have $\omega.\{e, a\} = \{e\}$. This implies that $\omega = e\&\omega.a = e$, which means that $a = e$ is again not possible. All these absurd conclusions arise from the assumption that μ is algebraically regular. Thus μ is not algebraically regular. □

Lemma 1 *Let $G = \{g : g^2 = e \quad or \quad g^2 = g\}$ and $H = \{h \in G : h^2 = e\}$. Then H is an abelian sub group of G.*

Proof Let $a, b \in H$. Then $a^2 = e = b^2$. If $(ab)^2 = ab$, then $a\,(ab.ab)\,b = aabb = e$. i.e. $ba = e \Longrightarrow b = a \Longrightarrow ab = e \in H$. This completes the proof. □

Proposition 2 *Let G be a locally compact Hausdorff topological semi-group such that $g \in G \Longrightarrow g^2 = e \quad or \quad g^2 = g$. If there is $a \in G : a \neq e\&a^2 = e$, then G is not algebraically regular.*

Proof By Lemma 1, if $\Omega = \{g \in G : g^2 = g \quad or \quad g^2 = e\}$ is an abelian group. It is known that [5] if $\mu = \frac{1}{2}\left(\delta_e + \delta_a\right)$ and if μ were algebraically regular with generalised inverse $\mu^{\dagger}$, then $S\left(\mu^{\dagger}\right) = \{e, a\}$ and therefore $\mu^{\dagger}$ is also a probability measure with support in Ω. However it is known [5] that μ is not algebraically regular as an element of $P\,(\Omega)$. Therefore μ is not algebraically regular in $P\,(G)$. Hence the proof. □

Remark 1 Let $\Omega = \{g \in G : g^2 = e\}$. Then Ω is an abelian group. Clearly g is the inverse of each $g \in \omega$. Let $a, b, \in \Omega$. By hypothesis,$ab.ab = ab$ or $(ab)^2\,ab$. $a.\,(ab.ab)\,b = a\,(ab)\,b = a^2b^2 = e \Longrightarrow ba = e \Longrightarrow b = a \Longrightarrow ab = e \in \Omega \Longrightarrow a = b = e$ or $(ab)^2 = e$. In either case, we have $ab \in \Omega$. In the latter case we ha

ve $ab.ab = e \Longrightarrow ba = ab$

Theorem 5 *Let G be a locally compact topological semi-group such that $g \in SG \Longrightarrow g^2 = e \quad or \quad g^2 = g$. Then $P\,(G)$ is not algebraically regular.*

Proof Let $g \in G, g \neq e$ and $\mu = \frac{1}{2}\left(\delta_e + \delta_g\right)$. We show that μ is not algebraically regular. Let if possible$\mu^{\dagger}$ be such that

$$\mu \star \mu^{\dagger} \star \mu = \mu, \tag{4}$$

and let $H = S\left(\mu \star \mu^{\dagger}\right)$. Then H is a completely simple semi-group. Now we have $H.\{e, g\} = \{e, g\}$ and therefore for $h \in H$,$h.\{e, g\} \subset \{e, g\}$. This leads to the following possibilities:

(i) $h.e = e$ *or* $h.e = g \to h = e$ *or* $h = g$.
(ii) $hg = e$ *or* $hg = g$.

Hence (i) implies that $H \subseteqq \{e, g\}$. Therefore we have three possibilities namely $H = \{e\} or H = \{g\} or H = \{e, g\}$. Since $\{g\}$ is a proper ideal of $\{e, g\}$, $H \neq$\{e,g\}. So we are left with two options namely $H = \{e\}$ or $H = \{g\}$. But $H.\{e, g\} = \{e, g\}$. Therefore we get that $H = \{e\}$. Let $\Omega = S\left(\mu^{\dagger}\right)$. Then we have $\{e, g\}.\Omega = \{e\} \Longrightarrow \{e, g\}\omega = \{e\}$ for every $\omega \in \Omega \Longrightarrow e.\omega = e = g.\omega \forall \omega \Longrightarrow \omega = e = g$. Therefore this absurd conclusion arises from the assumption that μ is algebraically regular. This completes the proof. □

To summarise the results proved above, we have the following.

Theorem 6 *Let G be a nontrivial, locally compact, Hausdorff topological semi-group with identity. i.e.,G is a monoid. Then $P(G)$ is not regular.*

4 Identification of Elements in $P(G)$ Which Are Algebraically Regular/Not Regular

In [5], obstructions for algebraic regularity were considered at least for special kinds of compact abelian groups. In this section, we consider a similar problem. Consider $P(G)$ where G is a locally compact Hausdorff topological semi-group with identity. Recall that the groups for which the analysis was carried out in [5] was for $G = \{g : g^2 = g \text{ } or \text{ } g^2 = e\}$ which reduced to $G = \{g : g^2 = e\}$. Even for semi-groups, the situation is the same by Lemma 4.7 above and hence we can apply the same results established in [5]. So the details are omitted.

Remark 2 The set $P(G)$ is a closed convex set under weak* topology of measures, and $\{\delta_g : g \in G\}$ is the set of extreme points of $P(G)$. Hence by Krein-Millmann theorem, the closed convex hull $\overline{conv}\{\delta_g : g \in G\} = P(G)$. In particular, if one considers the subsemi-group $conv\{\delta_g : g \in G\}$, it may be possible to locate all regular elements in it geometrically. This possibility is under investigation.

Remark 3 In a general semi-group Ω if $\omega \in \Omega$ has a generalised inverse, then it has a Moore-Penrose inverse: to be explicit, if $\omega g \omega = g$ for some $g \in \Omega$ then

$$\omega \omega^{\dagger} \omega = \omega \quad \& \tag{5}$$

$$\omega^{\dagger} \omega \omega^{\dagger} = \omega^{\dagger} \tag{6}$$

where $\omega^{\dagger} = \omega g \omega$.

So to characterise generalised invertibility, it will be enough to characterise Moore-Penrose invertibility.

5 Probability Measures on Stochastic/Doubly Stochastic Matrices

Let G be a locally compact, Hausdorff topological semi-group. Then we have

(1) Every invertible element and idempotent element are regular.
(2) It is easy to see that δ_g is regular if and only if g is regular.
(3) In a band G(an idempotent semi-group) δ_g is regular for all $g \in G$.
(4) However, detecting regular elements in $P(G)$ remains a task!
(5) An element $\mu \in P(G)$ is invertible in $P(G)$if and only if $\mu = \delta_g$ for some $g \in G$ & g invertible in G.

Proof We shall sketch a proof as follows. Let $\mu \in P(G)$ be invertible with inverse ν. Then, by support theorem we have

$$\{e\} = S(\mu \star \nu) = S(\nu \star \mu) = closure[S(\mu)\, S(\nu)] \tag{7}$$

Therefore we get $S(\mu)\, S(\nu) \subseteqq \{e\} \Longrightarrow g.h = e \forall g \in S(\mu)\, \& h \in S(\nu) \Longrightarrow g = h^{-1} \forall g \in S(\mu)\, \& h \in S(\nu)$. This implies that $S(\mu)\, \& S(\nu)$ are singleton sets. Thus $\mu = \delta_g \& \nu = \delta_{g^{-1}}$ for some $g \in G$. (6): Thus (6) implies that the only invertible elements of $P(G)$ are extreme points of $P(G)$. However, there can be algebraically regular elements in $P(G)$ other than these extreme points. Notice that even for compact groups it is not clear about this question!. □

Stochastc/Doubly Stochastic Matrices

A matrix A with non-negative entries is called stochastic if the row sum is 1 for all rows and is called doubly stochastic if it is stochastic and column sum is 1 for all columns. It is easy to see that the set $\Omega = \{A \in M(R) : A \quad stochastic\}$ is a compact semi-group under Euclidean topology and the set Ω_d of all doubly stochastic matrices is a compact subsemi-group of Ω. More details can be found in [3, 8, 9]. Here an attempt is made to detect algebraically regular elements in $\Omega \& \Omega_d$. The invertibility and generalised invertibility of stochastic and doubly stochastic matrices have been studied, and explicit characterisation has been obtained by Wall [9]. Moreover, an explicit characterisation of idempotents in $\Omega \& \Omega_d$ can also be found in [9]. So a partial picture of generalised invertible elements in $P(\Omega)\, \& P(\Omega_d)$ is more or less clear.

First, we consider $P(G)$ when G is the semi-group of all stochastic matrices of order n. It is easy (well known also) to see that Ω is a compact semi-group containing the compact semi-group Ω_d and both $\Omega \& \Omega_d$ are not regular semi-groups. Invertibility and generalised invertibility have been studied almost completely in [9]. We noted that [(3) above] that for $g \in G, \delta_g$ is invertible (generalised invertible respectively) if and only if g is invertible(respectively generalised invertible).

However, $P(G)$ is a closed convex set; therefore, an investigation is necessary to check the invertibility/generalised invertibility of probability measures such as convex combinations of Dirac delta measures.

Question; Let $\{g_1, g_2, \ldots., g_m\} \subset \Omega$ and $\mu = \Sigma_{k=1}^{m} \alpha_k \delta_{g_k}, 0 \le \alpha_k \le 1 \forall k, \& \Sigma_{k=1}^{m} \alpha_k = 1$. The problem is to know when is μ invertible/generalised invertible. However, (5) above gives a clear answer to the invertibility question. Recall that in 1955 Wendel proved the same assertion for compact, Hausdorff topological groups [4]. Therefore the problem remaining is to characterise the generalised invertibility. Case(1): Let $\mu = \left(\alpha_0 \delta_e + \alpha_1 \delta_{g_1}\right)$, a convex combination of $\delta_e \& \delta_g$, and $h \in S\left(\mu^{\dagger}\right)$. By product theorem we have

$$\{e, g_1\} S\left(\mu^{\dagger}\right)\{e, g_1\} = S(\mu) \tag{8}$$

and therefore $\{e, g_1\} h \{e, g_1\} \subseteqq S(\mu)$.

Proposition 3 *Let G be a locally compact semi-group with identity e and let $g \in G : g^2 = g \& g \neq e$. If $\mu = \alpha_1 \delta_e + \beta_1 \delta_g : \alpha_1, \beta_1 > 0, \alpha_1 + \beta_1 = 1$. Then μ is not algebraically regular.*

Proof Let, if possible,$\mu^{\dagger}$ is a generalised inverse of μ. We may assume that $\mu^{\dagger}$ is a semi inverse. Therefore we have

$$\mu \star \mu^{\dagger} \star \mu = \mu, \, and \tag{9}$$

$$\mu^{\dagger} \star \mu \star \mu^{\dagger} = \mu^{\dagger}. \tag{10}$$

By product theorem we have closure $S(\mu) S\left(\mu^{\dagger}\right) S(\mu) = S(\mu)$. This implies that $S(\mu) S\left(\mu^{\dagger}\right) S(\mu) \subseteqq S(\mu) \Longrightarrow \{e, g\} h \{e, g\} \subseteqq \{e, g\} \forall h \in S\left(\mu^{\dagger}\right)$. In particular we have $h \in \{e, g\} \forall h \in S\left(\mu^{\dagger}\right) \Longrightarrow S\left(\mu^{\dagger}\right) \subseteqq \{e, g\}$. Now similar calculations using (6.4) lead to the reverse inclusion. This completes the proof of claim. Therefore we have $\mu^{\dagger} = \alpha_1 \delta_e + \beta_1 \delta_g$ where $\alpha_2, \beta_2, \ge 0 \& \alpha_2 + \beta_2 = 1$.

Expanding equation (6.4), we get

$$\left(\alpha_1 \delta_e + \beta_1 \delta_g\right)\left(\alpha_2 \delta_e + \beta_2 \delta_g\right)\left(\alpha_1 \delta_e + \beta_1 \delta_g\right) = \alpha_1 \delta_e + \beta_1 \delta_g. \tag{11}$$

Using associativity and distributivity of the convolution $\star$ we find that

$$\alpha_1 \alpha_1 \alpha_2 \delta_e + [\alpha_1 \alpha_2 \beta 1 + \alpha_1 \beta_2 + \beta_1 \alpha_2 + \beta_1 \beta_2] \delta_g = \alpha_1 \delta_e + \beta_1 \delta_g. \tag{12}$$

Comparing coefficient of $\delta_e \& \delta_g$, we and by using the properties of $\alpha_1, \beta_1, \alpha_2 \& \beta_2$, we find that $\alpha_1 \alpha_2 = 1$ which is not possible. This is because, by assumption, all coefficients lie strictly between 0&1. □

Remark 4 (1): One can easily extend the above proposition 6.1 to a finite convex combination of the form $\mu = \Sigma_{k=0}^{n} \alpha_k \delta_{g_k}$ where $g_k^2 = g_k \neq e, k = 1, 2, \ldots, n$.
(2): The above remark applies to topological semi-groups containing identity, which are **BANDS**.

6 Concluding Remarks

It is interesting to note that Premchand, in his thesis [8], has studied the structure of the semi-groups Ω &Ω_d by computing the Biordered Sets of Nambooripad [6]. It would be interesting to obtain the biordered sets of $P(\Omega)$ & $P(\Omega_d)$, which is altogether a different problem. The study of semi-group of probability measures on topological semi-groups is to be treated separately. Bands are trivial for groups; it is nontrivial for semi-groups which are not groups.

Acknowledgements The author is thankful to KSCSTE, Government of Kerala, for financial support by awarding the Emeritus Scientist Fellowship, during which period a major part of this work was done. Also, a part of this research work was presented at the international conference, ICSAOT-22, 28-31 March 2022, held at the Department Of Mathematics, CUSAT, in honour of Prof. P.G. Romeo.

References

1. Parthasarathy, K.R.: Probability measures on metric spaces. In: Probability and Mathematical Statistics, No. 3. Academic Press, Inc., New York-London (1967)
2. Grenander, U.: Probabilities on Algebraic Structures. John Wiley & Sons Inc, New York-London (1963)
3. Mukherjea, A., Tserpes, N.A.: Measures on Topological Semigroups: Convolution Products and Random Walks, Lecture Notes in Mathematics, vol. 547. Springer, Berlin-New York (1976)
4. Wendel, J.G.: Haar measure and the semigroup of measures on a compact group. Proc. Am. Math. Soc. **5**, 923–929 (1954)
5. Namboodiri, M.N.N.: Regularity of Regular Probability Measures on Compact Hausdorff Topological Groups (preprint-2022)
6. Nambooripad, K.S.S.: Structure of regular semigroups. I, Mem. Am. Math. Soc. **22**, vii+119 (1979)
7. Mukherjea, A.: The convolution equation $\sigma \star \mu = \mu$ on non-compact non-abelian semiroups. In: Proceedings of Indian Academy of Sciences (Mathematical Sciences), vol. 129 (2019)
8. Premchand, S.: Semigroup of Stochastic Matrices, Thesis Submitted for the Award of the Degree of Doctor of Philosophy. University of Kerala, Trivandrum (1985)
9. Wall, J.R.: Generalized inverses of stochastic matrices. Linear Algebra Appl. **10**, 147–154 (1975)

Compatible and Discrete Normal Categories

A. R. Rajan

Abstract The concept of the normal category was introduced by K. S. S. Nambooripad in his study of the structure of regular semigroups using cross connections. Normal categories are essentially the categories of principal left ideals and the categories of principal right ideals of regular semigroups with appropriate morphisms. A cross connection relates two normal categories $\mathcal{C}$ and $\mathcal{D}$ so that the pair produces a regular semigroup S such that one is the category of principal left ideals and the other the category of principal right ideals of the semigroup S. This article considers the question of compatibility of two normal categories in order that they can be connected by a cross connection. It is known that for any normal category $\mathcal{C}$ the normal dual of $\mathcal{C}$ is a normal category $\mathcal{D}$ such that $\mathcal{C}$ and $\mathcal{D}$ are compatible. We attempt at finding all normal categories $\mathcal{D}$ that are compatible with a given normal category $\mathcal{C}$. We describe it for the case of a particular class of normal categories in which the partial order on the vertex set is the identity relation. These categories are called discrete normal categories.

Keyword Regular semigroups · Normal categories · Cross connection · Compatible categories

1 Introduction

Normal categories have been introduced by Nambooripad [8] as basic objects for describing the structure of regular semigroups. This structure theory is known as cross connection theory. The concept of cross connections in describing the structure of regular semigroups was introduced by Grillet [4] in a series of three papers in Semigroup Forum.

A. R. Rajan (✉)
Department of Mathematics, Institute of Mathematics Research and Training (IMRT), University of Kerala, Thiruvananthapuram, India
e-mail: arrunivker@yahoo.com

A. A. Ambily and V. B. Kiran Kumar (eds.), *Semigroups, Algebras and Operator Theory*, Springer Proceedings in Mathematics & Statistics 436,
https://doi.org/10.1007/978-981-99-6349-2_6

Grillet used two partially ordered sets which are essentially the partially ordered sets of Green's $\mathcal{L}$-classes and $\mathcal{R}$-classes of a regular semigroup. Grillet's theory was limited to provide the structure of a special class of regular semigroups known as fundamental regular semigroups.

Nambooripad [8] generalized the cross connection theory to the general class of regular semigroups. Nambooripad's cross connection theory used two categories called normal categories in place of the partially ordered sets in Grillet's theory. These normal categories are essentially the category of principal left ideals and the category of principal right ideals of a regular semigroup.

Characterization of normal categories arising from various classes of semigroups has been a topic of interest (see [1, 2, 9–15, 17–19], etc.). In Nambooripad's cross connection theory [8] a normal category $\mathcal{C}$ gives rise to another normal category $\mathcal{N}^*\mathcal{C}$ called the normal dual of $\mathcal{C}$ and a regular semigroup $T\mathcal{C}$ which is the semigroup of all normal cones in $\mathcal{C}$. Further the category $\mathcal{L}(S)$ of principal left ideals of $S = T\mathcal{C}$ is isomorphic to $\mathcal{C}$ and the category $\mathcal{R}(S)$ of principal right ideals of $T\mathcal{C}$ is isomorphic to $\mathcal{N}^*\mathcal{C}$.

Given two normal categories $\mathcal{C}$ and $\mathcal{D}$, a cross connection can exist between them only when one of them is isomorphic to the category $\mathcal{L}(S)$ of principal left ideals and the other is isomorphic to the category $\mathcal{R}(S)$ of principal right ideals of some regular semigroup S. In this case we say that the categories $\mathcal{C}$ and $\mathcal{D}$ are compatible and that one is a compatible dual of the other.

It may be observed that for every normal category $\mathcal{C}$ the normal dual $\mathcal{N}^*\mathcal{C}$ is a compatible dual of $\mathcal{C}$. We consider the question of finding all compatible duals of a given normal category. A complete determination of compatibility in the case of a normal category with only one object is given in this article. All these duals are shown to be discrete normal categories (cf. [16]).

2 Preliminaries

Here we provide the definition and elementary properties relating to regular semigroups, normal categories, and cross connections. The main references are [3, 5, 6, 8, 15, 17].

2.1 Categories

We begin with some notations and terminology of category theory. We consider only small categories. That is, categories in which the class of objects is a set.

The set of objects of a small category $\mathcal{C}$ is denoted by $v\mathcal{C}$ and the set of morphisms is denoted by $Mor(\mathcal{C})$ or $\mathcal{C}$ itself.

For objects a, b in $\mathcal{C}$ the set of all morphisms from a to b is denoted by $Hom(a, b)$ or $[a, b]_{\mathcal{C}}$ or $[a, b]$ and we call it a homset or a morphism set. We often write

$f : a \to b$ to mean that $f \in [a, b]_{\mathcal{C}}$. In this case a is said to be the domain of f and b its codomain. For objects a, b, c and morphisms $f : a \to b,\ g : b \to c$ in a category $\mathcal{C}$ there is a composition fg such that $fg : a \to c$. For each $a \in v\mathcal{C},\ [a, a]_{\mathcal{C}}$ contains an identity and is denoted by 1_a. Here

$$1_a f = f \text{ and } f 1_b = f$$

for all $f : a \to b$. A category $\mathcal{C}$ is said to be connected if for all $a, b \in v\mathcal{C}$ the morphism set $[a, b]_{\mathcal{C}}$ is nonempty.

For each object a in a category $\mathcal{C}$ the Hom functor $Hom(a, -) : \mathcal{C} \to \mathcal{S}et$ is defined as follows.

For $b \in v\mathcal{C}$

$$Hom(a, -)(b) = Hom(a, b) = [a, b]_{\mathcal{C}}.$$

For convenience we write $H(a, -)$ in place of $Hom(a, -)$. For $g : b \to c$

$$H(a, -)(g) = H(a, g) : H(a, b) \to H(a, c) \text{ maps } f \mapsto fg$$

for all $f \in H(a, b)$.

For $f : b \to a$ in $\mathcal{C}$ the natural transformation $H(f, -) : H(a, -) \to H(b, -)$ is described as follows.

For $c \in v\mathcal{C}$, the component of $H(f, -)$ at c is denoted by $H(f, c)$ and

$$H(f, c) : H(a, c) \to H(b, c) \text{ is given by } g \mapsto fg$$

for all $g \in H(a, c)$.

That is the following diagram is commutative.

$$\begin{array}{ccc} H(a, c) & \xrightarrow{H(f, c)} & H(b, c) \\ {\scriptstyle H(a, h)} \downarrow & & \downarrow {\scriptstyle H(b, h)} \\ H(a, d) & \xrightarrow[H(f, d)]{} & H(b, d) \end{array}$$

for $h : c \to d$.

It is easy to see that every natural transformation from $Hom(a, -)$ to $Hom(b, -)$ is determined by a morphism $f : b \to a$ (cf. [6]).

2.2 Normal Categories

KSS Nambooripad introduced normal categories using the concept of category with subobjects. The concept of subobjects induces a partial order on the object set $v\mathcal{C}$ and an inclusion morphism $j : a \to b$ for $a \leq b$ in $v\mathcal{C}$. Here we consider the modified description as categories with a partial order on objects and associated inclusion morphisms (cf. [15–17]). First we define categories with normal factorization and then go over to define normal categories.

A *category with normal factorization* is a small category $\mathcal{C}$ with the following properties.

- The vertex set $v\mathcal{C}$ of $\mathcal{C}$ is a partially ordered set such that whenever $a \leq b$ in $v\mathcal{C}$, there is a monomorphism $j(a, b) : a \to b$ in $\mathcal{C}$. This morphism is called the inclusion from a to b.
- For each $a \in v\mathcal{C}$, $j(a, a) = 1_a$.
- If $a \leq b \leq c$ then $j(a, b)j(b, c) = j(a, c)$.
- If $f : a \to b$ and $a, b \leq c$ and if

$$fj(b, c) = j(a, c) \text{ then } a \leq b \text{ and } f = j(a, b).$$

- Every inclusion $j(a, b)$ has a right inverse. That is, there is $q : b \to a$ such that $jq = 1_a$.
- Every morphism f in $\mathcal{C}$ has a factorization

$$f = quj$$

where q is a retraction, u is an isomorphism and j is an inclusion. Such a factorization is called normal factorization in $\mathcal{C}$.

Normal cones in categories with normal factorizations are defined as follows.

Definition 1 Let $\mathcal{C}$ be a category with normal factorization. A normal cone is a function γ from $v\mathcal{C}$ to $\mathcal{C}$ satisfying the following.

(i) There is an object $c = c_\gamma$ in $v\mathcal{C}$ called the vertex of γ such that for every $a \in v\mathcal{C}$

$$\gamma(a) : a \to c.$$

(ii) Whenever $a \leq b$, $\gamma(a) = j(a, b)\gamma(b)$ where $j(a, b)$ is the inclusion morphism from a to b.

(iii) There is an object $b \in v\mathcal{C}$ such that $\gamma(b)$ is an isomorphism.

The following is the diagram of a normal cone.

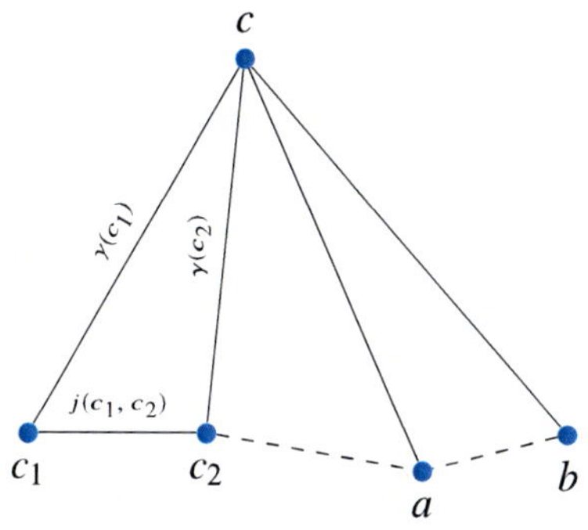

Remark 1 The M-set of a normal cone γ is defined as

$$M\gamma = \{a \in v\mathcal{C} : \gamma(a) \text{ is an isomorphism}\}.$$

Note that $M\gamma$ is nonempty for all normal cones γ.

Definition 2 A normal category is a small category $\mathcal{C}$ with normal factorization such that for each $a \in v\mathcal{C}$, there exists a normal cone γ with $\gamma(a) = 1_a$.

In the following we use 1 to denote the identity morphism on any object. Also for $f : a \to b$ writing $f = 1$ means that $a = b$ and $f = 1_a$.

Proposition 1 ([8]) *Let $\mathcal{C}$ be a normal category and $f : a \to b$ be a morphism in $\mathcal{C}$. Let $f = quj$ be a normal factorization of f. Then we have the following.*

(i) If f is an isomorphism and an inclusion then $f = 1$.
(ii) If f is an isomorphism and a retraction then $f = 1$.
(iii) If $f = q_1u_1j_1$ is another normal factorization of f then $qu = q_1u_1$ and $j = j_1$.
(iv) If $f = quj$ is a monomorphism then $q = 1$.
(v) If $f = quj$ is an epimorphism then $j = 1$.

Remark 2 In view of the above proposition every morphism $f = quj : a \to b$ in $\mathcal{C}$ can be written as $f = f^\circ j$ where $f^\circ = qu$.
It is easy to see that f° is an epimorphism. f° is called the epimorphic part of f and codomain of f° is called the image of f. We often write $Im\ f$ for image of f.

Theorem 1 ([8]) *Let $\mathcal{C}$ be a normal category and let $T\mathcal{C}$ be the set of all normal cones in $\mathcal{C}$. Then $T\mathcal{C}$ is a regular semigroup with product defined as follows. For $\gamma, \delta \in T\mathcal{C}$*

$$(\gamma\delta)(a) = \gamma(a)(\delta(c_\gamma))^\circ$$

for all $a \in v\mathcal{C}$. Further $T\mathcal{C}$ has the following properties.

1. *γ is an idempotent if and only if $\gamma(c) = 1_c$ where $c = c_\gamma$.*
2. *$\gamma \mathcal{L} \delta$ if and only if, $c_\gamma = c_\delta$.*
3. *If $\gamma \mathcal{R} \delta$ then $M\gamma = M\delta$ where*

$$M\gamma = \{a \in v\mathcal{C} : \gamma(a) \textit{ is an isomorphism }\}.$$

4. *γ $\mathcal{D}$ δ if and only if there is an isomorphism from c_γ to c_δ.*

Here $\mathcal{L}$, $\mathcal{R}$ and $\mathcal{D}$ are Greens relations.

2.3 Normal Duals and Cross Connections

Every normal category $\mathcal{C}$ has an associated normal category called the normal dual of $\mathcal{C}$ and is denoted by $\mathcal{N}^*\mathcal{C}$. A cross connection between two normal categories is a pair of functors connecting each with the normal dual of the other. The normal dual is a category of set valued functors described as follows.

The set of objects of $\mathcal{N}^*\mathcal{C}$ is the set of all functors $H(\gamma, -) : \mathcal{C} \to \mathcal{S}et$ where γ is a normal cone in $\mathcal{C}$ and for all $a \in v\mathcal{C}$

$$H(\gamma, a) = \{\gamma * f^o : f \in [c, a]_{\mathcal{C}}\}$$

where $c = c_\gamma$ is the vertex of γ. Also for a morphism $g : a \to b$, $H(\gamma, g) : H(\gamma, a) \to H(\gamma, b)$ is given by

$$\gamma * f^o \mapsto \gamma * (fg)^o.$$

Then $H(\gamma, -)$ is a functor from $\mathcal{C}$ to $\mathcal{S}et$ where $\mathcal{S}et$ is the category of sets and $\mathcal{N}^*\mathcal{C}$ is the category of all functors $H(\gamma, -) : \mathcal{C} \to \mathcal{S}et$ with all natural transformations between them as morphisms. These morphisms have a particularly simple description as follows. Let $\rho : H(\gamma, -) \to H(\delta, -)$ be a morphism in $\mathcal{N}^*\mathcal{C}$. Then there is a morphism $g : c_\delta \to c_\gamma$ such that $\rho = H(g, -)$ where $H(g, -) : H(\gamma, -) \to H(\delta, -)$ is the natural transformation with components

$$H(g, a) : \gamma * f^o \mapsto \delta * (gf)^o$$

for $\gamma * f^o \in H(\gamma, a)$.

The partial order in $v\mathcal{N}^*\mathcal{C}$ is given by

$$H(\gamma, -) \leq H(\delta, -) \text{ if and only if } H(\gamma, c) \subseteq H(\delta, c)$$

for all $c \in v\mathcal{C}$. It can be seen that $\mathcal{N}^*\mathcal{C}$ is also a normal category.

The following properties of the normal dual $\mathcal{N}^*\mathcal{C}$ of $\mathcal{C}$ are needed in the description of cross connections.

Proposition 2 ([8]) *Let $\mathcal{C}$ be a normal category and $\mathcal{N}^*\mathcal{C}$ be the normal dual of $\mathcal{C}$. Then we have the following.*

1. *For every $\gamma \in T\mathcal{C}$ there is an idempotent normal cone ε such that $H(\gamma, -) = H(\varepsilon, -)$.*
2. *For $H(\gamma, -), H(\delta, -) \in v\mathcal{N}^*\mathcal{C}$,*

$$H(\gamma, -) = H(\delta, -) \textit{ if and only if } \gamma \mathcal{R} \delta$$

in $T\mathcal{C}$ where $T\mathcal{C}$ is the semigroup of normal cones in $\mathcal{C}$.

3. *If $H(\gamma, -) = H(\delta, -)$ then $M\gamma = M\delta$ and so we denote $M\gamma$ by $MH(\gamma, -)$ also.*

Now we proceed to define cross connection. Here we use a special type of isomorphisms of normal categories called local isomorphisms.

Definition 3 Let $\mathcal{C}$, $\mathcal{D}$ be normal categories. A local isomorphism from $\mathcal{C}$ to $\mathcal{D}$ is a functor $F : \mathcal{C} \to \mathcal{D}$ satisfying the following.

1. F is inclusion preserving.
2. F is full and faithful.
3. $F|\langle c \rangle : \langle c \rangle \to \langle F(c) \rangle$ is an isomorphism of categories.

Here $\langle c \rangle$ if the full subcategory of $\mathcal{C}$ with object set

$$v\langle c \rangle = \{a \in v\mathcal{C} : a \leq c\}.$$

Definition 4 Let $\mathcal{C}$, $\mathcal{D}$ be normal categories. A cross connection between $\mathcal{C}$ and $\mathcal{D}$ is a pair (Γ, Δ) of functors where $\Gamma : \mathcal{D} \to \mathcal{N}^*\mathcal{C}$ and $\Delta : \mathcal{C} \to \mathcal{N}^*\mathcal{D}$ are local isomorphisms such that

$$c \in M\Gamma(d) \text{ if and only if } d \in M\Delta(c).$$

The local isomorphisms $\Gamma : \mathcal{D} \to \mathcal{N}^*\mathcal{C}$ and $\Delta : \mathcal{C} \to \mathcal{N}^*\mathcal{D}$ can be considered as functors from the product category $\mathcal{C} \times \mathcal{D}$ to $\mathcal{S}et$ as follows.

$$\Gamma(c, d) = (\Gamma(d))(c)$$

and

$$\Delta(c, d) = (\Delta(c))(d).$$

Then the cross connection (Γ, Δ) induces a natural isomorphism $\chi : \Gamma \to \Delta$ where $\chi(c, d) : \Gamma(c, d) \to \Delta(c, d)$ is given as follows.

Let $\Gamma(c, d) = H(\varepsilon, c)$ and $f : c_\varepsilon \to c$ so that $\varepsilon * f^o \in \Gamma(c, d)$. Also let $\Delta(c, d) = H(\tau, d)$ where $c_\tau = k \in M\Delta(c)$. Then by the property of cross connection $c \in M\Gamma(k)$ and so $\Gamma(k) = H(\sigma, -)$ for some normal cone σ in $\mathcal{C}$ with $c_\sigma = c$. Let $g : k \to d$ be such that $\Gamma(g) = H(f, -)$. Then

$$\chi(c, d) : \varepsilon * f^o \mapsto \tau * g^o.$$

The cross connection semigroup $S(\Gamma, \Delta)$ is described as follows.

Definition 5 Let (Γ, Δ) be a cross connection between normal categories $\mathcal{C}$ and $\mathcal{D}$. Then the cross connection semigroup is

$$S(\Gamma, \Delta) = \{(\gamma, \delta) \in \Gamma(c, d) \times \Delta(c, d) : \chi(c, d)(\gamma) = \delta\}$$

for $(c, d) \in v\mathcal{C} \times v\mathcal{D}$. The product is given by

$$(\gamma, \delta)(\gamma', \delta') = (\gamma\gamma', \delta'\delta).$$

2.4 Category of Principal Left Ideals of a Regular Semigroup

Let S be a regular semigroup. The category $\mathcal{L}(S)$ of principal left ideals of S is defined as follows.

$$v\mathcal{L}(S) = \{Se : e \in S \text{ and } e^2 = e\}$$

A morphism from Se to Sf is a partial right translation $\rho(e, u, f)$ defined by $x \mapsto xu$, where $u \in eSf$.

While describing properties of the normal category $\mathcal{L}(S)$ we make use of the Green's relations $\mathcal{L}$, $\mathcal{R}$ and $\mathcal{D}$ on the semigroup S [5]. We also use the biorder relations ω^l, ω^r and ω on the set $E(S)$ of idempotents of S [7]. Here ω^l and ω^r are quasi orders defined by

$$e\ \omega^l\ f \text{ if } ef = e \text{ and } e\ \omega^r\ f \text{ if } fe = e.$$

Further $\omega = \omega^l \cap \omega^r$ is a partial order on $E(S)$. We sometimes denote ω by $\leq$. The ideal $\omega^l(e)$ is defined by $\omega^l(e) = \{f : f\ \omega^l\ e\}$. Similarly $\omega^r(e)$ and $\omega(e)$ are defined.

The following theorem describes the properties of $\mathcal{L}(S)$ which we use in later discussions.

Theorem 2 ([8]) *Let $\mathcal{L}(S)$ be the category of principal left ideals of a regular semigroup S. Then we have the following.*

1. *$\mathcal{L}(S)$ is a normal category.*
2. *$e\rho(e, u, f) = u$.*
3. *$\rho(e, u, f)\rho(f, v, g) = \rho(e, uv, g)$.*
4. *$\rho(e, u, f) = \rho(e', v, f')$ if and only if $e\mathcal{L}e'$, $f\mathcal{L}f'$ and $v = e'u$.*
5. *$\rho(e, u, f)$ is an inclusion if and only if $\rho(e, u, f) = \rho(e, e, f)$.*
6. *$\rho(e, u, f)$ is a retraction if and only if $Sf \subseteq Se$ and $\rho(e, u, f) = \rho(e, g, g)$ for some idempotent g such that $Sg = Sf$.*
7. *$\rho(e, u, f)$ is an isomorphism if and only if $e\ \mathcal{R}\ u\ \mathcal{L}\ f$.*

3 Compatible and Discrete Normal Categories

Here we introduce the notion of compatible normal categories. The notion arises from the following question.

Given two normal categories $\mathcal{C}$ and $\mathcal{D}$ when are they related in such a way that one is the category of principal left ideals and the other the category of principal right ideals of a regular semigroup.

Pairs of categories related as above are said to be *compatible*. When $\mathcal{C}$ and $\mathcal{D}$ are compatible we say that each of them is a compatible dual of the other. When two normal categories are related by a cross connection one of them becomes the category of principal left ideals and the other the category of principal right ideals of a regular semigroup.

Definition 6 Normal categories $\mathcal{C}$ and $\mathcal{D}$ are said to be compatible if there exists a cross connection between them.

Remark 3 If S is a regular semigroup then $\mathcal{L}(S)$ and $\mathcal{R}(S)$ are compatible normal categories [8]. Consequently $\mathcal{R}(S)$ is a normal category compatible with $\mathcal{L}(S)$. Also it is shown in [8] that every normal category $\mathcal{C}$ is isomorphic to a category $\mathcal{L}(S)$ for a regular semigroup S. Thus every normal category has a compatible dual.

Now we consider the question of describing all normal categories $\mathcal{D}$ corresponding to a given normal category $\mathcal{C}$ such that $\mathcal{C}$ and $\mathcal{D}$ are compatible. We provide this description in the case of discrete normal categories which are defined as follows.

Definition 7 ([16]) A normal category $\mathcal{C}$ is said to be a discrete normal category if the partial order on $v\mathcal{C}$ is the identity relation. That is for $a, b \in v\mathcal{C}$,

$$a \leq b \text{ if and only if } a = b.$$

The following theorem describes the basic properties of discrete normal categories.

Theorem 3 ([16]) *Let $\mathcal{C}$ be a discrete normal category. Then we have the following.*

1. *Every homset $[a, b]_{\mathcal{C}}$ is nonempty for $a, b \in v\mathcal{C}$. That is $\mathcal{C}$ is connected.*
2. *Every morphism in $\mathcal{C}$ is an isomorphism and so for each $a \in v\mathcal{C}$ the homset $[a, a]_{\mathcal{C}}$ is a group under composition of morphisms.*
3. *The groups $[a, a]_{\mathcal{C}}$ and $[b, b]_{\mathcal{C}}$ are isomorphic for all $a, b \in v\mathcal{C}$.*
4. *For every morphism $f : a \to b$ the map*
$$u \mapsto f^{-1}uf$$
is a group isomorphism from $[a, a]$ to $[b, b]$.
5. *For every morphism $f : a \to b$ the map*
$$u \mapsto uf$$
is a bijection from $[a, a]$ to $[a, b]$ and the map
$$v \mapsto fv$$
is a bijection from $[b, b]$ to $[a, b]$.

Remark 4 The groups $[a, a]_{\mathcal{C}}$ in a discrete normal category $\mathcal{C}$ are called the maximal subgroups of $\mathcal{C}$. Since no two different objects are comparable in the partial order on $v\mathcal{C}$, every normal cone γ with a given vertex c is simply a collection of morphisms

$$\{\gamma(a) : a \to c \text{ for } a \in v\mathcal{C}\}.$$

The major properties of the semigroup $T\mathcal{C}$ of normal cones in this case are given in the following theorem.

Theorem 4 ([16]) *Let $\mathcal{C}$ be a discrete normal category. Let γ, δ be normal cones in $\mathcal{C}$ with vertices c and d respectively. Then we have the following.*

1. *$M\gamma = v\mathcal{C}$.*
2. *$\gamma \; \mathcal{R} \; \delta$ if and only if $\delta = \gamma * f$ for some $f : c \to d$.*
3. *For any normal cone γ, the cone $\sigma = \gamma * (\gamma(a))^{-1}$ is an idempotent normal cone with vertex a and $\gamma \; \mathcal{R} \; \sigma$.*
4. *If γ is an idempotent normal cone with vertex c then the $\mathcal{H}$-class of γ in $T\mathcal{C}$ is given by*
$$H_\gamma = \{\gamma * u : u \in [c, c]\}.$$
Further H_γ is a group isomorphic to the maximal subgroup $[c, c]$ of $\mathcal{C}$.
5. *The semigroup $T\mathcal{C}$ is a union of groups with each group isomorphic to the maximal subgroup $[c, c]$.*

3.1 Compatible Duals

We proceed to describe compatible duals of discrete normal categories. We begin with a single object category. Let $v\mathcal{C} = \{c\}$. Since $\mathcal{C}$ is a normal category, all morphisms in $\mathcal{C}$ are isomorphisms. So the category $\mathcal{C}$ in this case is a group. Let us denote this group by G. Now G is the group of all morphisms of $\mathcal{C}$.

Every normal cone in this case is given by a single morphism and the semigroup $T\mathcal{C}$ of normal cones coincides with the group G. Further we have the following.

Theorem 5 *Let $\mathcal{C}$ be a normal category with one object and $S = T\mathcal{C}$. Then the normal category $\mathcal{L}(S)$ is also a category with one object and is isomorphic to $\mathcal{C}$.*

Proof Since S is a group, $\mathcal{L}(S)$ has only one object. Now each morphism in $\mathcal{L}(S)$ is of the form $\rho(1, u, 1)$ where 1 is the identity in the group S and $u \in S$. Further the product in $\mathcal{L}(S)$ is given by

$$\rho(1, u, 1)\rho(1, v, 1) = \rho(1, uv, 1).$$

Hence $\mathcal{L}(S)$ is isomorphic to S and so is isomorphic to $\mathcal{C}$. □

The normal dual $\mathcal{N}^*\mathcal{C}$ in this case is also a category with one object since

$$v\mathcal{N}^*\mathcal{C} = \{H(\varepsilon, -) : \varepsilon \text{ is an idempotent normal cone }\}.$$

Moreover we have the following theorem.

Theorem 6 *Let $\mathcal{C}$ be a normal category with one object with group of morphisms G. Then the normal dual $\mathcal{N}^*\mathcal{C}$ is also a category with one object and its group of morphisms is anti isomorphic to G.*

Proof Let ε denote the idempotent normal cone in $\mathcal{C}$ so that $\varepsilon = \{1_c\}$ where $v\mathcal{C} = \{c\}$. Then $H(\varepsilon, c) = G$ and if $h : c \to c$ so that $h \in G$ we have

$$H(\varepsilon, h) : u \mapsto uh.$$

Also every natural transformation $\eta : H(\varepsilon, -) \to H(\varepsilon, -)$ is given by $H(f, -)$ so that

$$H(f, c) : G \to G \text{ maps } u \mapsto fu.$$

That is the following diagram is commutative.

$$\begin{array}{ccc} H\varepsilon,c & \xrightarrow{\;Hf,c\;} & H\varepsilon,c \\ {\scriptstyle H\varepsilon,h}\downarrow & & \downarrow{\scriptstyle H\varepsilon,h} \\ H\varepsilon,c & \xrightarrow[\;Hf,c\;]{} & H\varepsilon,c \end{array}$$

It follows that the product of two natural transformations $H(f, -)$ and $H(g, -)$ is given by

$$H(f, -)H(g, -) = H(gf, -).$$

Thus the group of morphisms of $\mathcal{N}^*\mathcal{C}$ is anti isomorphic to the group G. □

Now we have the following theorem on compatibility. For a category $\mathcal{C}$ we denote by $\mathcal{C}^{op}$ the category with the same objects as of $\mathcal{C}$ and morphisms given by

$$[a, b]_{\mathcal{C}^{op}} = [b, a]_{\mathcal{C}}$$

for all objects a, b.

Theorem 7 *Let $\mathcal{C}$ be a normal category with one object. Then the normal category $\mathcal{D} = \mathcal{C}^{op}$ is compatible with $\mathcal{C}$.*

Now we proceed to describe all compatible duals of the single object category. First we observe that if $\mathcal{D}$ is a compatible dual of a normal category with a single object then $\mathcal{D}$ is a discrete normal category. This is proved using the properties of cross connections.

Theorem 8 *Let $\mathcal{C}$ be a normal category with one object and $\mathcal{D}$ be a normal category compatible with $\mathcal{C}$. Then $\mathcal{D}$ is a discrete normal category and the maximal subgroups of $\mathcal{D}$ are anti isomorphic to G where G is the group of morphisms of $\mathcal{C}$.*

Proof Since $\mathcal{C}$ and $\mathcal{D}$ are compatible there is a cross connection between them. Let $\Gamma : \mathcal{D} \to \mathcal{N}^*\mathcal{C}$ and $\Delta : \mathcal{C} \to \mathcal{N}^*\mathcal{D}$ be local isomorphisms such that (Γ, Δ) is a cross connection. Now by Theorem 6 $\mathcal{N}^*\mathcal{C}$ is a category with one object. Since Γ is a local isomorphism each order ideal $\langle d \rangle$ of $v\mathcal{D}$ is singleton. It follows that $\mathcal{D}$ is discrete.

Let $H(\varepsilon, -)$ be the single object in $\mathcal{N}^*\mathcal{C}$. Then for every $d \in v\mathcal{D}$ we have $\Gamma(d) = H(\varepsilon, -)$. Also since Γ is a local isomorphism for $d \in v\mathcal{D}$, Γ restricted to $[d, d]_{\mathcal{D}}$ is an isomorphism onto $[H(\varepsilon, -), H(\varepsilon, -)]$. But by Theorem 6 the group of morphisms $[H(\varepsilon, -), H(\varepsilon, -)]$ is anti isomorphic to G. Thus $[d, d]_{\mathcal{D}}$ is anti isomorphic to G. Thus we see that all maximal subgroups of $\mathcal{D}$ are anti isomorphic to G. □

Now we show that all discrete normal categories with maximal subgroups anti isomorphic to G are compatible duals of the single object category $\mathcal{C}$ with group of morphisms G.

Theorem 9 *Let $\mathcal{C}$ be a normal category with one object and with group of morphisms G. Then every discrete normal category with maximal subgroups anti isomorphic to G are compatible duals of $\mathcal{C}$.*

Proof Let $\mathcal{D}$ be a discrete normal category as in the statement above. Since $\mathcal{C}$ has only one object the normal dual $\mathcal{N}^*\mathcal{C}$ is also a category with a single object. Let $H(\varepsilon, -)$ denote the single object of $\mathcal{N}^*\mathcal{C}$. Let G' be the group of morphisms of $\mathcal{N}^*\mathcal{C}$ so that G' is anti isomorphic to G. Since all maximal subgroups of $\mathcal{D}$ are anti isomorphic to G we see that for each $d \in v\mathcal{D}$ the group $[d, d]$ is isomorphic to G'.

Let $\Gamma : \mathcal{D} \to \mathcal{N}^*\mathcal{C}$ and $\Delta : \mathcal{C} \to \mathcal{N}^*\mathcal{D}$ be functors defined as follows.

$$\Gamma(d) = H(\varepsilon, -) \text{ for all } d \in v\mathcal{D}.$$

For defining Γ on morphisms we fix an object $d \in v\mathcal{D}$, a normal cone δ with vertex d and an isomorphism $\theta : [d, d] \to G'$. For morphisms $g \in [d, d]$ define

$$\Gamma(g) = H(u, -) \text{ where } H(u, -) = \theta(g)$$

and for morphisms $f : d \to k$ define

$$\Gamma(f) = H(v, -) \text{ where } H(v, -) = \theta(f\delta(k)).$$

By Theorem 3 $f \mapsto f\delta(k)$ is a bijection from $[d, k]$ to $[d, d]$ and so Γ is a local isomorphism.

The functor $\Delta : \mathcal{C} \to \mathcal{N}^*\mathcal{D}$ is defined by fixing an object d, an idempotent normal cone δ with vertex d of $\mathcal{D}$ and an anti isomorphism $\phi : G \to [d, d]$. Let $v\mathcal{C} = \{c\}$. Define

$$\Delta(c) = H(\delta, -)$$

and for $u : c \to c$ in $\mathcal{C}$

$$\Delta(u) = H(\phi(u), -).$$

Clearly Δ is also a local isomorphism. Now

$$M\Gamma(d) = MH(\varepsilon, -) = \{c\}$$

for all $d \in v\mathcal{D}$ and

$$M\Delta(c) = MH(\delta, -) = v\mathcal{D}$$

by Theorem 4. So it follows that (Γ, Δ) is a cross connection. So $\mathcal{C}$ and $\mathcal{D}$ are compatible. □

References

1. Azeef Mohammed, P.A., Rajan, A.R.: Cross connections of completely simple semigroups. Asian Eur. J. Math. **9**(3), 16500534 (2016)
2. Azeef Mohammed, P.A., Romeo, P.G., Nambooripad, K.S.S.: Cross connection structure of concordant semigroups. Int. J. Algbra Comput. **30**(1), 181–216 (2020)
3. Clifford, A.H., Preston, G.B.: Algebraic Theory of Semigroups, vol. I. American Mathematical Society (1961)
4. Grillet, P.A.: Structure of regular semigroups I: a representation, II: cross connections, III: the reduced case. Semigroup Forum **8**(177–183), 254–265 (1974)
5. Howie, J.M.: Fundamentals of Semigroup Theory. Academic, New York (1995)
6. Mac Lane, S.: Categories for the Working Mathematician. Springer, New York (1979)
7. Nambooripad, K.S.S.: Structure of regular semigroups I. Mem. Am. Math. Soc. **22**(224) (1979). Centre for Mathematical Sciences, Trivandrum (1989)
8. Nambooripad, K.S.S.: Theory of cross connections, Pub. No. 28. Centre for Mathematical Sciences, Trivandrum (1994)
9. Preenu, C.S., Rajan, A.R., Zeenath, K.S.: Category of principal left ideals of normal bands. In: Romeo, P.G., Volkov, M., Rajan, A.R. (eds.) Semigroups, Algebras and Operator Theory. Springer (2022)
10. Rajan, A.R.: Certain categories derived from normal categories. In: Romeo, P.G., Meakin, J.C., Rajan, A.R. (eds.) Semigroups, Algebras and Operator Theory. Springer (2015)
11. Rajan, A.R.: Normal categories of inverse semigroups. East West J. Math. (2015)
12. Rajan, A.R., Azeef Mohammed, P.A.: Normal categories of partitions of a set. South East Asian Bull. Math. **41**(5) (2015)
13. Rajan, A.R., Azeef Mohammed, P.A.: Normal categories from the transformation semigroup. Bull. Allahabad Math. Soc. **30**(1), 61–74 (2015)
14. Rajan, A.R.: Structure theory of regular semigroups using categories. In: Rizvi, S., Ali, A., Fillippis, V. (eds.) Algebras and its Applications, Springer Proceedings in Mathematics and Statistics, vol. 174. Springer (2016)

15. Rajan, A.R.: Inductive groupoids and normal categories of regular semigroups, pp. 193–200. In: Proceedings of the International Conference at Aligargh Muslim University, DeGruyter (2018)
16. Rajan, A.R.: Discrete normal categories and cross connections. Bull. Southeast Asian Math. Soc. **45**, 1–14 (2021)
17. Rajan, A.R., Preenu, C.S., Zeenath, K.S.: Cross connections for normal bands and partial order on morphisms. Semigroup Forum **104**, 125–147 (2022)
18. Rajan, A.R., Mathew, N.S., Bingjun, Y.: Normal categories of strongly unit regular semigroups. Bull. Southeast Asian Math. Soc. (to appear)
19. Lukose, S., Rajan, A.R.: Ring of normal cones. Indian J. Pure Appl. Math. **41**, 663–681 (2010)

A Range of the Multidimensional Permanent on (0, 1)-Matrices

I. M. Evseev and A. E. Guterman

Abstract In this paper we consider the multidimensional matrices of zeros and ones. Certain estimates for consecutive values of the permanent function on such matrices are known. We investigate new values of the permanent function appearing while the dimension increases. Using these investigations, we obtain new estimates for consecutive values of the permanent and show that for a series of matrix parameters they are better than the estimates known previously.

Keywords (0, 1)-matrices · Multidimensional matrices · Permanent

1 Introduction and Main Definitions

Let n be a positive integer and M_n be the set of $n \times n$ matrices over a field $\mathbb{F}$. The *permanent* of a matrix $A = (a_{ij}) \in M_n$ is the function

$$per(A) = \sum_{\sigma \in S_n} a_{1\sigma(1)} \ldots a_{n\sigma(n)},$$

where S_n denotes the group of permutations on the set $\{1, \ldots, n\}$. The permanent was introduced later than the determinant and was actively studied during the past century, see [1, 2].

Many open problems of the theory of the permanents are already nontrivial for the class of (0, 1)-matrices, i.e., matrices of order n, all entries of which belong to

I. M. Evseev · A. E. Guterman (✉)
Lomonosov Moscow State University, Moscow 119991, Russia
e-mail: alexander.guterman@gmail.com

I. M. Evseev
e-mail: illlya.e@yandex.ru

Moscow Center of Fundamental and Applied Mathematics, Moscow 119991, Russia

A. E. Guterman
Technion - Israel Institute of Technology, 3200003 Haifa, Israel

A. A. Ambily and V. B. Kiran Kumar (eds.), *Semigroups, Algebras and Operator Theory*, Springer Proceedings in Mathematics & Statistics 436,
https://doi.org/10.1007/978-981-99-6349-2_7

the set $\{0, 1\}$. One of the important research directions related to the permanent of $(0, 1)$-matrices is the investigation of the set of values that permanent can achieve for a fixed n. For more details see [3, Sect. 1]. This problem has been the subject of active investigations over the years (see, for example, works [4, 5] and their bibliography).

There exists a generalization of a matrix, which is called a multidimensional matrix. Here and in the further text we assume that $\mathbb{N} = \{1, 2, \dots\}$.

Definition 1 Let $k, n \in \mathbb{N}$. A *k-dimensional matrix of order n* is an array $(a_I)_{I \in I_n^k}$, $a_I \in \mathbb{R}$, where $I_n^k = \{(\alpha_1, \dots, \alpha_k) \mid \alpha_i \in \{1, \dots, n\}\}$ is a set of *indices*. We denote by $\mathfrak{M}(n, k)$ the set of all k-dimensional matrices of order n.

Notice that in case $k = 2$ each index consists of two components, and we get ordinary square matrices.

Similarly to the case of plain (2-dimensional) matrices, a k-dimensional matrix with all entries from the set $\{0, 1\}$ is called a *k-dimensional* $(0, 1)$*-matrix*. Let us denote the set of all k-dimensional $(0, 1)$-matrices of order n by $\mathfrak{A}(n, k)$.

We denote matrices by capital latin letters, and their entries by corresponding small latin letters. If $A \in \mathfrak{M}(n, k)$, then a *diagonal* of A is the set of indices $\{I_1, \dots, I_n\}$, $I_j \in I_n^k$, where any two index vectors from this set differ in each coordinate. The n-tuple of entries $a_{I_1}, \dots, a_{I_n}$ of $A \in \mathfrak{M}(n, k)$ is called a *diagonal*, if the set of indices $\{I_1, \dots, I_n\}$ forms a diagonal. The set of all diagonals of A is denoted by $D(A)$.

A diagonal d of $A \in \mathfrak{A}(n, k)$ is said to be a *unit* diagonal if all its entries are 1.

Definition 2 The *permanent* of a matrix $A \in \mathfrak{M}(n, k)$ is the value

$$per(A) = \sum_{d \in D(A)} \prod_{I \in d} a_I.$$

The *permanent* of $A \in \mathfrak{A}(n, k)$ is equal to the number of its unit diagonals.

In the monographs [6, 7] the permanent of a multidimensional matrix was considered as a generalization of the determinant. The detailed and self-contained exposition of the research results on multidimensional permanents is presented in the papers [8, 9], the monographs [10, 11], and the survey paper [12], see also the references therein. In particular, there are many connections between the multidimensional permanents and the hypergraphs. The permanents of multidimensional matrices helped to solve several combinatorial problems, see the works [13, 14]. It is also relevant to a number of other applications, for example, transversals in latin squares, 1-factors of hypergraphs, tilings, MDR-codes, combinatorial designs, etc., see [12–16] and references therein.

The possible values of the permanent for 2-dimensional matrices have attracted a lot of attention. However, similar issues for multidimensional matrices almost have not been investigated. For the first time these problems were considered in [3]. Namely, the multidimensional analog of the Brualdi and Newman estimate [4] for the upper bound of the set of consecutive values of the permanent on $\mathfrak{A}(n, 2)$ and its generalizations from [5] were proved in [3].

Theorem 1 ([3, Theorem 7.15]) *For any integer* j, $0 \leq j \leq 2^{k-1} \cdot k^{n-2}$, *there exists a matrix* $A \in \mathfrak{A}(n, k)$ *such that* $per(A) = j$.

Besides that, a formula for the permanent of multidimensional (0, 1)-matrices was established. As a consequence, explicit formulas for the permanent of matrices with a small number of zeros were deduced. In addition, all values of the permanent for 3-dimensional (0, 1)-matrices of order 3 were computed.

The main aim of the present paper is to improve the upper bound from Theorem 1. Namely, we investigate the extension of the set of values attainable by the permanent function under increasing of the dimension of a matrix. In particular, we prove that if S is a certain subset of values of the permanent on $\mathfrak{A}(n, k)$ for some fixed n and k, then $\bigcup_{m=1}^{n} m!S$ is a subset of values of the permanent on $\mathfrak{A}(n, k+1)$. Then we apply this method for estimation of consecutive values of the permanent. Thus, we obtain a new upper bound which improves the upper bound from Theorem 1 for large values of the dimension of a matrix with respect to its order.

In the present paper we suppose that $k, n \in \mathbb{N}$ are the dimension and the order of a multidimensional matrix, respectively. We assume that $k \geq 2$ and $n \geq 2$ unless otherwise is stated.

Our paper is organized as follows. We introduce the remaining definitions and notations used in the paper in Sect. 2. In Sect. 3 we introduce and study notion of resembling matrices that is useful for the subsequent investigations. In Sect. 4 we compute the new values of the permanent function for matrices of increasing dimension. We obtain the main estimates for consecutive values of the permanent of matrices of large dimension in Sect. 5.

2 Preliminaries

Let us introduce concepts and notations needed in this paper following classical monographs [10, 11, 17].

For $d \in \{0, \ldots, k\}$, a *d-dimensional plane* of $A \in \mathfrak{M}(n, k)$ is the d-dimensional submatrix of A, obtained by fixing $(k-d)$ coordinates and letting the other d coordinates vary from 1 to n. If a d-dimensional plane P of $A \in \mathfrak{M}(n, k)$ is obtained by fixing coordinates with indices $i_1, \ldots, i_{k-d}$, $1 \leq i_1 < \ldots < i_{k-d} \leq k$, then we denote the vector $(\cdot, \ldots, \cdot, \alpha_{i_1}, \cdot, \ldots, \cdot, \alpha_{i_j}, \cdot, \ldots, \cdot, \alpha_{i_{k-d}}, \cdot, \ldots, \cdot) \equiv P$ of the length k, where α_{i_j}, $j = 1, \ldots, k-d$, are values of the fixed coordinates. Any $(k-1)$-dimensional plane of a k-dimensional matrix is said to be a *hyperplane*. If $k = 2$, then the hyperplane of a matrix is any its row or column. The *direction* of a d-dimensional plane A' of $A \in \mathfrak{M}(n, k)$ is the vector $\Theta = (\theta_1, \ldots, \theta_k)$, where $\theta_i = 0$ if and only if i-th coordinate of the plane A' is variable and $\theta_i = 1$ if it is constant.

Let $A \in \mathfrak{M}(n, k)$, a_I be a fixed entry of A. A matrix $(A|I) \in \mathfrak{M}(n-1, k)$ is said to be a submatrix of A, which is *complementary to the entry* a_I, if it is obtained from A by deleting all the entries a_J such that a_J lies with a_I in the same hyperplane.

Suppose that $n \geq 2$ and $A \in \mathfrak{M}(n, k)$. Then a *partial diagonal (of the length s)* is a subset $\{I_1, \ldots, I_s\}$ of a diagonal of A (i.e., all components of the indices $I_1, \ldots, I_s$ are distinct). We say that entries $a_{I_1}, \ldots, a_{I_s}$, $s \in \{1, \ldots, n\}$, form a *partial diagonal (of the length s)* if they constitute a part of a diagonal. Note that a single entry is always considered as a partial diagonal. We say that sets of entries $\{a_{I_1}, \ldots, a_{I_s}\}$, $s \in \{1, \ldots, n\}$, and $\{a_{J_1}, \ldots, a_{J_t}\}$, $t \in \{1, \ldots, n\}$, *lie diagonally*, if the entries $a_{I_1}, \ldots, a_{I_s}, a_{J_1}, \ldots, a_{J_t}$ constitute a partial diagonal.

Definition 3 Consider a matrix $A \in \mathfrak{M}(n, k)$. A *unit partial diagonal of the length s* is a partial diagonal $\{I_1, \ldots, I_s\}$ of A such that $a_{I_1} = \cdots = a_{I_s} = 1$.

The s-tuple of entries $a_{I_1}, \ldots, a_{I_s}$ of $A \in \mathfrak{M}(n, k)$ is called a *unit partial diagonal*, if the set of indices $\{I_1, \ldots, I_s\}$ forms a unit partial diagonal.

A matrix $B \in \mathfrak{M}(n, k)$ is obtained by a *transposition* from $A \in \mathfrak{M}(n, k)$, if there exists a permutation $\sigma \in S_k$ such that $\sigma \neq e$ and for any $\alpha = (\alpha_1, \ldots, \alpha_k) \in I_n^k$

$$b_{(\alpha_1,\ldots,\alpha_k)} = a_{(\sigma(\alpha_1),\ldots,\sigma(\alpha_k))},$$

see [12, Sect. 1].

We introduce equivalence transformations on $\mathfrak{M}(n, k)$ as follows: it is said that $A, B \in \mathfrak{M}(n, k)$ are *equivalent* if B can be obtained from A by a composition of transformations of the following two types:

(1) Transposition.
(2) Permutation of hyperplanes of $A \in \mathfrak{M}(n, k)$ having the same direction.

Corresponding transformations are called *equivalence transformations*.

Let P be a submatrix of $A \in \mathfrak{A}(n, k)$ and let $U \subset I_n^k$ be the set of indices of unit entries of P. Then we say that U is a *support* of P.

For the later use we observe the following elementary properties of the multidimensional permanent, see [12, 1.1, Properties 1–3].

Lemma 1 *The permanents of equivalent matrices are equal.*

Lemma 2 *Assume that matrices $A_1 = (a_{1,\alpha}), A_2 = (a_{2,\alpha}) \in \mathfrak{M}(n, k)$ coincide everywhere except, possibly, for one hyperplane Γ. Then $per(A_1) + per(A_2) = per(B)$, where $B = (b_\alpha) \in \mathfrak{M}(n, k)$ coincides with A_1 everywhere except, possibly, the hyperplane Γ, and any entry b_γ of Γ is equal to $a_{1,\gamma} + a_{2,\gamma}$.*

Lemma 3 (Laplace decomposition formula) *If Γ is a hyperplane of $A \in \mathfrak{M}(n, k)$, then*

$$per(A) = \sum_{a_\alpha \in \Gamma} a_\alpha per(A|\alpha).$$

3 Resembling Matrices

The definition of resembling matrices was introduced in [5] for 2-dimensional matrices. This notion is very useful to produce consecutive values of the permanent for plain matrices. We are going to extend the definition of resembling matrices to the multidimensional case. However, a more strict definition is convenient for multidimensional matrices, in particular, under increasing of the dimension. Therefore, in the present paper we define resembling matrices as follows.

Definition 4 Suppose that $m \in \mathbb{N}$, $m > 2$. We say that matrices $A_1, \dots, A_m \in \mathfrak{A}(n, k)$ are *resembling*, or, in other words, each of them *resembles* the others, if $A_1, \dots, A_m$ coincide everywhere except, possibly, for one hyperplane.

Notice that this definition does not coincide with the definition from [5] for $k = 2$. We consider some examples below.

The notion of resembling matrices from [5] considers matrices with accuracy to equivalence transformations. So, we name such matrices as *resembling up to equivalence* in this text.

Definition 5 We say that two matrices $A, B \in \mathfrak{A}(n, k)$ are *resembling up to equivalence*, or, in other words, one of them *resembles up to equivalence* the other, if there exists an equivalent to B matrix B' such that A and B' are resembling.

This generalization completely coincides with the definition from [5] for $k = 2$.

Remark 1 If the matrices $A, B \in \mathfrak{A}(n, k)$ are resembling, then A, B are resembling up to equivalence. But the opposite implication is not always true.

We will use Remark 1 further to produce consecutive values of the multidimensional permanent from known values of the permanent on 2-dimensional matrices.

Example

Let us illustrate the notions of resembling and resembling up to equivalence matrices on the examples of matrices from $\mathfrak{A}(2, 3)$ and $\mathfrak{A}(3, 3)$. As in [3], we describe 3-dimensional matrices by enumerating all their hyperplanes of the direction $(1, 0, 0)$ separating them with symbols "$\vdots$". Matrices from $\mathfrak{A}(2, 3)$ we write in the form:

$$\begin{pmatrix} a_{111} \ a_{112} & \vdots & a_{211} \ a_{212} \\ a_{121} \ a_{122} & \vdots & a_{221} \ a_{222} \end{pmatrix}.$$

The following matrices are not resembling, but they are resembling up to equivalence:

$$A_1 = \begin{pmatrix} 1 \ 1 & \vdots & 0 \ 1 \\ 1 \ 1 & \vdots & 1 \ 1 \end{pmatrix}, \quad A_2 = \begin{pmatrix} 1 \ 1 & \vdots & 1 \ 1 \\ 1 \ 0 & \vdots & 0 \ 0 \end{pmatrix}.$$

Indeed, there is no hyperplane Γ such that A_1 and A_2 coincide everywhere except, possibly, Γ. However, if we apply the transposition $(12) \in S_3$ to coordinates of A_2, then we obtain the matrix

$$A_3 = \begin{pmatrix} 1\ 1 & \vdots & 1\ 0 \\ 1\ 1 & \vdots & 0\ 0 \end{pmatrix},$$

which coincides with A_1 everywhere except the hyperplane $(2, \cdot, \cdot)$. Hence A_1, A_2 are resembling up to equivalence. Notice that A_1 and A_3 are resembling.

Matrices from $\mathfrak{A}(3, 3)$ are written in the following way:

$$\begin{pmatrix} a_{111}\ a_{112}\ a_{113} & \vdots & a_{211}\ a_{212}\ a_{213} & \vdots & a_{311}\ a_{312}\ a_{313} \\ a_{121}\ a_{122}\ a_{123} & \vdots & a_{221}\ a_{222}\ a_{223} & \vdots & a_{321}\ a_{322}\ a_{323} \\ a_{131}\ a_{132}\ a_{133} & \vdots & a_{231}\ a_{232}\ a_{233} & \vdots & a_{331}\ a_{332}\ a_{333} \end{pmatrix}.$$

Let us consider matrices

$$B_1 = \begin{pmatrix} 1\,1\,1 & \vdots & 1\,0\,0 & \vdots & 0\,1\,0 \\ 1\,1\,1 & \vdots & 0\,1\,0 & \vdots & 1\,1\,1 \\ 1\,1\,1 & \vdots & 0\,0\,0 & \vdots & 0\,1\,0 \end{pmatrix}, \quad B_2 = \begin{pmatrix} 1\,1\,1 & \vdots & 0\,0\,1 & \vdots & 1\,0\,0 \\ 1\,1\,1 & \vdots & 0\,0\,1 & \vdots & 0\,1\,0 \\ 1\,1\,1 & \vdots & 1\,1\,1 & \vdots & 0\,0\,0 \end{pmatrix}.$$

If we rearrange the hyperplanes $(2, \cdot, \cdot)$ and $(3, \cdot, \cdot)$ of B_2, then we get the matrix

$$\begin{pmatrix} 1\,1\,1 & \vdots & 1\,0\,0 & \vdots & 0\,0\,1 \\ 1\,1\,1 & \vdots & 0\,1\,0 & \vdots & 0\,0\,1 \\ 1\,1\,1 & \vdots & 0\,0\,0 & \vdots & 1\,1\,1 \end{pmatrix},$$

which coincides with B_1 everywhere except the hyperplane $(3, \cdot, \cdot)$. So, B_1 and B_2 are resembling up to equivalence matrices.

Example

Let us compute the permanents of the matrices A_1, A_2, B_1, and B_2 from the previous example. Using the definition of the permanent, we have:

$$per(A_1) = a_{111}a_{222} + a_{112}a_{221} + a_{121}a_{212} + a_{122}a_{211} = 1 + 1 + 1 + 0 = 3,$$

$$per(A_2) = a_{111}a_{222} + a_{112}a_{221} + a_{121}a_{212} + a_{122}a_{211} = 0 + 0 + 1 + 0 = 1.$$

Further, by the Laplace decomposition formula (see [12, 1.1: Property 3]), we obtain

$$per(B_1) = \sum_{i_2,i_3=1}^{3} b_{(2,i_2,i_3)} per(B|(2, i_2, i_3)) = per(B|(2, 1, 1)) + per(B|(2, 2, 2)) =$$

$$per\begin{pmatrix} 1\,1 & \vdots & 1\,1 \\ 1\,1 & \vdots & 1\,0 \end{pmatrix} + per\begin{pmatrix} 1\,1 & \vdots & 0\,0 \\ 1\,1 & \vdots & 0\,0 \end{pmatrix} = 3 + 0 = 3,$$

$$per(B_2) = \sum_{i_2,i_3=1}^{3} b_{(3,i_2,i_3)} per(B|(3, i_2, i_3)) = per(B|(3, 1, 1)) + per(B|(3, 2, 2)) =$$

$$per\begin{pmatrix} 1\,1 & \vdots & 0\,1 \\ 1\,1 & \vdots & 1\,1 \end{pmatrix} + per\begin{pmatrix} 1\,1 & \vdots & 0\,1 \\ 1\,1 & \vdots & 1\,1 \end{pmatrix} = 3 + 3 = 6.$$

Example

Consider matrices from $\mathfrak{A}(2, 4)$. We write them by enumerating all their hyperplanes of the direction $(1, 0, 0, 0)$, which are 3-dimensional matrices of order 2:

$$\begin{pmatrix} a_{1111}\ a_{1112} & \vdots & a_{1211}\ a_{1212} \\ a_{1121}\ a_{1122} & \vdots & a_{1221}\ a_{1222} \end{pmatrix}, \quad \begin{pmatrix} a_{2111}\ a_{2112} & \vdots & a_{2211}\ a_{2212} \\ a_{2121}\ a_{2122} & \vdots & a_{2221}\ a_{2222} \end{pmatrix}.$$

Here the first hyperplane is $(1, \cdot, \cdot, \cdot)$, the second hyperplane is $(2, \cdot, \cdot, \cdot)$.

Let $A, B \in \mathfrak{A}(2, 4)$ such that

$$\Gamma_1^A, \Gamma_2^A \in A, \quad \Gamma_1^A = \begin{pmatrix} 0\,1 & \vdots & 1\,0 \\ 1\,0 & \vdots & 0\,1 \end{pmatrix}, \quad \Gamma_2^A = \begin{pmatrix} 0\,1 & \vdots & 0\,0 \\ 0\,0 & \vdots & 0\,1 \end{pmatrix}$$

— hyperplanes of the direction $(1, 0, 0, 0)$ of A,

$$\Gamma_1^B, \Gamma_2^B \in B, \quad \Gamma_1^B = \begin{pmatrix} 1\,0 & \vdots & 0\,1 \\ 0\,1 & \vdots & 1\,0 \end{pmatrix}, \quad \Gamma_2^B = \begin{pmatrix} 1\,1 & \vdots & 1\,1 \\ 1\,1 & \vdots & 0\,1 \end{pmatrix}$$

— hyperplanes of the direction $(1, 0, 0, 0)$ of B.

We claim that A and B are resembling up to equivalence matrices. Indeed, rearrangement of the hyperplanes $(\cdot, \cdot, 1, \cdot)$ and $(\cdot, \cdot, 2, \cdot)$ of A provides the following matrix A':

$$\Gamma_1^{A'} = \begin{pmatrix} 1\,0 & \vdots & 0\,1 \\ 0\,1 & \vdots & 1\,0 \end{pmatrix}, \quad \Gamma_2^{A'} = \begin{pmatrix} 0\,0 & \vdots & 0\,1 \\ 0\,1 & \vdots & 0\,0 \end{pmatrix}.$$

This matrix coincides with B everywhere except the hyperplane $(2,\cdot,\cdot,\cdot)$. Hence A' and B are resembling and A, B are resembling up to equivalence.

The permanents of A and B can be computed by the definition:

$$per(A) = a_{1111}a_{2222} + a_{1112}a_{2221} + a_{1121}a_{2212} + a_{1122}a_{2211} + a_{1211}a_{2122} +$$

$$a_{1212}a_{2121} + a_{1221}a_{2112} + a_{1222}a_{2111} = 0+0+0+0+0+0+0+0 = 0,$$

$$per(B) = b_{1111}b_{2222} + b_{1112}b_{2221} + b_{1121}b_{2212} + b_{1122}b_{2211} + b_{1211}b_{2122} +$$

$$b_{1212}b_{2121} + b_{1221}b_{2112} + b_{1222}b_{2111} = 1+0+0+1+0+1+1+0 = 4.$$

4 Increasing the Dimension

Definition 6 Let $\alpha \in \mathbb{N}$, $S \subset \mathbb{N}$. Then we denote $\alpha S \equiv \{\alpha s | s \in S\}$.

In this section we investigate the extension of the set of values of the permanent function under increasing of the dimension. In particular, we prove that if S is a subset of values of the permanent on $\mathfrak{A}(n,k)$ for some fixed n and k, then $\bigcup_{m=1}^{n} m!S$ is a subset of values of the permanent on $\mathfrak{A}(n, k+1)$. We illustrate these results on the sets $\mathfrak{A}(3,4)$ and $\mathfrak{A}(4,3)$.

Before proving the results of this work, we will formulate and prove the following statement which shows how to construct the new values of the permanent function from the existing ones by increasing the dimension or the order of a matrix.

Lemma 4 ([3, Proposition 7.4]) *Let $A \in \mathfrak{A}(n,k_1)$ and $B \in \mathfrak{A}(n,k_2)$. Then there exists a matrix $C \in \mathfrak{A}(n, k_1+k_2-1)$ with $per(C) = per(A) \cdot per(B)$.*

Proof Let some permutation $\sigma \in S_n$ be fixed. Let C be the matrix such that the entry of C with index $\gamma = (i, \alpha_2, \ldots, \alpha_{k_1}, \beta_2, \ldots, \beta_{k_2}) = (i, \alpha, \beta)$ equals to the product of the entry of A with index $(i, \alpha_2, \ldots, \alpha_{k_1})$ by the entry of B with index $(\sigma(i), \beta_2, \ldots, \beta_{k_2})$.

Compute the permanent of C:

$$per(C) = \sum_{l \in D(C)} \prod_{\gamma \in l} c_\gamma = \sum_{\substack{l_A \in D(A) \\ l_B \in D(B)}} \prod_{(i,\alpha) \in l_A} a_{(i,\alpha)} \prod_{(\sigma(i),\beta) \in l_B} b_{(\sigma(i),\beta)} =$$

$$\left(\sum_{l_A \in D(A)} \prod_{(i,\alpha) \in l_A} a_{(i,\alpha)} \right) \left(\sum_{l_B \in D(B)} \prod_{(j,\beta) \in l_B} b_{(j,\beta)} \right) = per(A) \cdot per(B).$$

□

Applying Lemma 4, we have

Corollary 1 *Let $A \in \mathfrak{A}(n,k)$, $per(A) = x$, and $B \in \mathfrak{A}(n,2)$, $per(B) = y$. Then there exists a matrix $C \in \mathfrak{A}(n,k+1)$ with $per(C) = xy$.*

Lemma 5 *Assume that S is a subset of values of the permanent on the set $\mathfrak{A}(n,k)$. Then*

1. *$\bigcup_{m=1}^{n} m(n-1)!S$ is a subset of values of the permanent on the set $\mathfrak{A}(n,k+1)$.*
2. *$\bigcup_{m=1}^{n} m!S$ is a subset of values of the permanent on the set $\mathfrak{A}(n,k+1)$.*

Proof 1. Let us consider the case $1 \le m < n$. We construct a set of matrices

$$\{B^{(l)} = (b_{ij}^{(l)}) | l = 1, \dots, n-1\} \subset \mathfrak{A}(n,2),$$

in the following way: for each $l \in \{1, \dots, n-1\}$

$$b_{ij}^{(l)} = \begin{cases} 0, & i = 1, \dots, n-l, \ j = 1, \\ 1, & \text{otherwise.} \end{cases}$$

Compute the number of unit diagonals in $B^{(l)}$. There are exactly l ones in the first column of $B^{(l)}$. We choose one of these 1's. The remaining columns consist of unit entries. Thus, there are exactly $(n-1)!$ options to complete the unit entry selected above to a unit diagonal. Hence the number of unit diagonals in $B^{(l)}$ equals $l \cdot (n-1)!$, i.e., $per(B^{(l)}) = l \cdot (n-1)!$.

In case $m = n$ we consider the matrix $B^{(n)} \in \mathfrak{A}(n,2)$ such that $b_{ij}^{(n)} = 1$ for $i, j = 1, \dots, n$. By the definition of the permanent, $per(B^{(n)}) = n!$.

Let $x \in S$ and $l \in \{1, \dots, n\}$. By the definition of S, there exists a matrix $A \in \mathfrak{A}(n,k)$ such that $per(A) = x$. Applying Corollary 1 to A and $B^{(l)}$, we obtain that there exists a matrix $C_x^{(l)} \in \mathfrak{A}(n,k+1)$ such that $per(C_x^{(l)}) = l \cdot (n-1)! \cdot x$.

Since $l \in \{1, \dots, n\}$ and $x \in S$ are arbitrary, $\bigcup_{m=1}^{n} m(n-1)!S$ is a subset of values of the permanent on the set $\mathfrak{A}(n,k+1)$.

2. Let $1 \le m < n-1$. For each $l \in \mathbb{N}$, $1 < l \le n-1$, we consider the following matrix $B^{(l)} \in \mathfrak{A}(n,2)$:

$$b_{ij}^{(l)} = \begin{cases} 0, & i+j \le n,\ j \le l, \\ 1, & \text{otherwise.} \end{cases}$$

Compute the permanent of $B^{(l)}$. There is exactly one unit entry in the first column of $B^{(l)}$. Let us choose this entry. Further, each column with index j, $2 \le j \le l$, contains one more unit entry than the column with index $j-1$. There is a unique non-zero entry of the column with index j that lies diagonally with each already chosen unit entry. Thus, there is a unique choice of a unit partial diagonal in the columns with indices $1, \dots, l$. In the remaining $n-l$ columns all entries are ones. Then there are exactly $(n-l)!$ options to complete the already chosen unit partial diagonal to a unit diagonal. Consequently, $per(B^{(l)}) = (n-l)!$.

Let $x \in S$ and $l \in \{2, \dots, n-1\}$. By the definition of S, there exists a matrix $A \in \mathfrak{A}(n,k)$ such that $per(A) = x$. By Corollary 1 applied to A and $B^{(l)}$, there exists a matrix $\tilde{C}_x^{(l)} \in \mathfrak{A}(n,k+1)$ with $per(\tilde{C}_x^{(l)}) = (n-l)! \cdot x$. If we replace $n-l$ by m, we obtain that for any $m \in \{1, \dots, n-2\}$ there exists a matrix $\tilde{C}_x^{(n-m)} \in \mathfrak{A}(n,k+1)$ with $per(\tilde{C}_x^{(n-m)}) = m! \cdot x$.

By clause 1 of the lemma, there exist matrices $C_x^{(1)}$, $C_x^{(n)} \in \mathfrak{A}(n,k+1)$ with $per(C_x^{(1)}) = (n-1)! \cdot x$, $per(C_x^{(n)}) = n! \cdot x$.

Since $x \in S$ is arbitrary, $\bigcup_{m=1}^{n} m!S$ is a subset of values of the permanent on the set $\mathfrak{A}(n,k+1)$. □

In the following lemma we construct 2-dimensional matrices of a specific form.

Lemma 6 *Assume that S is a subset of values of the permanent on the set $\mathfrak{A}(n,k)$. Then for any $m \in \mathbb{N}$, $1 < m \le n$, $i_2, \dots, i_m \in \mathbb{Z}$, $i_j \ge 0$, $i_2 + \dots + i_m \le n-m$, the set $2^{i_2} \cdot \ldots \cdot m^{i_m} m!S$ is a subset of values of the permanent on the set $\mathfrak{A}(n,k+1)$.*

Proof Let m and $I = (i_2, \dots, i_m)$ be given.

If $i_2 + \dots + i_m = 0$, then $i_2 = \ldots = i_m = 0$ by the condition $i_j \ge 0$. We can use clause 2 of Lemma 5 and conclude the proof.

Further we assume that $i_2 + \dots + i_m > 0$. Let us construct a matrix $B \in \mathfrak{A}(n,2)$ such that $per(B) = 2^{i_2} \cdot \ldots \cdot m^{i_m} m!$.

Let us denote $\lambda := n - m - (i_2 + \dots + i_m)$, $\alpha := |\{u \in \mathbb{N} | i_u > 0\}|$. Firstly, let all the entries of B be equal to 1. If $i_2 + \dots + i_m < n-m$, then we put

$$b_{jl} = 0,\ l = 1, \dots, \lambda,\ j = 1, \dots, n-l.$$

Now let $d_1 = \min\{u \in \mathbb{N} | i_u > 0\}$. We set

$$b_{jl} = 0,\ l = \lambda+1, \dots, \lambda+i_{d_1},\ j = 1, \dots, n-d_1-(l-1).$$

Further, let $d_2 = \min\{u \in \mathbb{N}, u > d_1 | i_u > 0\}$. Then we put

$$b_{jl} = 0,\ l = \lambda + i_{d_1} + 1, \ldots, \lambda + i_{d_1} + i_{d_2},\ j = 1, \ldots, n - d_2 - (l - 1).$$

Let us continue the process described in this paragraph by induction. At step t, $t \leq \alpha$, we consider $d_t = \min\{u \in \mathbb{N}, u > d_{t-1} | i_u > 0\}$ and set

$$b_{jl} = 0,\ l = \lambda + i_{d_1} + \cdots + i_{d_{t-1}} + 1, \ldots, \lambda + i_{d_1} + \cdots + i_{d_t},$$

$$j = 1, \ldots, n - d_t - (l - 1).$$

In this way, we will come to the column with index $n - m$, since $\lambda + i_2 + \cdots + i_m = n - m$.

Let us find the number of unit diagonals in B.

- If $\lambda > 0$, then there exists a unique unit partial diagonal of the length λ in the first λ columns. By construction, these entries have the indices $(n, 1)$, $(n - 1, 2) \ldots$, $(n - \lambda + 1, \lambda)$. Let us fix them.
- If $\lambda > 0$, then there exist exactly d_1 unit entries in the column with index $\lambda + 1$ that lie diagonally with the already fixed entries. If $\lambda = 0$, there are exactly d_1 unit entries in the first column. Let us choose one of d_1 unit entries in both cases. If $i_{d_1} > 1$, then, by construction, there exist exactly $d_1 + \lambda + 1$ unit entries in the column of B with index $\lambda + 2$. There are d_1 entries of them that lie diagonally with the already selected entries. We again choose one of them. In each of the remaining columns with indices less or equal $\lambda + i_{d_1}$ we can select one of d_1 unit entries similarly. Therefore, we have $d_1^{i_{d_1}}$ options to choose the following i_{d_1} entries of the unit diagonal. A similar conclusion is true for all d_t, $1 < t \leq \alpha$.
- We have fixed $\lambda + i_2 + \cdots + i_m = n - m$ entries. It is required to select m ones in the remaining m columns. All the entries in these columns are ones. So, we have $m!$ variants.

Hence the number of unit diagonals in B equals $d_1^{i_{d_1}} \cdot \ldots \cdot d_\alpha^{i_{d_\alpha}} m! = 2^{i_2} \cdot \ldots \cdot m^{i_m} m!$. The last equality follows from the definition of the numbers d_t, $t = 1, \ldots, \alpha$, and i_j, $j = 2, \ldots, m$. Since the number of unit diagonals in B equals $per(B)$, we obtain $B \in \mathfrak{A}(n, 2)$ such that $per(B) = 2^{i_2} \cdot \ldots \cdot m^{i_m} m!$.

Let $x \in S$. By the definition of S, there exists a matrix $A \in \mathfrak{A}(n, k)$ such that $per(A) = x$. Corollary 1 applied to A and B provides a matrix $C_x \in \mathfrak{A}(n, k + 1)$ with $per(C_x) = 2^{i_2} \cdot \ldots \cdot m^{i_m} m! \cdot x$.

Since $x \in S$ is arbitrary, $2^{i_2} \cdot \ldots \cdot m^{i_m} m! S$ is a subset of values of the permanent on the set $\mathfrak{A}(n, k + 1)$. □

Example

In case $n = 6$, $m = 3$, $i_2 = 1$, $i_3 = 1$ Lemma 6 provides us the following matrix B:

$$B = \begin{pmatrix} 0&0&0&1&1&1 \\ 0&0&1&1&1&1 \\ 0&0&1&1&1&1 \\ 0&1&1&1&1&1 \\ 0&1&1&1&1&1 \\ 1&1&1&1&1&1 \end{pmatrix}.$$

Here $\lambda = 1$.

Lemma 7 *If S is a subset of values of the permanent on the set $\mathfrak{A}(n, k)$, then $3 \cdot S$ is a subset of values of the permanent on the set $\mathfrak{A}(n, k + 1)$.*

Proof Let us construct a matrix $B \in \mathfrak{A}(n, 2)$ such that $per(B) = 3$.

Consider the following matrix of order 3:

$$B^{(3)} = \begin{pmatrix} 1&0&1 \\ 0&1&1 \\ 1&1&1 \end{pmatrix}.$$

It can be directly proved that the permanent of $B^{(3)}$ is equal to 3. If $n = 3$, then we can set $B = B^{(3)}$. If $n > 3$, we immerse $B^{(3)}$ to B as follows:

$$b_{ij} = \begin{cases} b_{ij}^{(3)}, & i, j = 1, 2, 3, \\ 1, & i = j,\ i > 3, \\ 0, & \text{otherwise.} \end{cases}$$

Each unit diagonal of $B^{(3)}$ may be completed to the unit diagonal of B by entries with indices $(4, 4), \ldots, (n, n)$. So, every unit diagonal of the matrix $B^{(3)}$ corresponds to a unique unit diagonal of B. Therefore, the permanent of B equals 3.

Let $x \in S$. By the definition of S, there exists a matrix $A \in \mathfrak{A}(n, k)$ such that $per(A) = x$. Corollary 1 applied to A and B provides a matrix $C_x \in \mathfrak{A}(n, k + 1)$ with $per(C_x) = 3 \cdot x$.

Since $x \in S$ is arbitrary, $3 \cdot S$ is a subset of values of the permanent on the set $\mathfrak{A}(n, k + 1)$. □

Example

In case $n = 4$ we obtain the following matrix B in Lemma 7:

$$B = \begin{pmatrix} 1 & 0 & 1 & 0 \\ 0 & 1 & 1 & 0 \\ 1 & 1 & 1 & 0 \\ 0 & 0 & 0 & 1 \end{pmatrix}.$$

Let us illustrate the obtained results by the sets $\mathfrak{A}(3, 4)$ and $\mathfrak{A}(4, 3)$.

Example

Let us apply Lemma 5 in case $n = 3$, $k = 3$. Here $n! = 6$, $(n-1)! = 2$, $(n-1) \cdot (n-1)! = 4$. From [3, Theorem 8.1] we know that the permanent attains on the set $\mathfrak{A}(3, 3)$ values $0, 1, \ldots, 26, 28, 29, 32, 36$. Hence, by Lemma 5(1), the permanent achieves the following values on $\mathfrak{A}(3, 4)$: $0, 6, 12, \ldots, 156, 168, 174, 192, 216$; $0, 2, 4, \ldots, 52, 56, 58, 64, 72$, and $0, 4, 8, \ldots, 104, 112, 116, 128, 144$.

Example

By [5, Corollary 5.11], the permanent on $\mathfrak{A}(4, 2)$ attains values $0, 1, 2, 3, \ldots, 12, 14, 18, 24$. Since $4! = 24$, $(4-1)! = 6$, $2 \cdot (4-1)! = 12$, $3 \cdot (4-1)! = 18$, by Lemma 5(1), the permanent achieves the following values on $\mathfrak{A}(4, 3)$: $0, 24, 48, \ldots, 288, 336, 432, 576$; $0, 6, \ldots, 72, 84, 108, 144$; $12, \ldots, 144, 168, 216, 288$, and $18, \ldots, 216, 252, 324, 432$.

Now let us apply Lemma 5(2) for $m = 2$. In this case we have $(n - m)! = (4 - 2)! = 2$. Therefore, by the lemma, there exist matrices from $\mathfrak{A}(4, 3)$ with permanents $2, 4, \ldots, 24, 28, 36, 48$.

Example

By [5, Corollary 5.11], the permanent on $\mathfrak{A}(4, 2)$ attains values $0, 1, 2, 3, \ldots, 12, 14, 18, 24$. By Lemma 6, the permanent on $\mathfrak{A}(4, 3)$ attains, in particular, values $2^2 \cdot 2! \cdot per(A) = 8 \cdot per(A)$, $A \in \mathfrak{A}(4, 2)$. Let us list them: $0, 8 \ldots, 96, 112, 144, 192$.

Example

By [3, Theorem 8.1], we know that the permanent on $\mathfrak{A}(3,3)$ attains values $0, 1, \ldots, 26, 28, 29, 32, 36$. Hence, by Lemma 7, the permanent on $\mathfrak{A}(3,4)$ achieves values $0, 3, \ldots, 78, 84, 87, 96, 108$.

By [5, Corollary 5.11], the permanent on $\mathfrak{A}(4,2)$ attains values $0, 1, 2, 3, \ldots, 12, 14, 18, 24$. Thus, by Lemma 7, the permanent achieves the following values on $\mathfrak{A}(4,3)$: $0, 3, \ldots, 36, 42, 54, 72$.

5 Increasing of Dimension and Consecutive Values of Permanent

In this section we construct new estimates for consecutive values of the permanent based on increasing of the dimension of the original matrix.

Before proving the main result of this work, we will formulate the following theorems for the permanent of 2-dimensional matrices. We use the notion of resembling up to equivalence matrices here, see more details in Sect. 3.

Theorem 2 ([5, Corollary 2.4]) *Let $(n, N) \in \mathbb{N}^2$. Assume that for any integer x, $0 \le x \le N-1$ there exists a pair of resembling up to equivalence matrices $A_x, B_x \in \mathfrak{A}(n,2)$ such that $per(A_x) = x$, $per(B_x) = x+1$. Then for any integer z, $0 \le z \le 2N-1$ there exists a pair of resembling up to equivalence matrices $C_z, D_z \in \mathfrak{A}(n+1,2)$ such that $per(C_z) = z$, $per(D_z) = z+1$.*

Theorem 3 ([5, Lemma 4.1]) *Conditions of Theorem 2 are satisfied for $n = 6$, $N = 67$, i.e., for any integer x, $0 \le x \le 66$ there exists a pair of resembling up to equivalence matrices $A_x, B_x \in \mathfrak{A}(6,2)$ such that $per(A_x) = x$, $per(B_x) = x+1$.*

Lemma 8 *Let $m_1, m_2 \in \mathbb{N} \setminus \{1\}$, $A_1, \ldots, A_{m_1} \in \mathfrak{A}(n,k)$, and $B_1, \ldots, B_{m_2} \in \mathfrak{A}(n,2)$. Assume that $A_1, \ldots, A_{m_1}$ are resembling and $B_1, \ldots, B_{m_2}$ are resembling. Then there exist resembling matrices $C_{i,j} \in \mathfrak{A}(n,k+1)$ with*

$$per(C_{i,j}) = per(A_i) \cdot per(B_j),\ i = 1, \ldots, m_1,\ j = 1, \ldots, m_2,$$

and such that $C_{i,j}$ coincide everywhere except, possibly, the hyperplane $\Gamma = (1, \cdot, \ldots, \cdot)$.

Proof Application of the same equivalence transformations to the matrices $A_1, \ldots, A_{m_1}$ and Lemma 1 allow us to assume that $A_1, \ldots, A_{m_1}$ coincide everywhere except, possibly, the hyperplane $\Gamma_1 = (1, \cdot, \ldots, \cdot)$ of the dimension $k-1$. Similarly, we may assume that $B_1, \ldots, B_{m_2}$ coincide everywhere except, possibly, the first row $L = (1, \cdot)$.

Then we put $A = B_j$, $B = A_i$ and use (for $i = 1, \ldots, m_1$, $j = 1, \ldots, m_2$) Lemma 4 choosing $\sigma = id \in S_n$ in the proof. Thus, we get $C_{i,j} \in \mathfrak{A}(n, k+1)$ with

$$per(C_{i,j}) = per(A_i) \cdot per(B_j).$$

Now we claim that $C_{i,j}$ coincide everywhere except, possibly, the hyperplane $\Gamma = (1, \cdot, \ldots, \cdot)$ of the dimension k. Indeed, by the proof of Lemma 4, for any $i \in \{1, \ldots, m_1\}$, $j \in \{1, \ldots, m_2\}$, and $l \in \{1, \ldots, n\}$

$$c_{i,j,(l,\beta_2,\alpha_2,\ldots,\alpha_k)} = b_{j,(l,\beta_2)} a_{i,(l,\alpha_2,\ldots,\alpha_k)},$$

since $\sigma(l) = l$. The matrices $B_1, \ldots, B_{m_2}$ coincide everywhere except, possibly, the first row $L = (1, \cdot)$ and $A_1, \ldots, A_{m_1}$ coincide everywhere except, possibly, the hyperplane $\Gamma_1 = (1, \cdot, \ldots, \cdot)$. So, for any $i_1, i_2, j_1, j_2, \beta_2, \alpha_2, \ldots, \alpha_k$ and $l > 1$, we have

$$c_{i_1,j_1,(l,\beta_2,\alpha_2,\ldots,\alpha_k)} = c_{i_2,j_2,(l,\beta_2,\alpha_2,\ldots,\alpha_k)}.$$

Hence $C_{i,j}$ coincide everywhere except, possibly, Γ, i.e., $C_{i,j}$ are resembling matrices. □

Lemma 9 *Let $t, m \in \mathbb{N} \setminus \{1\}$, $G_1, \ldots, G_t \in \mathfrak{A}(n, k)$. Assume that $G_1, \ldots, G_t$ are resembling. Denote $\mathcal{G} = \{G_1, \ldots, G_t\}$. Let $A_1, \ldots, A_{2m} \in \mathcal{G}$ and $B_1, \ldots, B_m \in \mathfrak{A}(n, 2)$ satisfy the following properties:*

1. *$B_1, \ldots, B_m$ are resembling, i.e., there exists a hyperplane (a row or a column) L such that $B_1, \ldots, B_m$ coincide everywhere except L.*
2. *The supports of L do not intersect for any different B_i and B_j, i.e., ones in the rows (resp., columns) L of B_i and B_j are located in non-intersecting sets of columns (resp., rows).*

Then there exists a pair of resembling matrices $D_1, D_2 \in \mathfrak{A}(n, k+1)$ with

$$per(D_1) = per(B_1) \cdot per(A_1) + \cdots + per(B_m) \cdot per(A_m),$$

$$per(D_2) = per(B_1) \cdot per(A_{m+1}) + \cdots + per(B_m) \cdot per(A_{2m}).$$

Proof Applying unique equivalence transformations, i.e., transpositions and permutations of hyperplanes having the same direction, to $B_1, \ldots, B_m$ and using Lemma 1, we may assume that $L = (1, \cdot)$. Notice that property 2) holds true for new matrices $B_1, \ldots, B_m$.

1. Since $A_1, \ldots, A_{2m}$ and $B_1, \ldots, B_m, B_1, \ldots, B_m$ satisfy the conditions of Lemma 8 for $m_1 = m_2 = 2m$, there exist matrices $C_{i,i} \in \mathfrak{A}(n, k+1)$, $i = 1, \ldots, 2m$ that coincide everywhere except, possibly, the hyperplane $\Gamma = (1, \cdot, \ldots, \cdot)$ and satisfy the conditions

$$per(C_{i,i}) = per(B_i) \cdot per(A_i),\ i = 1\ldots, m,$$

$$per(C_{i,i}) = per(B_{i-m}) \cdot per(A_i),\ i = m+1\ldots, 2m.$$

2. Let us construct $D_1 \in \mathfrak{A}(n, k+1)$.

Consider $i \in \{1, \ldots, m\}$ and the hyperplane L of B_i. By construction of Lemma 8,

$$c_{i,i,(1,\beta_2,\alpha_2,\ldots,\alpha_k)} = b_{i,(1,\beta_2)} a_{i,(1,\alpha_2,\ldots,\alpha_k)}.$$

So, if $\beta_2 \notin pr_2(supp(L))$, where $pr_2(\{(i_1, j_1), \ldots, (i_v, j_v)\}) = \{j_1, \ldots, j_v\}$, then $c_{i,i,(1,\beta_2,\alpha_2,\ldots,\alpha_k)} = 0$. Therefore, the support $supp(\Gamma)$ of the hyperplane Γ of $C_{i,i}$ is a subset of

$$\cup_{\beta_2 \in pr_2(supp(L))} \cup_{\alpha_2,\ldots,\alpha_k \in \{1,\ldots,n\}} (1, \beta_2, \alpha_2, \ldots, \alpha_k).$$

By property (2) of the matrices $B_1, \ldots, B_m$, for any different B_i and B_j the sets $pr_2(supp(L))$ do not intersect. Thus, the supports $supp(\Gamma)$ of any distinct matrices $C_{i,i}$ and $C_{j,j}$ do not intersect.

Let $D_1 \in \mathfrak{A}(n, k+1)$ be the matrix such that for any $\gamma_2, \ldots, \gamma_{k+1}$

$$d_{1,(\gamma_1,\gamma_2,\ldots,\gamma_{k+1})} = \begin{cases} c_{1,1,(\gamma_1,\gamma_2,\ldots,\gamma_{k+1})}, & \gamma_1 > 1, \\ \sum_{l=1}^{m} c_{l,l,(\gamma_1,\gamma_2,\ldots,\gamma_{k+1})}, & \gamma_1 = 1. \end{cases}$$

For any $\gamma_2, \ldots, \gamma_{k+1}$ there is at most one non-zero summand in $\sum_{l=1}^{m} c_{l,l,(1,\gamma_2,\ldots,\gamma_{k+1})}$, since the supports $supp(\Gamma)$ of any different $C_{i,i}$ and $C_{j,j}$ do not intersect. It implies that the matrix D_1 consists of zeros and ones. By Lemma 2,

$$per(D_1) = per(B_1) \cdot per(A_1) + \cdots + per(B_m) \cdot per(A_m).$$

In the same way we get $D_2 \in \mathfrak{A}(n, k+1)$ with the required value of the permanent using matrices $C_{m+1,m+1}, \ldots, C_{2m,2m}$.

3. Now we claim that D_1 resembles D_2.

Indeed, by clause 1 of the proof, $C_{1,1}$ and $C_{m+1,m+1}$ coincide everywhere except, possibly, the hyperplane Γ. The matrices D_1, $C_{1,1}$ coincide anywhere except, possibly, Γ by clause 2. The same is true for D_2 and $C_{m+1,m+1}$. Therefore, the matrices D_1, D_2 coincide everywhere except, possibly, Γ. Hence D_1 and D_2 are resembling. □

Lemma 10 *Consider $n \geq 3$, $t \in \mathbb{N} \setminus \{1\}$, and $G_1, \ldots, G_t \in \mathfrak{A}(n, k)$. Assume that $G_1, \ldots, G_t$ are resembling. Denote $\mathcal{G} = \{G_1, \ldots, G_t\}$. Let $A_i, A'_i \in \mathcal{G}$ for $i = 1, 2, 3$ and $per(A_i) = x_i$, $per(A'_i) = y_i$. Then there exists a pair of resembling matrices $D_1, D_2 \in \mathfrak{A}(n, k+1)$ with $per(D_1) = 2x_1 + x_2 + x_3$, $per(D_2) = 2y_1 + y_2 + y_3$.*

Proof It is sufficient to construct matrices $B_1, B_2, B_3 \in \mathfrak{A}(n, 2)$ satisfying conditions of Lemma 9 and such that $per(B_1) = 2$, $per(B_2) = 1$, $per(B_3) = 1$.

Consider matrices

$$\tilde{B}^{(1)} = \begin{pmatrix} 1 & 0 & 1 \\ 0 & 1 & 1 \\ 0 & 1 & 1 \end{pmatrix}, \quad \tilde{B}^{(2)} = \begin{pmatrix} 0 & 0 & 1 \\ 1 & 1 & 1 \\ 0 & 1 & 1 \end{pmatrix}, \quad \tilde{B}^{(3)} = \begin{pmatrix} 0 & 0 & 1 \\ 0 & 1 & 1 \\ 1 & 1 & 1 \end{pmatrix}.$$

It can be directly checked (for instance, apply the Laplace decomposition formula by the first column) that $per(\tilde{B}^{(1)}) = 2$, $per(\tilde{B}^{(2)}) = 1$, $per(\tilde{B}^{(3)}) = 1$. The matrices $\tilde{B}^{(1)}, \tilde{B}^{(2)}, \tilde{B}^{(3)}$ coincide everywhere except the first column and the supports of their first columns do not intersect.

If $n = 3$, we set $B^{(i)} = \tilde{B}^{(i)}$ for $i = 1, 2, 3$. If $n > 3$, for each $i \in \{1, 2, 3\}$ we construct $B^{(i)} \in \mathfrak{A}(n, 2)$ as follows:

$$b^{(i)}_{jl} = \begin{cases} \tilde{b}^{(i)}_{jl}, & j, l = 1, 2, 3, \\ 1, & j = l, \ j > 3, \\ 0, & \text{otherwise.} \end{cases}$$

The matrices $B^{(1)}, B^{(2)}, B^{(3)}$ also coincide everywhere except the first column, the supports of their first columns do not intersect, and $per(B^{(1)}) = 2$, $per(B^{(2)}) = 1$, $per(B^{(3)}) = 1$.

Then we set $B_i = B^{(i)}$ and apply Lemma 9 for t and $m = 3$. □

Remark 2 Assume that $G_1, G_2, G_3 \in \mathfrak{A}(n, k)$ satisfy the conditions of Lemma 10 for $t = 3$. By definition of resembling matrices, there exists a hyperplane Γ such that G_1, G_2, G_3 coincide everywhere except, possibly, Γ. Suppose that the hyperplane Γ of G_3 consists of zeros and $A_i, A'_i \in \mathcal{G} = \{G_1, G_2, G_3\}$ for $i = 1, 2, 3$. Therefore, by Lemma 10, there exists a pair of resembling matrices $D_1, D_2 \in \mathfrak{A}(n, k+1)$ with $per(D_1) = 2x_1 + x_2 + 0 = 2x_1 + x_2$, $per(D_2) = 2y_1 + y_2 + 0 = 2y_1 + y_2$.

Lemma 11 *Assume that for some $n \geq 3$, $N \in \mathbb{N}$ the following is true: for any integer x, $0 \leq x \leq N - 1$, there exists a pair of resembling up to equivalence matrices $A_x, B_x \in \mathfrak{A}(n, k)$ such that $per(A_x) = x$, $per(B_x) = x + 1$. Then for any integer y, $0 \leq y \leq 4N - 1$, there exists a pair of resembling up to equivalence matrices $C_y, D_y \in \mathfrak{A}(n, k+1)$ such that $per(C_y) = y$, $per(D_y) = y + 1$.*

Proof By Definition 5 of resembling up to equivalence matrices, for any integer $x \in [0, N - 1]$ there exist an equivalent to B_x matrix B'_x and a hyperplane Γ_x such that B'_x coincides with A_x everywhere except, possibly, Γ_x. At the same time, $per(B'_x) = per(B_x)$, since B'_x and B_x are equivalent. The matrix B'_x differs from A_x in the hyperplane Γ_x, since $per(B'_x) \neq per(A_x)$. So, replacing B_x with B'_x, we can assume that for any integer $x \in [0, N - 1]$ there exists a hyperplane Γ_x such that the matrices A_x and B_x coincide everywhere except Γ_x, i.e., A_x and B_x are resembling.

Let $x \in \{0, \ldots, N - 1\}$ and $\mathcal{G} = \{A_x, B_x\}$. Firstly, we put $A_1 = A_2 = A_3 = A_x$, $A'_1 = A'_2 = A_x$, $A'_3 = B_x$ and apply Lemma 10 for $t = 2$. We receive a pair of resembling matrices $C^{(0)}_x, E^{(1)}_x \in \mathfrak{A}(n, k+1)$ with

$$per(C_x^{(0)}) = 4x, \; per(E_x^{(1)}) = 4x + 1.$$

Then we set $A_1 = A_2 = A_x$, $A_3 = B_x$, $A'_1 = A_x$, and $A'_2 = A'_3 = B_x$. So, from Lemma 10 for $t = 2$ we get resembling matrices $C_x^{(1)}, E_x^{(2)} \in \mathfrak{A}(n, k+1)$ such that

$$per(C_x^{(1)}) = 4x + 1, \; per(E_x^{(2)}) = 4x + 2.$$

If we select $A_1 = A_x$, $A_2 = A_3 = B_x$, $A'_1 = A'_2 = B_x$, and $A'_3 = A_x$, then we obtain a pair of resembling matrices $C_x^{(2)}, E_x^{(3)} \in \mathfrak{A}(n, k+1)$ with

$$per(C_x^{(2)}) = 4x + 2, \; per(E_x^{(3)}) = 4x + 3.$$

Choice $A_1 = A_2 = B_x$, $A_3 = A_x$, $A'_1 = A'_2 = A'_3 = B_x$ provides resembling matrices $C_x^{(3)}, E_x^{(4)} \in \mathfrak{A}(n, k+1)$ such that

$$per(C_x^{(3)}) = 4x + 3, \; per(E_x^{(4)}) = 4x + 4.$$

Thus, for any $x = 0, \ldots, N-1$ and $j = 0, \ldots, 3$ we have the pair of resembling matrices $C_x^{(j)}, E_x^{(j+1)} \in \mathfrak{A}(n, k+1)$ with $per(C_x^{(j)}) = 4x + j$, $per(E_x^{(j+1)}) = 4x + j + 1$. Hence for any integer $y \in [0, 4N-1]$ there exists a pair of resembling matrices $C_y, D_y \in \mathfrak{A}(n, k+1)$ such that $per(C_y) = y$, $per(D_y) = y + 1$. By Remark 1, the statement of the lemma holds. □

Corollary 2 *Assume that the conditions of the previous lemma hold. Then for any nonnegative integer j the permanent function attains on the set $\mathfrak{A}(n, k+j)$ all integer values from the interval $[0, 4^j N]$.*

Proof Apply Lemma 11 j times subsequently. □

Theorem 4 *Let $n \geq 3$. For any integer l, $0 \leq l \leq 2^{n+2k-5} - 1$, there exists a pair of resembling up to equivalence matrices $A_l, B_l \in \mathfrak{A}(n, k)$ such that $per(A_l) = l$, $per(B_l) = l + 1$.*

Proof Consider the following matrices from $\mathfrak{A}(3, 2)$:

$$G_0 = \begin{pmatrix} 0 & 0 & 1 \\ 0 & 1 & 1 \\ 0 & 1 & 1 \end{pmatrix}, \quad G_1 = \begin{pmatrix} 0 & 0 & 1 \\ 0 & 1 & 1 \\ 1 & 1 & 1 \end{pmatrix}, \quad G_2 = \begin{pmatrix} 1 & 0 & 1 \\ 0 & 1 & 1 \\ 0 & 1 & 1 \end{pmatrix},$$

$$G_3 = \begin{pmatrix} 1 & 0 & 1 \\ 0 & 1 & 1 \\ 1 & 1 & 1 \end{pmatrix}, \quad G_4 = \begin{pmatrix} 1 & 1 & 1 \\ 0 & 1 & 1 \\ 1 & 1 & 1 \end{pmatrix}.$$

It can be proved by a direct calculation that $per(G_i) = i$. In addition, G_i and G_{i+1} differ only in the first column for $i = 0, 1, 2$, and G_3, G_4 differ only in the first row. Hence for each $i \in \{0, \ldots, 3\}$ the matrices G_i and G_{i+1} are resembling up to equivalence. Then we apply m times subsequently Theorem 2 for $n = 3$, $N = 4$. We obtain that for any integers $m \geq 3$ and x, $0 \leq x \leq 4 \cdot 2^{m-3} - 1$, there exists a pair of resembling up to equivalence matrices $A_x, B_x \in \mathfrak{A}(m, 2)$ such that $per(A_x) = x$, $per(B_x) = x + 1$.

Now we set $m = n$ and apply Corollary 2 for $k = 2, n$, and $N = 4 \cdot 2^{n-3} = 2^{n-1}$. We receive that for any integer $j \geq 0$ the permanent attains on $\mathfrak{A}(n, 2 + j)$ all integer values from $[0, 4^j N]$. Substituting $2 + j$ with k, we have

$$4^j N = 2^{n+2k-5}.$$

Since Corollary 2 is obtained by consecutive application of Lemma 11, we see that for any $l \in \mathbb{Z}$, $0 \leq l \leq 2^{n+2k-5} - 1$, there exists a pair of resembling up to equivalence matrices $X, Y \in \mathfrak{A}(n, k)$ with $per(A_l) = l$, $per(B_l) = l + 1$. □

The following corollary is implied by Theorem 4 immediately.

Corollary 3 *Let $n \geq 3$. For any integer l, $0 \leq l \leq 2^{n+2k-5}$, there exists a matrix $A \in \mathfrak{A}(n, k)$ such that $per(A) = l$.*

Let us compare the upper bound B_2 obtained in Corollary 3 with the upper bound B_1 from Theorem 1.

Lemma 12 *Let $B_1(n, k) = 2^{k-1}k^{n-2}$, $B_2(n, k) = 2^{n+2k-5}$. Then*

1.

$$\lim_{k \to \infty} \frac{B_1(n, k)}{B_2(n, k)} = 0.$$

2. If $k > 2$, then

$$\lim_{n \to \infty} \frac{B_2(n, k)}{B_1(n, k)} = 0.$$

3. The number $B_1(n, k)$ coincides with $B_2(n, k)$ if either $k = 2$ or

$$n = \frac{k - 2}{\log_2 k - 1} + 2, \; k > 2.$$

Proof Consider the ratio of B_1 to B_2:

$$\frac{B_1(n, k)}{B_2(n, k)} = \left(\frac{k}{2}\right)^{n-2} \cdot 2^{-k+2}.$$

1. We have

$$\lim_{k \to \infty} \frac{B_1(n, k)}{B_2(n, k)} = \lim_{k \to \infty} \left(\left(\frac{k}{2}\right)^{n-2} \cdot 2^{-k+2} \right) = 0.$$

2. Let $k > 2$. Since $0 < 2/k < 1$, we obtain

$$\lim_{n\to\infty} \frac{B_2(n,k)}{B_1(n,k)} = \lim_{n\to\infty} \left(\left(\frac{2}{k}\right)^{n-2} \cdot 2^{k-2} \right) = 0.$$

3. It can be checked that for $k > 2$ we have $B_1(n,k) = B_2(n,k)$ if and only if $n = \frac{k-2}{\log_2 k-1} + 2$. □

Actually clause 1 of Lemma 12 asserts that the bound $B_2(n,k)$ provides more values of the permanent on $\mathfrak{A}(n,k)$ than the bound $B_1(n,k)$ as $k \to \infty$. By clause 2 of Lemma 12, $B_1(n,k)$ provides more values of the permanent on $\mathfrak{A}(n,k)$ than $B_2(n,k)$ as $n \to \infty$.

Example

Consider the set of matrices $\mathfrak{A}(3,4)$. In this case the estimate of Theorem 1 has the form $B_1(3,4) = 2^{4-1} \cdot 4^{3-2} = 32$. From Corollary 3 we obtain the bound $B_2(3,4) = 2^{3+2\cdot 4-5} = 64$.

On the set $\mathfrak{A}(4,3)$ we have $B_1(4,3) = 2^{3-1} \cdot 3^{4-2} = 36$ and $B_2(4,3) = 2^{4+2\cdot 3-5} = 32$.

For $\mathfrak{A}(3,4)$ the second estimate allows to obtain more values of the permanent and for $\mathfrak{A}(4,3)$ the first estimate allows to obtain more values.

Theorem 5 *Let $n \geq 6$. Then for any integer l, $0 \leq l \leq 66 \cdot 2^{n-6} \cdot 4^{k-2}$, there exists a matrix $A \in \mathfrak{A}(n,k)$ such that $per(A) = l$.*

Proof By [5, Lemma 4.1] (see Theorem 3 from this text), for $n = 6$ and for any integer x, $0 \leq x \leq 66$, there exists a pair of resembling up to equivalence matrices $A_x, B_x \in \mathfrak{A}(6,2)$ with $per(A_x) = x$, $per(B_x) = x+1$. Applying $n-6$ times subsequently Theorem 2 for $n = 6$ and $N = 66$, we get that for any integer x, $0 \leq x \leq 66 \cdot 2^{n-6} - 1$, there exists a pair of resembling up to equivalence matrices $A_x, B_x \in \mathfrak{A}(n,2)$ such that $per(A_x) = x$, $per(B_x) = x+1$. Therefore, by Corollary 2 for $N = 66 \cdot 2^{n-6}$, $j = k-2$, we obtain that the permanent attains on $\mathfrak{A}(n,k)$ all integer values from $[0, 66 \cdot 2^{n-6} \cdot 4^{k-2}]$. □

Acknowledgements The authors are grateful to the referee for careful reading and valuable suggestions on the presentation of the results. The work of the second author was funded by the European Union (ERC, GENERALIZATION, 101039692). Views and opinions expressed are however those of the author(s) only and do not necessarily reflect those of the European Union or the European Research Council Executive Agency. Neither the European Union nor the granting authority can be held responsible for them.

References

1. Minc, H.: Permanents. Addison-Wesley, Englewood Cliffs (1978)
2. Cheon, G.-S., Wanless, I.M.: An update on Minc's survey of open problems involving permanents. Linear Algebra Appl. **403**, 314–342 (2005)
3. Guterman, A.E., Evseev, I.M., Taranenko, A.A.: Values of the permanent function on multidimensional (0,1)-matrices. Sib. Math. J. **63**, 262–276 (2022)
4. Brualdi, R.A., Newman, M.: Some theorems on the permanent. J. Res. Nat. Bur. Stand. **69B**(3), 159–163 (1965)
5. Guterman, A.E., Taranin, K.A.: On the values of the permanent of (0,1)-matrices. Linear Algebra Appl. **552**, 256–276 (2018)
6. Muir, T.: A Treatise on the Theory of Determinants. Macmillan Co., London (1933)
7. Oldenburger, R.: Higher dimensional determinants. Am. Math. Mon. **47**, 25–33 (1940)
8. Dow, S.J., Gibson, P.M.: Permanents of d-dimensional matrices. Linear Algebra Appl. **90**, 133–145 (1987)
9. Dow, S.J., Gibson, P.M.: An upper bound for the permanent of a 3-dimensional (0,1)-matrix. Proc. Am. Math. Soc. **99**(1), 29–34 (1987)
10. Sokolov, N.P.: Introduction to the Theory of Multidimensional Matrices. Naukova Dumka, Kiev (1972)
11. Sokolov, N.P.: Space Matrices and Their Applications. Fizmatlit, Moscow (1960)
12. Taranenko, A.A.: Permanents of multidimensional matrices: properties and applications. J. Appl. Ind. Math. **10**, 567–604 (2016)
13. Linial, N., Luria, Z.: An upper bound on the number of high-dimensional permutations. Combinatorica **34**(4), 471–486 (2014)
14. Taranenko, A.A.: Multidimensional permanents and an upper bound on the number of transversals in Latin squares. J. Combin. Des. **23**, 305–320 (2015)
15. Avgustinovich, S.V.: Multidimensional permanents in enumeration problems. Diskretn. Anal. Issled. Oper. **15**(5), 3–5 (2008)
16. Taranenko, A.A.: Upper bounds on the numbers of 1-factors and 1-factorizations of hypergraphs. Electron. Notes Discret. Math. **49**, 85–92 (2015)
17. Gelfand, I.M., Kapranov, M.M., Zelevinsky, A.V.: Discriminants, Resultants, and Multidimensional Determinants. Springer, New York (1994)

Generalized Essential Submodule Graph of an R-module

Rajani Salvankar, Babushri Srinivas Kedukodi, Harikrishnan Panackal, and Syam Prasad Kuncham

Abstract Let M be a finitely generated left module over a ring R with 1. We introduce generalized essential submodule graph, denoted as $g\text{-}\mathcal{E}_M$. We show that every maximal submodule is a universal vertex of $g\text{-}\mathcal{E}_M$; consequently, we prove that the graph is connected with diameter at most 2. Furthermore, we define a proper g-essential submodule graph and prove the existence of a path between every two small submodules. We compute the g-essential submodule graph of the $\mathbb{Z}$-module $\mathbb{Z}_n$ for different values of n and obtain some properties.

Keywords Essential ideal · Module · Superfluous ideal

1 Introduction and Preliminaries

The idea of a graph constructed from a ring structure was initiated from the concept of a zero-divisor graph (see [4, 6]). Later, based on a given ring structure, various types of graphs such as annihilator essential graph (see [13]), essential graph (see [12]), co-maximal ideal graph (see [15]), total graph (see [3]), prime graph (see [9]), and hypercubes of dimension n (see [7]) were studied.

The authors (see [12]) considered the essential graph of a ring with multiplication operation is commutative, and found that the zero-divisor graph is a subgraph of the essential graph. In commutative rings, the authors (see [1, 11]) studied the basic

R. Salvankar · B. S. Kedukodi · H. Panackal · S. P. Kuncham (✉)
Department of Mathematics, Manipal Institute of Technology, Manipal Academy of Higher Education, Manipal, KA, India
e-mail: syamprasad.k@manipal.edu

R. Salvankar
e-mail: rajanisalvankar@gmail.com

B. S. Kedukodi
e-mail: babushrisrinivas.k@manipal.edu

H. Panackal
e-mail: pk.harikrishnan@manipal.edu

A. A. Ambily and V. B. Kiran Kumar (eds.), *Semigroups, Algebras and Operator Theory*, Springer Proceedings in Mathematics & Statistics 436,
https://doi.org/10.1007/978-981-99-6349-2_8

properties of the essential ideal graph (also known as, sum-essential graph) and characterized rings (resp. module over rings) based on the different types of graphs. They considered the set of all non-trivial ideals as the vertex set in a commutative ring (resp. submodules in case of modules) and an edge is defined if the sum of two ideals (resp. submodules) is essential in the underlying ring (resp. module).

Throughout, R denotes a ring with identity and M denotes a finitely generated unitary left R-module. In this paper, we define a generalized essential submodule graph of a module over a commutative ring. We notice that the essential ideal graph of a commutative ring and sum-essential graph of a module are its subgraphs. We derive some properties such as diameter and completeness of this graph based on module properties. We prove that a maximal submodule of M is a universal vertex, and hence the graph is connected with diameter not more than 2. We define g-complement of a submodule and show that every non-zero non-g-essential submodule is adjacent to its g-complement. Further, we define the proper g-essential submodule graph, which is induced by non-g-essential submodules of M, and show that there exists a path between every two small submodules. Finally, we list the types of graphs obtained for $\mathbb{Z}$-module $\mathbb{Z}_n$ for various values of n and obtain some results.

We denote $A \leq M$ if A is a submodule of M. A submodule A of M is essential if $A \cap B \neq (0)$ for any non-zero submodule B of M, and if every non-zero submodule of M is essential, then M is uniform. $A \leq M$ is small if $A + B = M$ where B is any submodule of M, which implies $B = M$, denoted by $A \ll M$. The authors [18] have introduced g-essential submodule as a generalization of an essential submodule. $A \leq M$ is g-essential if $A \cap B = (0)$ and $B \leq M$ is small, which implies $B = (0)$, denoted as $A \leq_{ge} M$. Moreover, if every non-zero submodule of M is g-essential, then M is g-uniform. We refer to [2, 5] for the notions of maximal submodule, Artinian module, and primitive submodule. The graphs considered in this paper are simple graphs. As usual, we denote the vertex set as V and $d(u, v)$ to represent the shortest $u - v$ path, while the eccentricity of a vertex u is defined as $e(u) = max\{d(u, v) : v \in V\}$. We use $u \sim v$ to denote the edge between u and v. The radius is the minimum eccentricity, and the diameter is the maximum eccentricity. A vertex is universal if it is adjacent to every other vertex of G.

For all other notations and definitions in graph theory, we refer to [8, 10].

2 Generalized Essential Submodule Graph

We introduce the notion of generalized essential submodule graph.

Definition 1 The g-essential submodule graph of M, denoted as g-$\mathcal{E}_M$, is a graph whose vertex set is the set of all non-trivial submodules of M, and two distinct vertices A, B are adjacent if $A + B \leq_{ge} M$.

Example 1

1. If M is simple, then g-$\mathcal{E}_M$ is a null graph.
2. Let x be a fixed prime number. Let $R = \{\frac{m}{x^n} \in \mathbb{Q} : m \in \mathbb{Z}, n \geq 0\}$ be the ring with usual addition and multiplication over $\mathbb{Q}$. Let $M = R$. Then each submodule of M is of the form $A_k = \{\frac{1}{x^k}, \frac{2}{x^k} \cdots\}$ where k is some non-negative integer, and M is Artinian having a unique minimal submodule $\mathbb{Z}$. In this case, the intersection of any two non-zero submodules is non-zero. Therefore, every non-zero submodule of M is g-essential. Hence, g-$\mathcal{E}_M$ is a complete graph.
3. Let $R = \mathbb{Z}_4$ and $M = \frac{\mathbb{Z}_4[x]}{\langle x^2 \rangle} = \{ax + b : a, b \in \mathbb{Z}_4\}$. The non-trivial submodules of M are $(2x), (2x+2), (2), (1), (2, 2x), (x), (x+1), (x+2), (x+3), (2x+1), (2, x+2), (x+1, 2), (2x, 1)$. Out of these, $(2x)$, $(2x+2)$, (2), and $(2, 2x)$ are small, and $(2, 2x), (2, x+2), (x+1, 2)$, and $(2x, 1)$ are g-essential. Hence, in the corresponding g-essential ideal graph, there exists an edge between every pair of vertices except the pairs $(2x) \sim (x)$, $(2x) \sim (x+2)$, $(2) \sim (2x+1)$, $(2) \sim (1)$, $(2x+2) \sim (x+1)$, and $(2x+2) \sim (x+3)$.

Note 1
1. If M is Artinian and contains only one minimal submodule, then g-$\mathcal{E}_M$ is a complete graph.
2. Let $A \in V(g\text{-}\mathcal{E}_M)$ be a universal vertex. If A is not g-essential in M, then A is a simple submodule of M.
3. If M has a unique non-zero small submodule, then it is a universal vertex in g-$\mathcal{E}_M$.

Notation The set of all maximal submodules of M is denoted by $Max(M)$.

Proposition 1 *[2] If M is semisimple, then* $rad(M) = (0)$.

Remark 1 Suppose M is not simple. If the radical of M is (0), then g-$\mathcal{E}_M$ is a complete graph. In particular, if M is semisimple, g-$\mathcal{E}_M$ is a complete graph.

Proof We have

$$\begin{aligned} rad(M) &= \cap\{K : K \in Max(M)\} \\ &= \sum\{A : A \ll M\}. \end{aligned}$$

Now $rad(M) = (0)$ implies M has no non-zero small submodules. Therefore, every non-zero submodule of M is g-essential. Hence, g-$\mathcal{E}_M$ is a complete graph. When M is semisimple, then by Proposition 1, the result follows. □

Theorem 1 *[2] Every proper submodule of M is contained in a maximal submodule. In particular, M has a maximal submodule.*

Remark 2 If $A \leq_{ge} M$, then $A + B \leq_{ge} M$ for any submodule B of M.

Proof Let A be a proper g-essential submodule of M. Let B be a submodule of M. To prove $A + B$ is g-essential, let K be a small submodule of M such that $(A + B) \cap K = (0)$. Now $A \cap K \subseteq (A + B) \cap K = (0)$, which implies $A \cap K = (0)$. Since $A \leq_{ge} M$, we get $K = (0)$. Therefore $A + B \leq_{ge} M$. □

Remark 3 Any proper g-essential submodule of M is a universal vertex in g-$\mathcal{E}_M$.

Proof Let A be a proper g-essential submodule of M. Let B be a non-zero proper submodule of M. We prove $A \sim B \in E(g\text{-}\mathcal{E}_M)$. Since $A \leq_{ge} M$, by Remark 2, we have $A + B \leq_{ge} M$. Therefore, $A \sim B \in E(g\text{-}\mathcal{E}_M)$. Since B is arbitrary, $A \sim B \in E(g\text{-}\mathcal{E}_M)$ for all $B \in V(\mathcal{E}_M)$. Hence, A is a universal vertex in g-$\mathcal{E}_M$. □

Remark 4 If $rad(M) \neq (0)$, then $rad(M)$ is a universal vertex in g-$\mathcal{E}_M$.

Proof Let $K \ll M$ be such that $K \cap rad(M) = (0)$. Since

$$rad(M) = \sum \{A : A \text{ is a small submodule of } M\},$$

we have $K \subseteq rad(M)$. Therefore $K = K \cap rad(M) = (0)$. Hence $rad(M)$ is g-essential. □

Proposition 2 *1. Every maximal submodule of M is a universal vertex in g-$\mathcal{E}_M$.*
2. The graph g-$\mathcal{E}_M[Max(M)]$ induced by $Max(M)$ is a clique.

Proof
1. Let K be a maximal submodule of M. Let L be a non-zero submodule of M such that $K \cap L = (0)$. We prove that L is not small. Since $L \neq (0)$ and $K \cap L = (0)$, we get $L \nsubseteq K$. Now $K \subsetneq K + L \subseteq M$ and K is a maximal submodule of M, which implies that $L + K = M$. Therefore, L is not small in M. Hence, K is g-essential and by Remark 3, K is a universal vertex in g-$\mathcal{E}_M$.
2. Follows directly from 1.

□

Proposition 3 *The subgraph of g-$\mathcal{E}_M$ induced by the set of all primitive submodules of M is a clique.*

Proof Let I and J be two primitive submodules of G.
Case (i): I and J are contained in the same maximal submodule K.
Since I and J are primitive, $I \nsubseteq J$ and $J \nsubseteq I$. Therefore $I + J \neq I$ and $I + J \neq J$. Now $I \subsetneq I + J \subseteq K$ and I is primitive, which imply $I + J = K$. By Proposition 2, we have K is g-essential, and hence $I \sim J \in E(g\text{-}\mathcal{E}_M)$.
Case (ii): I and J are contained in two distinct maximal submodules K_i and K_j. $I \subseteq K_i$ and $J \nsubseteq K_i$ imply $K_i \subsetneq I + J$. Since K_i is maximal, we get $I + J = M$, and therefore $I \sim J \in E(g\text{-}\mathcal{E}_M)$. □

Proposition 4 *g-$\mathcal{E}_M$ is an empty graph if and only if M has exactly one non-trivial submodule.*

Proof Suppose g-$\mathcal{E}_M$ is an empty graph. Then by Proposition 2, we get $|Max(M)| = 1$. Let K be the maximal submodule of M. Suppose there exists a non-trivial submodule $B \neq K$ of M. Then by Proposition 2, we get $I \sim K \in E(g\text{-}\mathcal{E}_M)$, a contradiction, since g-$\mathcal{E}_M$ is an empty graph. Conversely, if M has exactly one non-zero, non-g-essential ideal, then g-$\mathcal{E}_M = K_1$, an empty graph. □

Proposition 5 *g-$\mathcal{E}_M$ is a complete graph if and only if every non-zero, non-g-essential submodule is a minimal submodule.*

Proof Assume that every non-zero, non-g-essential submodule of M is a minimal submodule. To prove g-$\mathcal{E}_M$ is a complete graph, let A and B be two distinct vertices of g-$\mathcal{E}_M$. Then A and B are two non-zero proper submodules of M.
Case 1: Either A or B is g-essential.
In this case, by Remark 2, $A \sim B \in E(g\text{-}\mathcal{E}_M)$.
Case 2: Neither A nor B is g-essential.
Since A and B are non-zero, from the hypothesis A and B are minimal. Then $(0) \neq A \subsetneq A + B$, which implies that $A + B$ is not minimal. Again from the hypothesis, $A + B \leq_{ge} M$, and hence $A \sim B \in E(g\text{-}\mathcal{E}_M)$.
Conversely, suppose that g-$\mathcal{E}_M$ is a complete graph. We prove that every non-zero, non-g-essential submodule is a minimal submodule. On the contrary, let A be a non-zero, non-g-essential submodule, which is not minimal. Then, there exists a non-zero submodule B of M such that $B \subsetneq A$. Since g-$\mathcal{E}_M$ is complete, $A \sim B \in E(g\text{-}\mathcal{E}_M)$, which implies $A + B = A \leq_{ge} M$, a contradiction to the assumption. Thus, every non-zero, non-g-essential submodule of M is a minimal submodule. □

Corollary 1 *If every non-maximal submodule of M is a minimal submodule, then g-$\mathcal{E}_M$ is a complete graph.*

Proof The proof follows directly from Proposition 2 and 5. □

Definition 2 Let A be a submodule of M. A submodule B of M is said to be a g-complement of A in M if B is maximal with respect to $A \cap B = (0)$ and B is small in M.

Example 2

In the $\mathbb{Z}$-module $\mathbb{Z}_{36}$, the submodule $18\mathbb{Z}_{36}$ is a g-complement of $4\mathbb{Z}_{36}$, but it is not a complement, since $4\mathbb{Z}_{36} \cap 9\mathbb{Z}_{36} = (0)$ and $18\mathbb{Z}_{36} \subseteq 9\mathbb{Z}_{36}$.

Lemma 1 *Let A be a submodule of M and C be its g-complement. Then $A + C \leq_{ge} M$.*

Proof Let K be a small submodule of M such that $(A + C) \cap K = (0)$. Then $A \cap (C + K) = (0)$. We claim that $C + K$ is a small submodule of M. Let D be a submodule of M such that $(C + K) + D = M$. Since $C + (K + D) = M$ and C is small in M, we have $K + D = M$. Again since K is small in M, we get $D = M$. Therefore, $C + K$ is a small submodule of M. Since C is a g-complement of A and $C + K$ is a small submodule of M satisfying $A \cap (C + K) = (0)$, we get a contradiction. Therefore, $A + C \leq_{ge} M$. □

Remark 5 Every non-g-essential submodule of M is adjacent to its g-complement in g-$\mathcal{E}_M$.

Proof Let A be a non-g-essential submodule of M. Then there exists a non-zero proper small submodule B of M such that $A \cap B = (0)$. Now by Zorn's lemma, A has a non-zero g-complement, say $A^{'}$. Then by Lemma 1, $A + A^{'} \leq_{ge} M$. Therefore, $A \sim A^{'} \in E(g\text{-}\mathcal{E}_M)$. □

Lemma 2 *Let $B \in V(g\text{-}\mathcal{E}_M)$. If $deg_{g\text{-}\mathcal{E}_M} B = 1$, then either B is simple or $V(g\text{-}\mathcal{E}_M) = \{A, B\}$ where A is a simple submodule properly contained in B.*

Proof Suppose B is not simple. Then it has a proper submodule say, A. If $B \leq_{ge} M$, then B is a universal vertex. Since $deg_{g\text{-}\mathcal{E}_M} B = 1$, we get $V(g\text{-}\mathcal{E}_M) = \{A, B\}$. Suppose B is not g-essential in M, then it has a non-zero g-complement, say C. Now $B \oplus C \leq_{ge} M$ and also $B + (A \oplus C) \leq_{ge} M$. As B cannot be adjacent to two submodules, we get $A \oplus C$ is equal to either C or M. If $A \oplus C = C$, then $A = (0)$, a contradiction. If $A \oplus C = M$, then $B \oplus C = M$, and we get $A = B$, a contradiction. □

Proposition 6 *g-$\mathcal{E}_M$ is a connected graph of diameter less than 3.*

Proof Since M is finitely generated, M has a maximal submodule, say K. By Proposition 3, K is a universal vertex. Let A and B be two non-zero proper submodules of M. Then $K \sim A$ and $K \sim B$ are edges in g-$\mathcal{E}_M$, which give a path of length 2 from A to B. Therefore, $diam(g\text{-}\mathcal{E}_M)$ is less than or equal to 2. □

Proposition 7 *If g-$\mathcal{E}_M$ has exactly one universal vertex, then M has a unique proper essential submodule.*

Proof Suppose g-$\mathcal{E}_M$ has a unique universal vertex, say A. By Proposition 2 and from the hypothesis, it is clear that $Max(M) = \{A\}$. If A is not minimal, then there exists a submodule B such that $B \subset A$. Since A is a universal vertex, $A \sim B \in E(g\text{-}\mathcal{E}_M)$, which implies $A + B = A \leq_{ge} M$. If A is minimal, we prove that A is g-essential. On the contrary, suppose A is not g-essential, then there exists a small submodule L of M such that $J \cap L = (0)$. Since A is minimal, we have $L \nsubseteq A$. By Theorem 1, L is contained in a maximal submodule, say P. Now, since A and P are maximal, we get $A + P = M$. Also, since A is minimal, we get $A \cap P = (0)$. Then it can be easily verified that M is isomorphic to $\frac{M}{A} \times \frac{M}{P}$. Since A and P are maximal, $\frac{M}{A}$ and $\frac{M}{P}$ are simple. Hence, M has only two proper submodules, and g-$\mathcal{E}_M = K_2$, which has two universal vertices, a contradiction. Therefore, A is g-essential. □

Remark 6 Let A be a non-zero proper submodule of M and B be any submodule of M. If $A \sim B \in E(g\text{-}\mathcal{E}_M)$, then $A \sim K \in E(g\text{-}\mathcal{E}_M)$ for any proper submodule K of M which contains B.

Proof Let $(0) \neq K$ be a proper submodule of M such that $B \subseteq K$. Let P be a small submodule of M such that $(A + K) \cap P = (0)$. Then $(A + B) \cap P \subseteq (A + K) \cap P = (0)$. Since $A \sim B \in E(g\text{-}\mathcal{E}_M)$, $A + B \leq_{ge} M$, and we get $P = (0)$. Therefore $A + K \leq_{ge} M$. Hence, $A \sim K \in E(g\text{-}\mathcal{E}_M)$. □

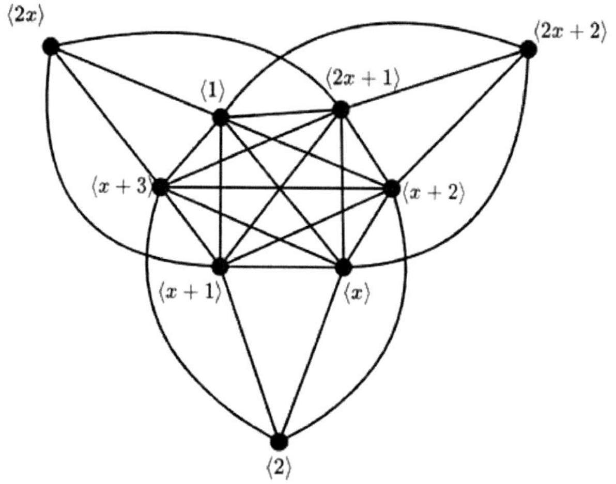

Fig. 1 g-essential ideal graph of $\mathbb{Z}_4$-module $\mathbb{Z}_4[x]/<x^2>$

Definition 3 The proper g-essential submodule graph of M denoted by g-$\mathcal{P}_M$ is a subgraph of g-$\mathcal{E}_M$, induced by the submodules which are non-g-essential in M.

Example 3

In Example 1(3), there are nine non-zero non-g-essential ideals. The corresponding proper g-essential ideal graph is given in Figure 1.

Definition 4 M is said to be g-uniform if every non-zero submodule of M is g-essential.

Remark 7 1. Every uniform module is g-uniform but a g-uniform module need not be uniform. For example, a semisimple module is g-uniform but not uniform.
2. g-$\mathcal{P}_M$ is a null graph if and only if M is g-uniform.

Proposition 8 *Let M be an R-module in which every non-maximal submodule is a minimal submodule. Then g-$\mathcal{P}_M$ is an empty graph if and only if M has only one non-zero and non-g-essential submodule.*

Proof Suppose M has exactly one non-trivial and non-g-essential submodule. Then g-$\mathcal{P}_M$ is an empty graph. Conversely, suppose that g-$\mathcal{P}_M$ is an empty graph. On the contrary, suppose that M has two proper non-g-essential submodules, say A and B. Then $A, B \in V(g\text{-}\mathcal{P}_M)$. Since every non-maximal submodule is a minimal submodule, by Corollary 1, it follows that g-$\mathcal{E}_M$ is a complete graph. Since g-$\mathcal{P}_M$ is an induced subgraph of g-$\mathcal{E}_M$, we get $A \sim B \in E(g\text{-}\mathcal{P}_M)$, a contradiction. Therefore, g-$\mathcal{P}_M$ is an empty graph. □

Proposition 9 *There exists a path between any two small submodules in g-$\mathcal{P}_M$.*

Proof Let I, J be two distinct small submodules in $V(g\text{-}\mathcal{P}_M)$. If $I+J \leq_{ge} M$, then $I \sim J \in E(g\text{-}\mathcal{P}_M)$. Assume that $I+J$ is not g-essential. Let S_I, S_J be g-complements of I and J, respectively. Since I and J are small submodules, we get S_I and S_J are not g-essential. Therefore $S_I, S_J \in V(g\text{-}\mathcal{P}_M)$. Also, by Lemma 1, we get $I+S_I \leq_{ge} M$ and $J+S_J \leq_{ge} M$. If $I+S_J$ or $J+S_I$ is g-essential, then either $I \sim S_J \sim J$ or $I \sim S_J \sim J$ is a path from I to J. If $J+S_I$ and $I+S_J$ are not g-essential, then $J+S_I, I+S_J \in V(g\text{-}\mathcal{P}_M)$ and hence $I \sim J+S_I \sim I+S_J \sim J$ is a path from I to J. □

Example 4

Consider the $\mathbb{Z}$-module $\mathbb{Z}_n$. The generalized essential graph of $\mathbb{Z}_n$ for different values of n are listed below.

1. For $n = 2^4$, $g\text{-}\mathcal{E}_M$ is a complete graph on 3 vertices and $g\text{-}\mathcal{P}_M$ is a null graph.
2. For $n = 2 \cdot 3$, $g\text{-}\mathcal{E}_M$ is a complete graph on 2 vertices and $g\text{-}\mathcal{P}_M$ is a null graph.
3. For $n = 2 \cdot 3 \cdot 5$, $g\text{-}\mathcal{E}_M$ is a complete graph and $g\text{-}\mathcal{P}_M$ is a null graph.
4. For $n = 2^2 \cdot 3$, $g\text{-}\mathcal{E}_M$ is a complete graph and $g\text{-}\mathcal{P}_M$ is an empty graph with only one vertex.
5. For $n = 2^2 \cdot 3^2$, in $g\text{-}\mathcal{E}_M$, there exists an edge between every pair except $(18\mathbb{Z}_{36}, \mathbb{Z}_{36})$ and $(\mathbb{Z}_{36}, \mathbb{Z}_{36})$ and $g\text{-}\mathcal{P}_M$ is a complete bipartite graph $K_{2,2}$.
6. For $n = 2^3 \cdot 3^2$, in $g\text{-}\mathcal{E}_M$, there exists an edge between every pair except $(8\mathbb{Z}_{72}, \mathbb{Z}_{72})$, $(9\mathbb{Z}_{72}, 36\mathbb{Z}_{72})$, $(9\mathbb{Z}_{72}, 18\mathbb{Z}_{72})$ and $(18\mathbb{Z}_{72}, 36\mathbb{Z}_{72})$, and $g\text{-}\mathcal{P}_M$ is a complete bipartite graph $K_{2,3}$.

Now we prove the following propositions.

Proposition 10 *If n is not prime and $n = \prod_{i=1}^{k} p_i$, $(1 \leq i \leq k)$ are distinct primes, then the generalized essential submodule graph of $\mathbb{Z}$-module $\mathbb{Z}_n$ is complete.*

Proof Since n is not prime, we get $k \geq 2$. Clearly, the maximal submodules are of the form $p_i\mathbb{Z}_n$, $1 \leq i \leq k$. We prove that every proper non-zero submodule of $\mathbb{Z}_n$ is not small. Let $J = \{p_1, p_2, \cdots p_n\}$. Let B be a non-zero submodule of $\mathbb{Z}_n$. Then $B = (\prod_{p_i \in A} p_i)\mathbb{Z}_n$, where A is a proper subset of J. Since A is a proper subset of J, there exists at least one $p_j \in J \setminus A$. Now, consider $p_j\mathbb{Z}_n$, which is a maximal submodule of $\mathbb{Z}_n$. Since $p_j \notin A$, we have p_j does not divide $\prod_{p_i \in A} p_i$ and hence $(\prod_{p_i \in A} p_i)\mathbb{Z}_n \nsubseteq p_j\mathbb{Z}_n$. Now, $p_j \subsetneq (\prod_{p_i \in A} p_i)\mathbb{Z}_n + P_j\mathbb{Z}_n$ and $p_j\mathbb{Z}_n$ is maximal together imply $p_j\mathbb{Z}_n + B = \mathbb{Z}_n$, and therefore B is non-small. Hence, there exists no non-zero small submodule in $\mathbb{Z}_n$ which implies every proper submodule of $\mathbb{Z}_n$ is g-essential, and this shows that $g\text{-}\mathcal{E}_M$ is complete. □

Proposition 11 *If n is not prime and $n = p^k q$ where p and q are distinct primes, then the generalized essential submodule graph of $\mathbb{Z}$-module $\mathbb{Z}_n$ is complete.*

Proof A submodule A of $\mathbb{Z}_n$ is small if and only if it is contained in both the maximal ideals $p\mathbb{Z}_n$ and $q\mathbb{Z}_n$ if and only if A is of the form $p^i q\mathbb{Z}_n$ for some $1 \leq i \leq k-1$. Also the small submodules form a decreasing chain $pq\mathbb{Z}_n \supseteq p^2q\mathbb{Z}_n \subseteq \cdots p^{k-1}q\mathbb{Z}_n$, and so it is clear that the intersection of any two non-zero small submodules is non-zero. Hence, every non-zero small submodules of M is g-essential.

Clearly, the remaining proper submodules of the form $p^i\mathbb{Z}_n$ where $1 \leq i \leq k, q\mathbb{Z}_n$ of M are non-small. As the set of all small submodules are in a chain and the module Z_n is Artinian, it has a minimal small submodule $p_{k-1}q\mathbb{Z}_n$. Now to prove that a submodule B is g-essential, it suffices to show that B has non-zero intersection with $p^{k-1}q\mathbb{Z}_n$. Since $p^{k-1}q\mathbb{Z}_n$ is contained in $q\mathbb{Z}_n$, the intersection of $q\mathbb{Z}_n$ with $p^{k-1}q\mathbb{Z}_n$ is non-zero. Now the intersection of $p^{k-1}q\mathbb{Z}_n$ with $p^i\mathbb{Z}_n$ is $p^{k-1}q\mathbb{Z}_n$ is non-zero, for each $1 \leq i \leq k-1$. Hence, $q\mathbb{Z}_n$, $p^i\mathbb{Z}_n$ where $1 \leq i \leq k-1$ are g-essential. The submodule $p^k\mathbb{Z}_n \cap p^{k-1}q\mathbb{Z}_n = p^kq\mathbb{Z}_n = n\mathbb{Z}_n = (0)$. Therefore, $p^i\mathbb{Z}_n$ is the only non-zero submodule which is not g-essential. Hence we get a complete graph. □

3 Conclusion

We have defined the g-essential submodule graph of a finitely generated module. We have shown that the maximal submodule is always a universal vertex, and hence the g-essential submodule graph of such modules is always connected with diameter not more than 2. We have obtained several properties of g-essential submodule graphs based on the notions of connectivity, completeness, etc. As future scope, we will explore the connections to study the properties of essential elements and superfluous elements in lattices as motivated by the authors in [14, 16, 17].

Acknowledgements The authors express their gratitude to the referee(s)/editor(s) for their valuable suggestions that have improved the paper. The first author acknowledges Dr. T. M. A. Pai Fellowship, MAHE, Manipal, for pursuing Ph.D. All the authors acknowledge the organizers of ICSAOT 2022, and the Manipal Institute of Technology, Manipal Academy of Higher Education, Manipal, for their kind encouragement.

References

1. Amjadi, J.: The essential ideal graph of a commutative ring. Asian Eur. J. Math. **11**, (2018). https://doi.org/10.1142/S1793557118500584
2. Anderson, F.W., Fuller, K.R.: Rings and Categories of Modules. Springer Verlag, New York, Inc. (1992). https://doi.org/10.1007/978-1-4612-4418-9
3. Anderson, D.F., Badavi, A.: The total graph of a commutative ring. J. Algebra. **320**, 2706–2719 (2008)
4. Anderson, D.F., Asir, T., Badavi, A., Chelvam, T.T.: Graphs from Rings. Springer Nature (2021)
5. Atiyah, M.F., Macdonald, I.G.: Introduction to Commutative Algebra. Wesley (1969)
6. Beck, I.: Colouring of commutative rings. J. Algebra. **116**, 208–216 (1988)
7. Bhavanari, S., Kuncham, S.P.: An isomorphism theorem on directed hypercubes of dimension n. Indian J. Pure Appl. Math. **34**, 1453–1457 (2003)
8. Bhavanari, S., Kuncham, S.P.: Discrete Mathematics and Graph Theory. PHI learning (2009)
9. Bhavanari, S., Kuncham, S.P., Nagaraju, D.: Prime graph of a ring. J. Comb. Inf. Syst. Sci. **35**, 27–42 (2010). ISSN 0250-9628
10. Bhavanari, S., Kuncham, S.P.: Nearrings, fuzzy ideals, and graph theory. Chapman and Hall, Taylor and Francis Group (London, New York) (2013). ISBN 13: 9781439873106
11. Matczuk, J., Majidinya, A.: Sum-essential graphs of modules. J. Algebra. Appl. **20**, (2021)
12. Nikmehr, M.J., Nikandish, R., Bakhtyiari, M.: On the essential graph of a commutative ring. J. Algebra. Appl. **16**, (2017)
13. Babaei, Sanikeh, Payrovi, Shiroyeh: Esra Sengelen Sevim: on the annihilator submodules and the annihilator essential graph. Acta Mathematica Vietnamica **44**, 905–914 (2019)
14. Rajani, S., Tapatee, S., Harikrishnan, P., Kuncham, S.P.: Superfluous ideals of N-groups. Rend. Circ. Mat. Palermo, II. Ser (2023). https://doi.org/10.1007/s12215-023-00888-2
15. Sharma, P.K., Bhatwadekar, S.M.: A note on graphical representation of rings. J. Algebra. **176**, 124–127 (1995)
16. Tapatee, S., Davvaz, B., Panackal, H., Kedukodi, B.S., Kuncham, S.P.: Relative essential ideals in N-groups. Tamkang J. Math. **54**, 69–82 (2023). https://doi.org/10.5556/j.tkjm.54.2023.4136
17. Tapatee, S., Harikrishnan, P.K., Kedukodi, B.S., Kuncham, S.P.: Graphs with respect to superfluous elements in a lattice. Miskolc Math. Notes. **23**, 929–945 (2022). https://doi.org/10.1142/S1793557120500230
18. Zhou, D.X., Zhang, X.R.: Small-essential submodules and Morita duality. Southeast Asian Bull. Math. **35**, 1051–1062 (2011)

Algebras

Algebra is a branch which deals with the algebraic structures called algebras. The second part of this volume deals with various algebras such as Lie algebras, Geometric algebras and generalized essential submodule graph of a module.

On Category of Lie Algebras

S. N. Arjun and P. G. Romeo

Abstract In this paper we describe certain Lie algebra of a group algebra which we call group Lie algebra. Further we introduce the category of Lie algebras with the aim to have a unified approach to various Lie algebras arising in various situations.

Keywords Lie algebra · Plesken Lie algebra · Group algebra · Category · Functor

1 Introduction

The Lie algebras were introduced by Sophus Lie to solve problems in Lie groups with some ease. W. Plseken and Arjeh M. Cohen introduced the construction of a Lie algebra of a finite group known as Plesken Lie algebra. One question that they encountered in this regard was for which groups the construction would produce a simple Lie algebra? In [2], Arjeh M. Cohen and D. E. Taylor studied the Plesken Lie algebras in detail and determined the groups for which the Plesken Lie algebra is simple and semisimple. In this paper we consider Lie algebra of a group algebra which we call a group Lie algebra and discuss its properties and its relation with Plesken Lie algebras. Further we introduce the category of Lie algebras provided several examples and explore some interesting categorical properties. Throughout this paper we consider only finite dimensional Lie algebras over a field of characteristic zero.

2 Preliminaries

In this section we recall definitions of Lie algebra, group algebra and Plesken Lie algebra with some examples and discuss some interesting properties of these structures.

S. N. Arjun (✉) · P. G. Romeo
Department of Mathematics, Cochin University of Science and Technology, Kochi, KL, India
e-mail: arjunsnmaths1996@gmail.com

A. A. Ambily and V. B. Kiran Kumar (eds.), *Semigroups, Algebras and Operator Theory*,
Springer Proceedings in Mathematics & Statistics 436,
https://doi.org/10.1007/978-981-99-6349-2_9

Definition 1 [3] A Lie algebra L over an arbitrary field $\mathbb{F}$ is a vector space over $\mathbb{F}$ endowed with an operation called Lie bracket satisfying the following properties :

1. Bilinearity : For $x, y, z \in L, a, b \in \mathbb{F}$

$$[ax + by, z] = [ax, z] + [by, z]$$
$$[x, ay + bz] = [ax, y] + [bx, z]$$

2. $[x, x] = 0$ for all $x \in L$
3. Jacobi identity :
$[x, [y, z]] + [y, [z, x]] + [z, [x, y]] = 0$ for all $x, y, z \in L$.

Example

$End(V)$, the set of all linear transformations on a finite dimensional vector space V over a field $\mathbb{F}$ is a Lie algebra with Lie bracket $[x, y] = xy - yx$ for $x, y \in End(V)$.

A subspace K of a Lie algebra L is called a *Lie subalgebra* if $[x, y] \in K$ whenever $x, y \in K$.

Lie algebras are closely related to Lie groups, which are groups that are also smooth manifolds: any Lie group gives rise to a Lie algebra, which is its tangent space at the identity. Here we are considering only the Lie algebra structure defined by a finite group.

Next we describe Lie algebra determined by a finite group called Plesken Lie algebra. Let G be a group and $\mathbb{F}G$ be its group algebra over $\mathbb{F}$, then for each $g \in G$, $g - g^{-1} \in \mathbb{F}G$, denote it by $\hat{g}$, then the linear span of $\hat{g}$ admits a Lie algebra structure as explained below.

Definition 2 [2] Plesken Lie algebra $\mathfrak{L}(G)$ of a group G over $\mathbb{F}$ is the linear span of elements $\hat{g} \in \mathbb{F}G$ together with the Lie bracket

$$[\hat{g}, \hat{h}] = \hat{g}\hat{h} - \hat{h}\hat{g}$$

That is, for any group $G = \{g_1, g_2, g_3, ..., g_n\}$, $\left\{\sum_{i=1}^{n} a_i \hat{g}_i : a_i \in \mathbb{F} \text{ for all } i\right\}$ together with the Lie bracket defined above is the Plesken Lie algebra $\mathfrak{L}(G)$.

Example

Consider the symmetric group S_3, then

$$\begin{aligned}\mathfrak{L}(S_3) &= \text{span}_{\mathbb{C}}\{\sigma - \sigma^{-1} : \sigma \in S_3\} \\ &= \{a_1((1) - (1)) + a_2((1\ 2) - (1\ 2)) + a_3((1\ 3) - (1\ 3)) + +a_4((2\ 3) - (2\ 3)) \\ &\quad + a_5((1\ 2\ 3) - (1\ 3\ 2)) + a_6((1\ 3\ 2) - (1\ 2\ 3)) : a_i \in \mathbb{C}\} \\ &= \{a((1\ 2\ 3) - (1\ 3\ 2)) : a \in \mathbb{C}\}\end{aligned}$$

is a one dimensional Plesken Lie algebra over $\mathbb{C}$ with Lie bracket

$$[a\widehat{(1\ 2\ 3)}, b\widehat{(1\ 2\ 3)}] = 0$$

Definition 3 Let G be a finite group and let $\mathbb{F}$ be $\mathbb{R}$ or $\mathbb{C}$. Define a vector space over $\mathbb{F}$ with elements of G as a basis, and denote it by $\mathbb{F}G$. That is; $\mathbb{F}G = \left\{\sum_{i=1}^{n} a_i g_i : a_i \in \mathbb{F}, g_i \in G\right\}$. The addition and scalar multiplication in $\mathbb{F}G$ are defined as follows: for $\alpha = \sum_{i=1}^{n} a_i g_i$ and $\beta = \sum_{i=1}^{n} b_i g_i$ in $\mathbb{F}G$ and $k \in \mathbb{F}$,

$$\alpha + \beta = \sum_{i=1}^{n} (a_i + b_i) g_i \text{ and } k\alpha = \sum_{i=1}^{n} (ka_i) g_i$$

The vector space $\mathbb{F}G$, with multiplication defined by

$$\left(\sum_{i=1}^{n} a_i g_i\right)\left(\sum_{j=1}^{n} b_j g_j\right) = \sum_{i,j=1}^{n} a_i b_j (g_i g_j), \quad a_i, b_j \in \mathbb{F}$$

is called the group algebra of G over $\mathbb{F}$.

The group algebra of a finite group G is a vector space of dimension $|G|$ which also carries extra structure involving the the product operation on G.

Example

$G = C_3 = < a : a^3 = e >$ where e is the identity in G. Then

$$\mathbb{C}G = \{\lambda_1 e + \lambda_2 a + \lambda_3 a^2 : \lambda_i \in \mathbb{C} \text{ for } i = 1, 2, 3\}$$

$\mathbb{C}G$ is a group algebra with usual addition and multiplication of serieses.

3 Group Lie Algebras and Plesken Lie Algebras

Let G be any finite group and $\mathbb{F}G$ the group algebra over the field $\mathbb{F}$. By defining the Lie bracket $[\ ,\ ] : \mathbb{F}G \times \mathbb{F}G \to \mathbb{F}G$ as follows :

$$[\alpha, \beta] = \alpha\beta - \beta\alpha \text{ for } \alpha = \sum_{i=1}^{n} a_i g_i \text{ and } \beta = \sum_{i=1}^{n} b_i g_i \text{ in } \mathbb{F}G$$

$\mathbb{F}G$ is a Lie algebra with respect to the given Lie bracket and is called the *group Lie algebra* denoted as $L_{\mathbb{F}G}$.

Let G and H be two groups whose Lie algebras are $L_{\mathbb{F}G}$ and $L_{\mathbb{F}H}$. Any linear map between $L_{\mathbb{F}G}$ and $L_{\mathbb{F}H}$ which preserves the Lie bracket is a *homomorphism* of group Lie algebras.

Proposition 1 *Let $f : G \to H$ be a group homomorphism. Then $\bar{f} : L_{\mathbb{F}G} \to L_{\mathbb{F}H}$ defined by*

$$\bar{f}(\alpha) = \bar{f}\left(\sum_{i=1}^{n} a_i g_i\right) = \sum_{i=1}^{n} a_i f(g_i)$$

is a homomorphism between group Lie algebras.

Next we describe the relation between Plesken Lie algebras and group Lie algebras.

Lemma 1 *The Plesken Lie algebra $\mathfrak{L}(G)$ over $\mathbb{F}$ is a Lie subalgebra of the group Lie algebra $L_{\mathbb{F}G}$.*

Proof Let $G = \{g_1, g_2, g_3, ..., g_n\}$ be a group. Then the group Lie algebra is $L_{\mathbb{F}G} = \left\{\sum_{i=1}^{n} \lambda_i g_i : a_i \in \mathbb{F} \text{ for all} i\right\}$ and the Plesken Lie algebra is $\mathfrak{L}(G) = \left\{\sum_{i=1}^{n} a_i \hat{g}_i : a_i \in \mathbb{F} \text{ for all } i\right\}$. For any $x \in \mathfrak{L}(G)$,

$$x = \sum_{i=1}^{n} a_i \hat{g}_i = \sum_{i=1}^{n} a_i g_i - \sum_{i=1}^{n} a_i {g_i}^{-1} \in L_{\mathbb{F}G}.$$

Thus $\mathfrak{L}(G) \subseteq L_{\mathbb{F}G}$. Let $\hat{\alpha} = \sum_{i=1}^{n} a_i \hat{g}_i$ and $\hat{\beta} = \sum_{i=1}^{n} b_i \hat{g}_i$ in $\mathfrak{L}(G)$, then

$$\hat{\alpha} + \hat{\beta} = \sum_{i=1}^{n} a_i \hat{g}_i + \sum_{i=1}^{n} b_i \hat{g}_i = \sum_{i=1}^{n} (a_i + b_i)\hat{g}_i \in \mathfrak{L}(G)$$

and $k(\hat{\alpha}) = k \sum_{i=1}^{n} a_i \hat{g}_i = \sum_{i=1}^{n} (ka_i)\hat{g}_i \in \mathfrak{L}(G)$. Thus, $\mathfrak{L}(G)$ is a subspace of $L_{\mathbb{F}G}$. For $\hat{g}, \hat{h} \in \mathfrak{L}(G)$,

$$\begin{aligned} \hat{g}, \hat{h}t &= \hat{g}\hat{h} - \hat{h}\hat{g} \\ &= (g - g^{-1})(h - h^{-1}) - (h - h^{-1})(g - g^{-1}) \\ &= \widehat{gh} - \widehat{gh^{-1}} - \widehat{g^{-1}h} + \widehat{g^{-1}h^{-1}}. \end{aligned}$$

Thus Lie bracket is closed in $\mathfrak{L}(G)$ and hence $\mathfrak{L}(G)$ is a Lie subalgebra of $L_{\mathbb{F}G}$. □

Any linear map between two Plesken Lie algebras $\mathfrak{L}(G)$ and $\mathfrak{L}(H)$ is a *Plesken Lie algebra homomorphism* if it preserves the Lie bracket.

Proposition 2 *Let* $f : G \to H$ *be a group homomorphism. Then* $\hat{f} : \mathfrak{L}(G) \to \mathfrak{L}(H)$ *defined by*

$$\hat{f}\left(\sum_{i=1}^{n} a_i \hat{g}_i\right) = \sum_{i=1}^{n} a_i \widehat{f(g_i)}$$

is a Plesken Lie algebra homomorphism.

Proof For $\sum_{i=1}^{n} a_i \hat{g}_i, \sum_{i=1}^{n} b_i \hat{g}_i \in \mathfrak{L}(G)$ and $k \in \mathbb{F}$,

$$\begin{aligned}
\hat{f}\left(\sum_{i=1}^{n} a_i \hat{g}_i + k \sum_{i=1}^{n} b_i \hat{g}_i\right) &= \hat{f}\left(\sum_{i=1}^{n} (a_i + kb_i)\hat{g}_i\right) \\
&= \sum_{i=1}^{n} (a_i + kb_i)\widehat{f(g_i)} \\
&= \hat{f}\left(\sum_{i=1}^{n} a_i \hat{g}_i\right) + k\hat{f}\left(\sum_{i=1}^{n} b_i \hat{g}_i\right).
\end{aligned}$$

Thus $\hat{f}$ is linear. For $\sum_{i=1}^{n} a_i \hat{g}_i, \sum_{j=1}^{n} b_j \hat{g}_j \in \mathfrak{L}(G)$,

$$\begin{aligned}
\hat{f}\left(\left[\sum_{i=1}^{n} a_i \hat{g}_i, \sum_{j=1}^{n} b_j \hat{g}_j\right]\right) &= \hat{f}\left(\sum_{i,j=1}^{n} a_i b_j (\hat{g}_i \hat{g}_j - \hat{g}_j \hat{g}_i)\right) \\
&= \hat{f}\left(\sum_{i,j=1}^{n} a_i b_j (\widehat{g_i g_j} - \widehat{g_i^{-1} g_j} - \widehat{g_i g_j^{-1}} + \widehat{g_i^{-1} g_j^{-1}})\right) \\
&= \sum_{i,j=1}^{n} a_i b_j (\widehat{f(g_i g_j)} - \widehat{f(g_i^{-1} g_j)} - \widehat{f(g_i g_j^{-1})} + \widehat{f(g_i^{-1} g_j^{-1})}) \\
&= \sum_{i,j=1}^{n} a_i b_j (\widehat{f(g_i) f(g_j)} - \widehat{f(g_i^{-1}) f(g_j)} \\
&\qquad - \widehat{f(g_i) f(g_j^{-1})} + \widehat{f(g_i^{-1}) f(g_j^{-1})}) \\
&= \left[\sum_{i=1}^{n} a_i \widehat{f(g_i)}, \sum_{j=1}^{n} a_j \widehat{f(g_j)}\right].
\end{aligned}$$

Hence $\hat{f}$ preserves Lie bracket and is a Plesken Lie algebra homomorphism. □

4 Category of Lie Algebras

Next we proceed to describe the category of Lie algebras inorder to obtain a unified account of various Lie algebras and their relations. First we recall some definitions and concepts in category theory, for a detailed discussion see [1].

Definition 4 (cf. [5]) A category $\mathcal{C}$ consists of the following data:

1. A class called the class of vertices or objects $\nu\mathcal{C}$.
2. A class of disjoint sets $\mathcal{C}(a, b)$ one for each pair $(a, b) \in \nu\mathcal{C} \times \nu\mathcal{C}$. An element $f \in \mathcal{C}$ is called a morphism from a to b, written $f : a \to b$; $a = dom\ f$ called the domain of f and $b = cod\ f$ called the codomain of f.
3. For $a, b, c, \in \nu\mathcal{C}$, a map
   $\circ : \mathcal{C}(a, b) \times \mathcal{C}(b, c) \to \mathcal{C}(a, c) \qquad (f, g) \to f \circ g$
   $\circ$ is called the *composition* of morphisms in $\mathcal{C}$.
4. For each $a \in \nu\mathcal{C}$, a unique $1_a \in \mathcal{C}(a, a)$ is called the identity morphism on a.

These must satisfy the following axioms :

- (cat1) The composition is associative : for $f \in \mathcal{C}(a, b), g \in \mathcal{C}(b, c)$ *and* $h \in \mathcal{C}(c, d)$, we have
$$f \circ (g \circ h) = (f \circ g) \circ h$$
- (cat 2) For each $a \in \nu\mathcal{C}$, $f \in \mathcal{C}\ (a, b)$ *and* $g \in \mathcal{C}(c, a)$,
$$1_a \circ f = f \quad \text{and} \quad g \circ 1_a = g.$$

Clearly $\nu\mathcal{C}$ can be identify as a subclass of $\mathcal{C}$ and with this identification we may regard a category in terms of the morphisms alone. A category $\mathcal{C}$ is a small category if the class $\mathcal{C}$ is a set. A morphism $f \in \mathcal{C}(a, b)$ is said to be an isomorphism if there exists $f^{-1} \in \mathcal{C}(b, a)$ such that $ff^{-1} = 1_a = e_a$, the domain identity and $f^{-1}f = 1_b = f_b$, the range identity.

Example

A group G can be regarded as a category $\mathcal{C}$ with the object set of $\mathcal{C}$ say $\nu C = \{G\}$, and morphisms $\mathcal{C}(G, G) = G$ and composition in $\mathcal{C}$ is the binary operation in G. Identity element in the group will be the identity morphism on the vertex G.

Definition 5 [1] For categories $\mathcal{C}$ and $\mathcal{D}$ a functor $T : \mathcal{C} \to \mathcal{D}$ with domain $\mathcal{C}$ and codomain $\mathcal{D}$ consists of two functions: the object function T, which assigns to each object c of $\mathcal{C}$ an object Tc of $\mathcal{D}$ and the arrow function which assigns to each arrow $f : c \to c'$ of $\mathcal{C}$ an arrow $Tf : Tc \to Tc'$ of $\mathcal{D}$, in such a way that $T(1_c) = 1_{Tc}$ and $T(g \circ f) = Tg \circ Tf$.

The category of Lie algebras is a category whose objects are Lie algebras and morphisms Lie algebra homomorphisms. A category whose objects are group Lie algebras $L_{\mathbb{F}G}$ and morphisms are appropriate algebra homomorphisms belongs to the category of Lie algebras. In the following example we describe the category whose objects are group Lie algebras of subgroups of the Heisenberg group $H(\mathbb{Z}/p\mathbb{Z})$.

Example

Consider the Heisenberg group $H(\mathbb{Z}/p\mathbb{Z}) = \{\begin{pmatrix} 1 & a & b \\ 0 & 1 & c \\ 0 & 0 & 1 \end{pmatrix} : a, b, c \in \mathbb{Z}/p\mathbb{Z}\}$ and its non-isomorphic subgroups :

$$H_1 = \{I_{3\times 3}, \text{ the idenity matrix }\},\ H_2 = \left\{ \begin{pmatrix} 1 & 0 & b \\ 0 & 1 & 0 \\ 0 & 0 & 1 \end{pmatrix} : b \in \mathbb{Z}/p\mathbb{Z} \right\}$$

$$H_3 = \left\{ \begin{pmatrix} 1 & 0 & b \\ 0 & 1 & c \\ 0 & 0 & 1 \end{pmatrix} : b, c \in \mathbb{Z}/p\mathbb{Z} \right\},\ H_4 = H(\mathbb{Z}/p\mathbb{Z})$$

The group algebras $\mathbb{F}H_i$ over $\mathbb{F}$ of each subgroups H_i of $H(\mathbb{Z}/p\mathbb{Z})$ together with a Lie bracket $[X, Y] = XY - YX$ where $X = \sum_{i=1}^{n} \lambda_i A_i$, $Y = \sum_{i=1}^{n} \mu_i B_i \in \mathbb{F}H_i$ is the Lie algebra $L_{\mathbb{F}H_i}$ of the group algebra $\mathbb{F}H_i$. Consider $L_{\mathbb{F}H(\mathbb{Z}/p\mathbb{Z})}$ whose objects are

$$L_{\mathbb{F}H_1} = \{\lambda I : \lambda \in \mathbb{F}\} = \text{the Lie algebra of scalar matrices}$$

$$L_{\mathbb{F}H_2} = \left\{ \sum_{i=1}^{p} \lambda_i A_i : A_i \in H_2, \lambda_i \in \mathbb{F} \right\} = \left\{ \begin{pmatrix} \sum_{i=1}^{p} \lambda_i & 0 & \sum_{i=1}^{p} \lambda_i b_i \\ 0 & \sum_{i=1}^{p} \lambda_i & 0 \\ 0 & 0 & \sum_{i=1}^{p} \lambda_i \end{pmatrix} : \lambda_i \in \mathbb{F} \right\}$$

$$L_{\mathbb{F}H_3} = \left\{ \sum_{i=1}^{p^2} \lambda_i A_i : A_i \in H_3, \lambda_i \in \mathbb{F} \right\} = \left\{ \begin{pmatrix} \sum_{i=1}^{p^2} \lambda_i & 0 & \sum_{i=1}^{p^2} \lambda_i b_i \\ 0 & \sum_{i=1}^{p^2} \lambda_i & \sum_{i=1}^{p^2} \lambda_i c_i \\ 0 & 0 & \sum_{i=1}^{p^2} \lambda_i \end{pmatrix} : \lambda_i \in \mathbb{F} \right\}$$

$$L_{\mathbb{F}H_4} = \left\{ \sum_{i=1}^{p^3} \lambda_i A_i : A_i \in H_4, \lambda_i \in \mathbb{F} \right\} = \left\{ \begin{pmatrix} \sum_{i=1}^{p^3} \lambda_i & \sum_{i} \lambda_i a_i & \sum_{i=1}^{p^3} \lambda_i b_i \\ 0 & \sum_{i=1}^{p^3} \lambda_i & \sum_{i=1}^{p^3} \lambda_i c_i \\ 0 & 0 & \sum_{i=1}^{p^3} \lambda_i \end{pmatrix} : \lambda_i \in \mathbb{F} \right\}$$

and morphisms : $\hom(L_{\mathbb{F}H_i}, L_{\mathbb{F}H_j}) = \{\bar{f}_{ij} : L_{\mathbb{F}H_i} \to L_{\mathbb{F}H_j} | \bar{f}\left(\sum_{i=1}^{n} a_i g_i\right) = \sum_{i=1}^{n} a_i f(g_i)$ where $f : H_i \to H_j$ is the group homomorphism$\}$ is the category $L_{\mathbb{F}(H(\mathbb{Z}/p\mathbb{Z}))}$ of Lie algebras of the group algebras of subgroups of $H(\mathbb{Z}/p\mathbb{Z})$.

Next we consider the category whose objects are Plesken Lie algebras $\mathfrak{L}(G)$ over a field $\mathbb{F}$ and morphisms are Plesken Lie algebra homomorphisms which we denote it by $\mathfrak{C}_{PLG}$. The following is an example for category of Plesken Lie algebras where we consider the objects as Plesken Lie algebras of subgroups of $H(\mathbb{Z}/p\mathbb{Z})$.

Example

Consider the Heisenberg group $H(\mathbb{Z}/p\mathbb{Z}) = \left\{ \begin{pmatrix} 1 & a & b \\ 0 & 1 & c \\ 0 & 0 & 1 \end{pmatrix} : a, b, c \in \mathbb{Z}/p\mathbb{Z} \right\}$ and its non-isomorphic subgroups H_1, H_2, H_3 and H_4 as given in previous example. For each $A = \begin{pmatrix} 1 & a & b \\ 0 & 1 & c \\ 0 & 0 & 1 \end{pmatrix} \in H_i$,

$$\hat{A} = A - A^{-1} = \begin{pmatrix} 1 & a & b \\ 0 & 1 & c \\ 0 & 0 & 1 \end{pmatrix} - \begin{pmatrix} 1 & -a & ac - b \\ 0 & 1 & -c \\ 0 & 0 & 1 \end{pmatrix} = \begin{pmatrix} 0 & 2a & 2b - ac \\ 0 & 0 & 2c \\ 0 & 0 & 0 \end{pmatrix}$$

Thus $\mathfrak{L}(H_i) = \left\{ \sum_{i=1}^{n} \lambda_i \hat{A}_i : A_i \in H_i, \lambda_i \in \mathbb{F} \right\}$ and so

$$\mathfrak{L}(H_1) = \{0\}, \quad \mathfrak{L}(H_2) = \left\{ \begin{pmatrix} 0 & 0 & 2\sum_{i=1}^{p} \lambda_i b_i \\ 0 & 0 & 0 \\ 0 & 0 & 0 \end{pmatrix} : b_i \in \mathbb{Z}/p\mathbb{Z}, \lambda_i \in \mathbb{F} \right\}$$

$$\mathcal{L}(H_3) = \left\{ \begin{pmatrix} 0 & 0 & 2\sum_{i=1}^{p^2} \lambda_i b_i \\ 0 & 0 & 2\sum_{i=1}^{p^2} \lambda_i c_i \\ 0 & 0 & 0 \end{pmatrix} : b_i, c_i \in \mathbb{Z}/p\mathbb{Z}, \lambda_i \in \mathbb{F} \right\},$$

$$\mathcal{L}(H_4) = \left\{ \begin{pmatrix} 0 & 2\sum_{i=1}^{p^3} \lambda_i a_i & \sum_{i=1}^{p^3} \lambda_i (2b_i - a_i c_i) \\ 0 & 0 & 2\sum_{i=1}^{p^3} \lambda_i c_i \\ 0 & 0 & 0 \end{pmatrix} : a_i, b_i, c_i \in \mathbb{Z}/p\mathbb{Z}, \lambda_i \in \mathbb{F} \right\}$$

The category whose objects are $\mathcal{L}(H_i), i = 1, 2, 3, 4$ and morphisms are $hom(\mathcal{L}(H_i), \mathcal{L}(H_j)) = \{\hat{f} : \mathcal{L}(H_i) \to \mathcal{L}(H_j) \mid \hat{f}\left(\sum_{i=1}^{n} \lambda_i \hat{A}_i\right) = \sum_{i=1}^{n} \lambda_i \widehat{f(A_i)}$ where $f : H_i \to H_j$ is the group homomorphism $\}$ is the category of Plesken Lie algebras of subgroups of $H(\mathbb{Z}/p\mathbb{Z})$.

Plesken Lie algebra of a group and group Lie algebras are two Lie algebras constructed from a group. The following theorem define a functor between the category of group Lie algebras and the category of Plesken Lie algebras.

Theorem 1 *If $L_{\mathbb{F}G}$ is the category of group Lie algebras and $\mathcal{C}_{PLG}$ is the category of Plesken Lie algebras, then there is functor T such that*

$$T : L_{\mathbb{F}G} \to \mathcal{C}_{PLG}.$$

Proof Define $T : L_{\mathbb{F}G} \to \mathcal{C}_{PLG}$ such that $\nu T : L_{\mathbb{F}G_i} \to \mathcal{L}(G_i)$ is defined by

$$\nu T\left(\sum_{i=1}^{n} a_i g_i\right) = \sum_{i=1}^{n} a_i (g_i - g_i{}^{-1})$$

and for a group homomorphism $f : G_i \to G_j$, $\bar{f} : L_{\mathbb{F}G_i} \to L_{\mathbb{F}G_j}$ is given by

$$\bar{f}\left(\sum_{i=1}^{n} a_i g_i\right) = \sum_{i=1}^{n} a_i f(g_i)$$

and $T\bar{f} : T(L_{\mathbb{F}G_i}) \to T(L_{\mathbb{F}G_j})$ is defined by

$$T\bar{f} = \hat{f}$$

where $\hat{f} : \mathcal{L}(G_i) \to \mathcal{L}(G_j)$ is given by

$$\hat{f}\left(\sum_{i=1}^{n} a_i(g_i - {g_i}^{-1})\right) = \sum_{i=1}^{n} a_i(f(g_i) - f(g_i)^{-1})$$

is a Plesken Lie algebra homomorphism (Proposition 2). Let $\bar{f_1} : L_{\mathbb{F}G_i} \to L_{\mathbb{F}G_j}$ and $\bar{f_2} : L_{\mathbb{F}G_j} \to L_{\mathbb{F}G_k}$ be two homomorphisms between group Lie algebras which are induced from the group homomorphisms $f_1 : G_i \to G_j$ and $f_2 : G_j \to G_k$. Then $T(\bar{f_1}) : T(L_{\mathbb{F}G_i}) \to T(L_{\mathbb{F}G_j})$ and $T(\bar{f_2}) : T(L_{\mathbb{F}G_j}) \to T(L_{\mathbb{F}G_k})$ are Plesken Lie algebra homomorphisms and their composition $T(\bar{f_2}) \circ T(\bar{f_1}) : T(L_{\mathbb{F}G_i}) \to T(L_{\mathbb{F}G_k})$ is also a Plesken Lie algebra homomorphism. Moreover,

$$\begin{aligned}
(T(\bar{f_2}) \circ T(\bar{f_1}))\left(\sum_{i=1}^{n} a_i(g_i - {g_i}^{-1})\right) &= (\hat{f_2} \circ \hat{f_1})\left(\sum_{i=1}^{n} a_i(g_i - {g_i}^{-1}))\right) \\
&= \hat{f_2}\left(\sum_{i=1}^{n} a_i(f_1(g_i) - f_1({g_i}^{-1}))\right) \qquad (1) \\
&= \sum_{i=1}^{n} a_i((f_2 \circ f_1)(g_i) - ((f_2 \circ f_1)(g_i))^{-1})
\end{aligned}$$

and

$$\begin{aligned}
T(\bar{f_2} \circ \bar{f_1})\left(\sum_{i=1}^{n} a_i(g_i - {g_i}^{-1})\right) &= (\widehat{f_2 \circ f_1})\left(\sum_{i=1}^{n} a_i(g_i - {g_i}^{-1})\right) \\
&= \sum_{i=1}^{n} a_i((f_2 \circ f_1)(g_i) - ((f_2 \circ f_1)(g_i))^{-1})
\end{aligned} \qquad (2)$$

(1) and (2) shows that $T(\bar{f_2} \circ \bar{f_1}) = T(\bar{f_2}) \circ T(\bar{f_1})$.
The identity Lie algebra homomorphism, $1_{L_{\mathbb{F}G_i}} : L_{\mathbb{F}G_i} \to L_{\mathbb{F}G_i}$ is induced from the identity group homomorphism $1_{G_i} : G_i \to G_i$. Then $T(1_{L_{\mathbb{F}G_i}}) : T(L_{\mathbb{F}G_i}) \to T(L_{\mathbb{F}G_i})$ is a Plesken Lie algebra homomorphism and

$$\begin{aligned}
T(1_{L_{\mathbb{F}G_i}})\left(\sum_{i=1}^{n} a_i(g_i - {g_i}^{-1})\right) &= \sum_{i=1}^{n} a_i(1_{G_i}(g_i) - (1_{G_i}(g_i))^{-1}) \\
&= \sum_{i=1}^{n} a_i(g_i - {g_i}^{-1}) \\
&= 1_{T(L_{\mathbb{F}G_i})}\left(\sum_{i=1}^{n} a_i(g_i - {g_i}^{-1})\right)
\end{aligned}$$

That is, $T(1_{L_{\mathbb{F}G_i}}) = 1_{T(L_{\mathbb{F}G_i})}$, hence T is a functor. □

Corollary 1 *The functor $T : L_{\mathbb{F}G} \to \mathcal{C}_{PLG}$ in Theorem 1 is a full functor.*

Proof For the functor $T : L_{\mathbb{F}G} \to \mathcal{C}_{PLG}$ define a map $T_{L_{\mathbb{F}G_i}, L_{\mathbb{F}G_j}} : hom(L_{\mathbb{F}G_i}, L_{\mathbb{F}G_j}) \to hom(TL_{\mathbb{F}G_i}, TL_{\mathbb{F}G_j})$ by

$$T_{L_{\mathbb{F}G_i}, L_{\mathbb{F}G_j}}(\bar{f}) = \hat{f}$$

Now it is enough to prove that $T_{L_{\mathbb{F}G_i}, L_{\mathbb{F}G_j}}$ is surjective. For $\hat{f} \in hom(TL_{\mathbb{F}G_i}, TL_{\mathbb{F}G_j})$, a Plesken Lie algebra homomorphism, define $\bar{f} : L_{\mathbb{F}G_i} \to L_{\mathbb{F}G_j}$ by

$$\bar{f}\left(\sum_{i=1}^{n} a_i g_i\right) = \sum_{i=1}^{n} a_i f(g_i)$$

is a homomorphism between Lie algebras of group algebras and $T\bar{f} : TL_{\mathbb{F}G_i} \to TL_{\mathbb{F}G_j}$ given by,

$$\begin{aligned}
T\bar{f}\left(\sum_{i=1}^{n} a_i \hat{g}_i\right) &= T\bar{f}\left(\sum_{i=1}^{n} a_i g_i - \sum_{i=1}^{n} a_i {g_i}^{-1}\right) \\
&= \sum_{i=1}^{n} a_i f(g_i) - \sum_{i=1}^{n} a_i f(g_i)^{-1} \\
&= \sum_{i=1}^{n} a_i \widehat{f(g_i)} \\
&= \hat{f}\left(\sum_{i=1}^{n} a_i \hat{g}_i\right)
\end{aligned}$$

That is, $T\bar{f} = \hat{f}$. Hence T is full. □

However, it should be noted that the functor defined above need not be faithful as illustrated in the following example.

Example

Let $K_4 = \{e, a, b, c\}$ be the Klein 4- group and $L_{\mathbb{F}K_4}$ Lie algebra of the group algebra $\mathbb{F}K_4$. Consider the identity map $1_{L_{\mathbb{F}K_4}}$ and $\bar{f} : L_{\mathbb{F}K_4} \to L_{\mathbb{F}K_4}$ given by

$$\bar{f}\left(\sum_{i=1}^{4} a_i g_i\right) = \sum_{i=1}^{4} a_i f(g_i)$$

where $g_i \in K_4$ and f is the trivial homomorphism. Clearly both $T\bar{f}$ and $T1_{L_{\mathbb{F}K_4}}$ are zeroes for the functor $T : L_{\mathbb{F}G} \to \mathcal{C}_{PLG}$.

References

1. Mac Lane, S.: Categories for the Working Mathematician, 2nd edn. Springer (1997)
2. Cohen, A.M., Taylor, D.E.: On a certain Lie algebra defined by a finite group. Am. Math. Mon., 633–639 (2007)
3. Humphreys, J.E.: Introduction to Lie Algebras and Representation Theory. Springer-Verlag (1972)
4. Hall, B.C.: Lie Groups, Lie Algebras, and Representations. An Elementary Introduction. Springer (2003)
5. Nambooripad, K.S.S.: Theory of Regular Semigroups. Sayahna Foundation, Trivandrum (2018)
6. James, G., Liebeck, M.: Representations and Characters of Groups, 2nd edn. Cambridge University Press (2001)

On Lattice Vector Spaces over a Distributive Lattice

Pallavi Panjarike, Kuncham Syam Prasad, Madeline Al-Tahan, Vadiraja Bhatta, and Harikrishnan Panackal

Abstract In this paper, we show a homomorphism between lattices that induces a lattice vector space over a distributive lattice. The notion of congruence relation on lattice vector space is introduced to study the quotient spaces. We obtain a correspondence between the set of matrices over distributive lattices and the set of linear transformations.

Keywords Lattice vector space · Lattice matrices · Distributive lattices

1 Introduction

Lattice vector spaces are commutative monoids together with a bounded distributive lattice. Several authors have studied certain basic properties of lattice vector spaces and linear transformation in it. In [2], Foster discussed the idea of Boolean vector representation. Later, Subrahmanyam [10] explored the notion of Boolean vector spaces (B-vector spaces), and gave a characterization for the existence of a basis in a normed B-vector space. Subrahmanyam [9] studied free mobility in B-vector spaces. Later, this notion was generalized to lattice vector spaces by Joy and

P. Panjarike · K. Syam Prasad · V. Bhatta · H. Panackal (✉)
Department of Mathematics, Manipal Institute of Technology, Manipal Academy of Higher Education, Manipal, KA, India
e-mail: pk.harikrishnan@manipal.edu

P. Panjarike
e-mail: pallavipanjrk@gmail.com

K. Syam Prasad
e-mail: syamprasad.k@manipal.edu

V. Bhatta
e-mail: vadiraja.bhatta@manipal.edu

M. Al-Tahan
Department of Mathematics and Statistics, Abu Dhabi University, Abu Dhabi, UAE
e-mail: altahan.madeleine@gmail.com

A. A. Ambily and V. B. Kiran Kumar (eds.), *Semigroups, Algebras and Operator Theory*, Springer Proceedings in Mathematics & Statistics 436,
https://doi.org/10.1007/978-981-99-6349-2_10

Thomas [6]. Concurrently, Give'on [3] explored the properties of square matrices over distributive lattices. Invertibility of lattice matrices were studied by Marenich and Kumarov [7]. Cui-Kui [1] investigated the solutions to lattice matrix equations over distributive lattices. Lattice matrix equations were used to study the transportation problem, medical diagnosis, etc. In [8] the authors studied some interesting applications of prime hyperideals in meet and join hyperlattices. In [11] the authors studied the combinatorial properties of lattices. We refer to [5] for the basic concepts and notations in classical vector spaces over a field.

In this paper, we prove homomorphism between lattices, which will give lattice vector space over a distributive lattice. We introduce the idea of congruence relation over lattice vector spaces and quotient spaces induced by this relation. Furthermore, we show the interrelation between linear transformation and lattice matrices, and study some properties.

2 Lattice Vector Spaces

Definition 1 [6] Let L be a distributive lattice with 0 and 1 and $(V, +)$ be an commutative monoid. We call V as a lattice vector space (or L-vector space or l.v.s.) if there exists a map $\cdot : L \times V \to V$ such that the following properties hold:

1. $u + v = 0_V$ if and only if $u = v = 0$, where 0_V is the identity element in $(V, +)$;
2. $l \cdot (u + v) = l \cdot u + l \cdot v$;
3. $(l_1 \vee l_2) \cdot v = l_1 \cdot v + l_2 \cdot v$;
4. $(l_1 \wedge l_2) \cdot v = l_1 \cdot (l_2 \cdot v)$;
5. $1 \cdot v = v$;
6. $0 \cdot v = 0_V$,

for all $l, l_1, l_2 \in L$, and $u, v \in V$.

Throughout the paper, L denotes a distributive lattice, and V stands for l.v.s. over L. For $l \in L$ and $u \in V$, we denote $lu = l \cdot u$.

Proposition 1 *[6] In an l.v.s. V,*

1. $l \cdot 0_V = 0_V$;
2. $v + v = v$,

for all $l \in L$ and $v \in V$.

Condition 2 of the Proposition 1 shows that $+$ satisfies the idempotent law and hence it forms a join semilattice with a partial order $\leq$ on V by $v \leq u$ if and only if $u + v = u$.

Remark 1 For $l \in L$ and $v \in V$, $lv \leq v$.

Definition 2 [4] Let $(L, \vee, \wedge)$, and $(L', \vee', \wedge')$ be two lattices. A map $\eta : L \to L'$ is called

1. a join homomorphism if $\eta(l_1 \vee l_2) = \eta(l_1) \vee' \eta(l_2)$;
2. a homomorphism if η is a join homomorphism and $\eta(l_1 \wedge l_2) = \eta(l_1) \wedge' \eta(l_2)$,

for all $l_1, l_2 \in L$.

Proposition 2 *Let $(L, \vee, \wedge)$ and $(L', \vee', \wedge')$ be distributive lattices with 0 and 1, and $\eta : L \to L'$ be a homomorphism with $\eta(0) = 0$ and $\eta(1) = 1$. Then $(L', \vee')$ is an l.v.s. over L with respect to the map $\cdot : L \times L' :\to L'$ by $l \cdot v' = \eta(l) \wedge' v'$, for all $(l, v') \in L \times L'$.*

Proof Since L' is a distributive lattice, for any $v_1, v_2 \in L'$ with $v_1 \vee' v_2 = 0$ implies $v_1 = v_2 = 0$. Now let $l \in L$ and $v_1, v_2 \in L'$. Then $l \cdot \left(v_1 \vee' v_2\right) = \eta(l) \wedge' \left(v_1 \vee' v_2\right)$. As L' is distributive, we get $l \cdot \left(v_1 \vee' v_2\right) = \left(\eta(l) \wedge' v_1\right) \vee' \left(\eta(l) \wedge' v_2\right) = l \cdot v_1 \vee' l \cdot v_2$. In a similar way, the other conditions will be followed from the fact that η is a homomorphism and L' is distributive. □

We use $\mathcal{I}(L)$ to denote the collection of all ideals of L, and $(l]$ to denote the set $\{x \in L : x \leq l\}$. $\mathcal{I}(L)$ is a lattice with respect to inclusion. Furthermore, $\mathcal{I}(L)$ is a distributive lattice (See [4]).

Proposition 3 *$\mathcal{I}(L)$ is an l.v.s. over L with respect to the map $\cdot : L \times \mathcal{I}(L) \to \mathcal{I}(L)$ defined by $l \cdot I = l \wedge I = \{l \wedge i : i \in I\}$ for all $(l, I) \in L \times \mathcal{I}(L)$.*

Proof The operation $+$ on $\mathcal{I}(L)$ defined by $I + J = \{x \in L : x \leq i \vee j,\ i \in I, j \in J\}$, for all $I, J \in \mathcal{I}(L)$, so that $(\mathcal{I}(L), +)$ is an commutative monoid. Further, $I + J = \{0\}$ implies that $\{x \in L : x \leq i \vee j,\ i \in I, j \in J\} = \{0\}$, which shows that $I = J = \{0\}$. Next, we have $l \wedge I = (l] \wedge I = (l] \cap I$ for all $(l, I) \in L \times \mathcal{I}(L)$. Let $l_1, l_2 \in L$ and $I_1, I_2 \in \mathcal{I}(L)$. Now,

$$\begin{aligned} l_1 \cdot (I_1 + I_2) &= l_1 \wedge (I_1 + I_2) \\ &= (l_1] \wedge (I_1 + I_2) \\ &= \left((l_1] \wedge I_1\right) + \left((l_1] \wedge I_2\right) (\text{since } \mathcal{I}(L) \text{ is distributive}) \\ &= \left(l_1 \wedge I_1\right) + \left(l_1 \wedge I_2\right) = l_1 \cdot I_1 + l_1 \cdot I_2. \end{aligned}$$

In a similar way, we can show that $(l_1 \vee l_2) \cdot I_1 = l_1 \cdot I_1 + l_2 \cdot I_2$, and $(l_1 \wedge l_2) \cdot I_1 = l_1 \cdot (l_2 \cdot I_1)$. Also, $1 \cdot I_1 = (1] \wedge I_1 = L \cap I_1 = I_1$, and $\{0\} \cdot I_1 = \{0\} \cap I_1 = \{0\}$. Thus, $\left(\mathcal{I}(L), +, \cdot\right)$ is a lattice vector space over L. □

Proposition 4 *Let $(L, \vee, \wedge)$ and $(L', \vee', \wedge')$ be distributive lattices with 0 and 1, and $\eta : L \to L'$ be a homomorphism with $\eta(0) = 0$ and $\eta(1) = 1$. Let*

$$\mathcal{H} = \left\{p : L \to L' : p \text{ is a join homomorphism}\right\}.$$

Define $+$ on $\mathcal{H}$ by

$$(p + q)(x) = p(x) \vee' q(x)$$

for all $x \in L$ and $p, q \in \mathcal{H}$. Then $(\mathcal{H}, +)$ is an l.v.s. over L with respect to the map $\circ : L \times \mathcal{H} :\to \mathcal{H}$ by $l \circ p = \eta(l) \wedge' p$, for all $(l, p) \in L \times \mathcal{H}$, where $(\eta(l) \wedge' p)(x) = \eta(l) \wedge' p(x)$ for all $x \in L$.

Proof First we show that $+$ is well defined. Clearly, for any $p, q \in \mathcal{H}$, $p+q$ is a join homomorphism. Now let $p_1, p_2, q_1, q_2 \in \mathcal{H}$ such that $p_1 = q_1$ and $p_2 = q_2$. For any $x \in L$,

$$\begin{aligned}(p_1 + p_2)(x) =& p_1(x) \vee' p_2(x)\\ =& q_1(x) \vee' q_2(x)\\ =& (q_1 + q_2)(x).\end{aligned}$$

We can easily verify that $(\mathcal{H}, +)$ is an commutative monoid with $\mathbf{0}(x) = 0$ for all $x \in L$, as $\mathbf{0}$ acts as the identity element of $\mathcal{H}$. Suppose that $p + q = q + p = \mathbf{0}$ for some $p, q \in \mathcal{H}$. Then $0 = \mathbf{0}(x) = (p+q)(x) = p(x) \vee' q(x) = q(x) \vee' p(x)$ implies that $p(x) = q(x) = 0$ for all $x \in L$, and hence, $p = q = \mathbf{0}$.

We first show $\circ$ is well defined.
Let $l_1, l_2 \in L$, and $p, q \in \mathcal{H}$ such that $l_1 = l_2$ and $p = q$. For any $x \in L$,

$$\begin{aligned}(l_1 \circ p)(x) =& \eta(l_1) \wedge' p(x)\\ =& \eta(l_2) \wedge' q(x)\\ =& (l_2 \circ q)(x).\end{aligned}$$

Therefore, $l_1 \circ p = l_2 \circ p$. Furthermore, $l \circ p$ is a join homomorphism for any $l \in L, p \in \mathcal{H}$. Let $x, y \in L$,

$$\begin{aligned}(l \circ p)(x \vee y) &= \eta(l) \wedge' (p(x \vee y))\\ &= \eta(l) \wedge' \big(p(x) \vee p(y)\big)\\ &= (\eta(l) \wedge' x) \vee' (\eta(l) \wedge' y)\\ &= (l \circ p)(x) \vee' (l \circ p)(y).\end{aligned}$$

Let $p, q \in \mathcal{H}, l \in L$. For any $x \in L$,

$$\begin{aligned}\big(l \circ (p+q)\big)(x) &= \eta(l) \wedge' \big((p+q)(x)\big)\\ &= \eta(l) \wedge' \big(p(x) \vee' q(x)\big)\\ &= \big(\eta(l) \wedge' p(x)\big) \vee' \big(\eta(l) \wedge' q(x)\big)\\ &= (l \circ p)(x) \vee' (l \circ q)(x).\end{aligned}$$

Let $l_1, l_2 \in L, p \in \mathcal{H}$. For any $x \in L$,

$$\begin{aligned}\big((l_1 \vee l_2) \circ (p)\big)(x) &= \big(\eta(l_1) \vee' \eta(l_2)\big) \wedge' p(x)\\ &= \big(\eta(l_1) \wedge' p(x)\big) \vee' \big(\eta(l_2) \wedge' p(x)\big)\\ &= (l_1 \circ p)(x) \vee' (l_2 \circ p)(x).\end{aligned}$$

So $(l_1 \vee l_2)(p) = l_1 \circ p + l_2 \circ p$. Also, for any $x \in L$,

$$\begin{aligned}\big((l_1 \wedge l_2) \circ p\big)(x) &= \eta(l_1 \wedge l_2) \wedge' p(x)\\ &= \eta(l_1) \wedge' \eta(l_2) \wedge' p(x)\\ &= \eta(l_1) \wedge' \big((l_2 \circ p)(x)\big)\\ &= l_1 \circ (l_2 \circ p)(x).\end{aligned}$$

And for any $p \in \mathcal{H}, x \in L$,

$$\begin{aligned}(1 \circ p)(x) &= \eta(1) \wedge' p(x) = p(x)\\ (0 \circ p)(x) &= \eta(0) \wedge' p(x) = 0 = \mathbf{0}(x).\end{aligned}$$

This shows that $1 \circ p = p$ and $0 \circ p = \mathbf{0}$. □

Proposition 5 *Let $(L, \vee, \wedge)$ and $(L', \vee', \wedge')$ be distributive lattices and let*

$$\mathcal{H} = \big\{p : L \to L' : p \text{ is a join homomorphism}\big\}.$$

Define + on $\mathcal{H}$ by

$$(p + q)(x) = p(x) \vee' q(x)$$

for all $x \in L$ and $p, q \in \mathcal{H}$. Then $(\mathcal{H}, +)$ is an l.v.s. over L' with respect to the map $\circ : L' \times \mathcal{H} :\to \mathcal{H}$ by $l' \circ q = l' \wedge' q$, for all $(l', q) \in L' \times \mathcal{H}$, where $(l' \wedge' q)(x) = l' \wedge' q(x)$ for all $x \in L$.

Proof Proof is analogous to the proof of Proposition 4. □

Example 1

Let $(V, +)$ be a commutative monoid in which all the elements are idempotents. Then $(V, +, \cdot)$ is an l.v.s. over $C_2 = \{0, 1 : 0 < 1\}$ where the scalar multiplication $\cdot : C_2 \times V \to V$ defined as $0 \cdot v = 0$ and $1 \cdot v = v$ for all $v \in V$.

Definition 3 [6] $\emptyset \neq V' \subseteq V$ is said to be a lattice subspace (or L-subspace) of V if V' itself is an l.v.s. over L under the restriction of $+$ and $\cdot$ to V'.

Proposition 6 *[6] $\emptyset \neq V' \subseteq V$ is a lattice subspace of V if and only if V' is closed under addition and scalar multiplication in V.*

Proposition 7 *Any arbitrary intersection of lattice subspaces is a lattice subspace.*

Unlike in vector spaces over a field, in an l.v.s., the union of the lattice subspaces W_1, W_2 can be a lattice subspace even though $W_1 \nsubseteq W_2$ and $W_2 \nsubseteq W_1$. We show this in Example 2.

Example 2

Let C_n denote a chain of length n. Then $V = C_2 \times C_3$ (where $C_3 = \{0, a, 1 : 0 < a < 1\}$) is an l.v.s. over C_2, with the scalar multiplication defined by $0 \cdot u = 0$ and $1 \cdot u = u$ for all $u \in V$. Consider the lattice subspaces $W_1 = \{(0, 0), (0, a)\}$ and $W_2 = \{(0, 0), (1, a)\}$. Here, $W_1 \not\subseteq W_2$ and $W_2 \not\subseteq W_1$, but $W_1 \cup W_2 = \{(0, 0), (0, a), (1, a)\}$ is a lattice subspace of V.

3 Congruence Relation and Linear Transformations on l.v.s.

We introduce the notion of congruence relation on lattice vector spaces and obtain its quotient structure.

Definition 4 [6] Let V and W be l.v.s over L. A map $T : V \to W$ is called a linear transformation if it satisfies the following conditions:

1. $T(v + v') = T(v) + T(v')$;
2. $T(l \cdot v) = l \cdot T(v)$,

for all $v, v' \in V$ and $l \in L$.

Definition 5 An equivalence relation $\sim$ on an l.v.s. V over L is called a congruence relation on V if it satisfy the following conditions:

1. $u \sim v, w \sim x$ implies $u + w \sim v + x$;
2. $u \sim v$ implies $l \cdot u \sim l \cdot v$,

for all $u, v, w, x \in V$ and $l \in L$.
For any $w \in V$ we denote the congruence class containing w by $[w] = \{v \in V : v \sim w\}$.

Example 3

Take $V = L^2$. We define a relation $\sim$ on V by $(l_1, l_2) \sim (l_3, l_4)$ if and only if $l_2 = l_4$. Then $\sim$ is a congruence relation on V.

Proposition 8 *Let $\sim$ be a congruence relation on a lvs V and $V/\!\sim\, = \{[v] : v \in V\}$ be the set of all congruence classes of V with respect to $\sim$. Define $\oplus$ on $V/\!\sim$ by $[u] \oplus [v] = [u + v]$, and the scalar multiplication $\odot : L \times V/\!\sim$ by $l \odot [v] = [lv]$. Then $(V/\!\sim, \oplus, \odot)$ is an l.v.s. over L.*

Proof Suppose $[u_1] = [u_2]$ and $[w_1] = [w_2]$. Then $u_1 \sim u_2$ and $w_1 \sim w_2$. As $\sim$ is a congruence relation, we get $u_1 + w_1 \sim u_2 + w_2$. This yields $[u_1 + w_1] = [u_2 + w_2]$, and hence $[u_1] \oplus [w_1] = [u_2] \oplus [w_2]$. Now we show $(V/\sim, \oplus)$ is an commutative monoid. Let $[u], [v], [w] \in V/\sim$. Then,

$$\begin{aligned}\big([u] \oplus [v]\big) \oplus [w] &= \big([u + v]\big) \oplus [w] \\ &= \big[(u + v) + w\big] = \big[u + (v + w)\big] \\ &= [u] \oplus \big([v + w]\big) = [u] \oplus \big([v] \oplus [w]\big).\end{aligned}$$

Also $[0] \oplus [v] = [0 + v] = [v] = [v + 0] = [v] \oplus [0]$, for all $[v] \in V/\sim$. For any $[u], [w] \in V/\sim$, $[u] \oplus [w] = [u + w] = [w + u] = [w] \oplus [u]$. Hence, $(V/\sim, \oplus)$ is an commutative monoid. Moreover, $[u] \oplus [w] = [0]$ gives $0 \sim u + w$, and $u + w \sim 0$. Then $u + 0 \sim u + u + w$. Since $u + u = u$ and $\sim$ is transitive, we get $u \sim 0$. In a similar way we can show $w \sim 0$. So, $[u] = [w] = [0]$. Now we show that $l \odot [u] = [l \cdot u]$ is well defined. Let $l = l' \in L$ and $[u] = [w] \in V/\sim$. Then $u \sim w$. Since $\sim$ is congruence relation on V, we get $l \cdot u \sim l \cdot w$, and hence $l \cdot u \sim l' \cdot w$, That is, $[l \cdot u] = [l' \cdot w]$. Let $l, l' \in L, u, w \in V$. Now $l \odot \big([u] \oplus [w]\big) = l \odot \big([u + w]\big) = \big[l \cdot (u + w)\big] = \big[l \cdot u + l \cdot w\big] = [l \cdot u] \oplus [l \cdot w] = l \odot [u] \oplus l \odot [w]$. Also, $(l \vee l') \odot [u] = \big[(l \vee l') \cdot u\big] = [l \cdot u + l' \cdot u] = l \odot [u] \oplus l' \odot [u]$, and $(l \wedge l') \odot [u] = \big[(l \wedge l') \cdot u\big] = \big[l \cdot (l' \cdot u)\big] = l \odot [l' \cdot u]$. Furthermore, $1 \odot [u] = [1 \cdot u] = [u]$ and $0 \odot [u] = [0 \cdot u] = [0]$. Thus $(V/\sim, \oplus, \odot)$ is an l.v.s. over L. □

Proposition 9 *Let $\sim$ be a congruence relation on an l.v.s. over L and $V/\sim$ be the corresponding quotient space. Then the map $\eta : V \to V/\sim$ defined by $\eta(v) = [v]$ for all $v \in V$ is an onto linear transformation.*

Proof Suppose that $u_1 = u_2 \in V$ with $u_1 = u_2$. Then $u_1 \sim u_2$ implies $[u_1] = [u_2]$, that is, $\eta(u_1) = \eta(u_2)$. So η is well defined. Clearly η is onto. Let $u, v \in V$ and $l \in L$. Then $\eta(u + v) = [u + v] = [u] \oplus [v] = \eta(u) \oplus \eta(v)$, and $\eta(lu) = [lu] = l \odot [u] = l \odot \eta(u)$, and hence η is a linear transformation. □

Proposition 10 *Let V and W be l.v.s. over L and $T : V \to W$ be a linear transformation. Then $T(V)$ is isomorphic to a suitable quotient space of V.*

Proof Define a relation $\sim_T$ on V by $v \sim_T v'$ if and only if $T(v) = T(v')$. Clearly $\sim_T$ is an equivalence relation. Suppose that $u \sim_T u'$ and $v \sim_T v'$, where $u, u', v, v' \in V$. Then $T(u) = T(v)$ and $T(v) = T(v')$. This gives $T(u) + T(v) = T(u') + T(v')$. As T is a linear transformation, $T(u + v) = T(u' + v')$, and hence, $u + v \sim_T u' + v'$. Also, for any $u, v \in V, l \in L$ with $T(u) = T(v)$, we have $lT(u) = lT(v)$. Again as T is a linear transformation, $T(ul) = T(vl)$, and hence, $ul \sim_T vl$. Thus, $\sim_T$ is a congruence relation. Let $V/\sim_T$ be the quotient space of V corresponding to the congruence relation $\sim_T$. Define $\phi : V/\sim_T \to T(V)$ by $\phi\big([v]\big) = T(v)$, for all $[v] \in V/\sim_T$. Let $[v], [w] \in V/\sim_T$. Then $[v] = [w]$ if and only if $v \sim_T w$ if and only if $T(v) = T(w)$ if and only if $\phi\big([v]\big) = \phi\big([w]\big)$. Thus, ϕ is both well defined and one-one. Now, let $w \in T(V)$. Then $w = T(v)$ for some $v \in V$. This implies $\phi\big([v]\big) = T(v) = w$. Therefore ϕ is onto. Let $u, v \in V$ and $l, l' \in L$. Then $\phi\big(l \odot [v] \oplus l' \odot [w]\big) = \phi\big([lv + l'w]\big) = T(lv + l'w) = lT(v) + l'T(w) = l\phi\big([v]\big) + l'\phi\big([w]\big)$. Thus, ϕ is an isomorphism. □

Definition 6 [6] Let V be an l.v.s. and $\emptyset \neq U \subseteq V$.

1. We say $v \in V$ is a linear combination of the vectors in U, if there exist $u_i (1 \leq i \leq n) \in U$ and $l_i (1 \leq i \leq n) \in L$ such that $v = \sum\limits_{1 \leq i \leq n} l_i u_i$.
2. We write Span $U = \left\{ \sum\limits_{1 \leq i \leq n} a_i u_i : u_i \in U, a_i \in L, n \in \mathbb{N} \right\}$. If Span $U = V$, then we call U as a spanning subset of V.
3. U is said to be linearly independent (in short, l.i.) if every element $v \in$ Span U can be expressed uniquely as a linear combination of elements of U. Otherwise, U is linearly dependent.
4. U is said to be a basis if S is l.i. and Span $U = V$. If $|U|$ is finite, then we say V is finite dimensional.

Proposition 11 *Let V and W be l.v.s. over L and $T : V \to W$ be a linear transformation and $S = \{v_i : 1 \leq i \leq n\}$ be a l.i. subset of V. Then $\{T(v_i) : 1 \leq i \leq n\}$ is a l.i. subset of $T(V)$ if and only if T is one-one.*

Proof Suppose that T is one-one. Take $B = \{T(v_i) : 1 \leq i \leq n\}$, and let $v \in$ Span B. Then $v = \sum\limits_{1 \leq i \leq n} l_i T(v_i) = T\left(\sum\limits_{1 \leq i \leq n} l_i v_i \right)$ for some $l_i (1 \leq i \leq n) \in L$. Suppose that v can also be expressed as $v = \sum\limits_{1 \leq i \leq n} l_i' T(v_i)$. Then $T\left(\sum\limits_{1 \leq i \leq n} l_i v_i \right) = v = T\left(\sum\limits_{1 \leq i \leq n} l_i' v_i \right)$. Since T is one-one, $\sum\limits_{1 \leq i \leq n} l_i v_i = \sum\limits_{1 \leq i \leq n} l_i' v_i$. As S is l.i., by uniqueness of elements of L, we must have $l_i = l_i'$, for all $1 \leq i \leq n$. Conversely suppose that B is a l.i. subset of $T(V)$. Let $v, v' \in V$ such that $T(v) = T(v')$. Since S is a l.i. subset of V, there exist $l_i, l_i' (1 \leq i \leq n)$ such that $v = \sum\limits_{1 \leq i \leq n} l_i v_i$ and $v' = \sum\limits_{1 \leq i \leq n} l_i' v_i$. Now $T(v) = T(v')$ implies that $T\left(\sum\limits_{1 \leq i \leq n} l_i v_i \right) = T\left(\sum\limits_{1 \leq i \leq n} l_i' v_i \right)$, and hence, $\sum\limits_{1 \leq i \leq n} l_i T(v_i) = \sum\limits_{1 \leq i \leq n} l_i' T(v_i)$. As B is l.i., we get $l_i = l_i'$ for all $1 \leq i \leq n$, which shows that $v = v'$. □

Proposition 12 *Let V and W be l.v.s. over L and $T : V \to W$ be a linear transformation, and $\{v_i : 1 \leq i \leq n\}$ be a spanning subset of V. Then $\{T(v_i) : 1 \leq i \leq n\}$ is a spanning subset of W if and only if T is onto.*

Proof Let $w \in W$. Suppose that T is onto. Then $T(v) = w$ for some $v \in V$, $v = \sum\limits_{1 \leq i \leq n} l_i v_i, l_i \in L$. Now $w = T(v) = T\left(\sum\limits_{1 \leq i \leq n} l_i v_i \right) = \sum\limits_{1 \leq i \leq n} l_i T(v_i)$. Conversely, suppose that $\{T(v_i) : 1 \leq i \leq n\}$ is a spanning set of W. Now let $w \in W$. Then $w = \sum\limits_{1 \leq i \leq n} l_i T(v_i) = T\left(\sum\limits_{1 \leq i \leq n} l_i v_i \right)$ shows that T is onto. □

For the succeeding results, we consider only finite dimensional l.v.s.

Proposition 13 *Let V and W be l.v.s. over L, $\{v_i : 1 \le i \le n\}$ be a basis of V and $\{w_i : 1 \le i \le n\}$ be a subset of W. Then there exists a unique linear transformation $T : V \to W$ such that $T(v_i) = w_i$.*

Proof Let $B = \{v_i : 1 \le i \le n\}$. To prove the existence, let $x \in V = \text{Span } B$. Then $x = \sum_{1\le i\le n} l_i v_i$, where $l_i (1 \le i \le n) \in L$. Define $T : V \to W$ by $T(x) = \sum_{1\le i\le n} l_i w_i$. Clearly T is well defined. Let $v, v' \in V$ with $v = \sum_{1\le i\le n} l_i v_i$, $v' = \sum_{1\le i\le n} l'_i v_i$, where $l, l', l_i, l'_i (1 \le i \le n) \in L$. Then $T(lv + l'v') = T\Big(l \sum_{1\le i\le n} l_i v_i + l' \sum_{1\le i\le n} l'_i v_i\Big) = T\Big(\sum_{1\le i\le n} (ll_i + l'l'_i) v_i\Big) = \sum_{1\le i\le n} (ll_i + ll'_i) w_i = l \sum_{1\le i\le n} l_i w_i + l' \sum_{1\le i\le n} l'_i w_i = lT(v) + l'T(v')$. Now we prove the uniqueness part. Suppose $T' : V \to W$ is a linear transformation such that $T'(v_i) = w_i$ for all $1 \le i \le n$. Then for any $v = \sum_{1\le i\le n} l_i v_i$, where $l_i (1 \le i \le n) \in L$, $T'(v) = T'\Big(\sum_{1\le i\le n} l_i v_i\Big) = \sum_{1\le i\le n} l_i T'(v_i) = \sum_{1\le i\le n} l_i w_i = \sum_{1\le i\le n} l_i T(v_i) = T\Big(\sum_{1\le i\le n} l_i v_i\Big) = T(v)$. □

We show that there is a correspondence between the linear transformations on l.v.s. with finite dimension, and the set of all matrices over a distributive lattice. Let $A_{m\times n} = (a_{ij})_{m\times n}$, $B_{m\times n} = (b_{ij})_{m\times n}$, $C_{n\times p} = (c_{ij})_{n\times p}$ be lattice matrices over L. We have the following matrix operation: $A + B = (a_{ij} + b_{ij})_{m\times n}$ and $P = AC = (p_{ij})_{m\times p}$, where $p_{ij} = \bigvee_{1\le k\le n} (a_{ik} \wedge c_{kj})$. Also, for $l \in L$, $Q = lA = (q_{ij})_{m\times n}$, where $q_{ij} = (l \wedge a_{ij})$ (see [3]).

Theorem 1 *Let $A \in M_{n\times m}$, the set of $n \times m$ matrices over L. Then the map $T : L^m \to L^n$ defined by $T(X) = AX$ for all $X \in L^m$ is a linear transformation. Conversely, let $T : L^n \to L^m$ be a linear transformation. Then there exists $A \in M_{m\times n}$ such that $T(X) = AX$ for all $X \in L^n$.*

Proof Let $X, Y \in L^m, l, l' \in L$. Then $T(lX + l'Y) = A(lX + l'Y) = A(l'X) + A(l'Y) = l(AX) + l'(AY) = lT(X) + l'T(Y)$, showing T is linear. Conversely, Let $e_i = (0, 0, \cdots, 1, \cdots, 0)^t$, $1 \le i \le n$. Then $B = \{e_i : 1 \le i \le n\}$ is a basis for L^n. Take $A = \big[T(e_1)\ T(e_2) \ldots T(e_n)\big]$. We can verify that $T(e_i) = Ae_i$ for all e_i, $1 \le i \le n$. Then by Proposition 13, we get that $T(X) = AX$ for all $X \in L^n$. □

Definition 7 [7] Let $A, B \in M_{n\times n}$. Then B is said to be

1. a right inverse of A if $AB = i_d$
2. a left inverse of A if $BA = i_d$.
3. an inverse of A if $AB = BA = i_d$.

Remark 2 Let $T : V \to W$, $T' : W \to U$ be linear transformations, and A and B be matrix corresponding to T and T' respectively. Then BA is the matrix corresponding to $T' \circ T : V \to U$.

Theorem 2 *Let $T : L^n \to L^n$ be a linear transformation and A be the matrix w.r.t. T. Then*

1. *T is one-one if and only if A has a left inverse.*
2. *T is onto if and only if A has a right inverse.*
3. *T is bijective if and only if A has an inverse.*

Proof (1) Suppose that A has a left inverse say B, and $X, Y \in L^n$ such that $T(X) = T(Y)$. Then $AX = AY$, and so $B(AX) = B(AY)$. As L is distributive, we can write $(BA)X = (BA)Y$, gives $X = Y$. Conversely suppose that T is one-one. Take $f_i = T(e_i) = Ae_i, (1 \leq i \leq n)$. By Proposition 11, $\{f_i : 1 \leq i \leq n\}$ is a l.i. subset of L^n. By Proposition 13, there exists a linear transformation T' such that $T'(f_i) = e_i$ $(1 \leq i \leq n)$. Let B be the matrix of linear transformation T'. Now $e_i = T'(f_i) = T'(T(e_i)) = T'T(e_i)$, and hence, $T'T$ is the identity linear transformation and so, $BA = i_d$.
(2) Suppose that B is a right inverse of A and $Y \in L^n$. Take $X = BY$. Then $T(X) = AX = A(BY) = (AB)Y = Y$. Hence T is onto. Conversely, suppose that T is onto. Then for each $e_i (1 \leq i \leq n)$, there exists $X_i \in L^n$ such that $T(X_i) = e_i$. Now take $B = \begin{bmatrix} X_1 & X_2 & \cdots & X_n \end{bmatrix}$. Let T' be the linear transformation defined by $T'(Y) = BY$. Then $TT'(e_i) = e_i$ and hence $AB = i_d$.
(3) Follows from (1) and (2). □

4 Conclusion

We have established that the collection of join homomorphisms between distributive lattices form an l.v.s. over L. Using Propositions 2 to 5, we can construct several examples for l.v.s. over a distributive lattice to understand the structure better. Using Theorem 2, we can observe the linear transformations between lattice vector spaces as matrices and vice versa.

Acknowledgements The authors express their gratitude to the referee(s) and the editor(s) for their careful reading of the manuscript, and the valuable suggestions to improve the paper. Author[a] acknowledges Manipal Academy of Higher Education, Manipal for providing the scholarship under Dr T M A Pai fellowship. The authors[a,b,d,*] thank Manipal Institute of Technology, Manipal Academy of Higher Education, Manipal for the kind encouragement. Author[c] thank Abu Dhabi University for their support. The author[a] thank the organizers of ICSAOT-2022 for providing an opportunity to present the research work.

References

1. Cui-Kui, Z.: On matrix equations in a class of complete and completely distributive lattices. Fuzzy Sets Syst. **22**, 303–320 (1987)
2. Foster, A.L.: p-Rings and their Boolean-vector representation. Acta Math. **84**, 231–261 (1951)
3. Give'on, Y.: Lattice matrices. Inf. Control. **7**, 477–484 (1964)
4. Grätzer, G.: Lattice Theory: Foundation. Springer Science & Business Media (2011)
5. Hoffman, K., Kunze, R.: Linear Algebra. Englewood Cliffs, New Jersey (1971)
6. Joy, G., Thomas, K.V.: Lattice vector spaces and linear transformations. Asian Eur. J. Math. **12**, 1–12 (2019)
7. Marenich, E.E., Kumarov, V.G.: Inversion of matrices over a pseudocomplemented lattice. J. Math. Sci. **144**, 3968–3979 (2007)
8. Pallavi, P., Kuncham, S.P., Vadiraja, G.R.B., Harikrishnan, P.K.: Computation of prime hyperideals in meet hyperlattices. Bull. Comput. Appl. Math. **10**, 23–58 (2022)
9. Subrahmanyam, N.V.: Boolean vector spaces II. Math. Zeitschr. **87**, 401–419 (1965)
10. Subrahmanyam, N.V.: Boolean vector spaces I. Math. Zeitschr **83**, 422–433 (1964)
11. Tapatee, S., Harikrishnan, P.K., Kedukodi, B.S., Kuncham, S.P.: Graph with respect to superfluous elements in a lattice. Miskolc Math. Notes. **23**, 929–945 (2022)

Operator Theory

Operator theory is a branch of functional analysis, which plays a major role in many important areas, such as differential equations, mathematical physics, etc. The third part of this volume deals with operator theory and related areas.

Application of Geometric Algebra to Koga's Work on Quantum Mechanics

K. V. Didimos

Abstract One of the people who has offered alternatives to the Copenhagen interpretation of quantum mechanics is Toyoki Koga. Some of the important equations in quantum mechanics are the Schrödinger equation, Dirac equation, and Pauli equation; however, Koga only focused on the solutions of the first two. This article briefly introduces geometric algebra to study Koga's works, especially on the Dirac equation and how to get a translation of Koga's solution of the Dirac equation using geometric algebra.

Keywords Geometric algebra · Space · Time algebra · Dirac equation · Koga's solution

1 Introduction

Many scholars assumed at the turn of the twentieth century that the structure and behaviour of electrons were the product of an electromagnetic origin. With the introduction of quantum mechanics, however, everyone ceased considering this. Nevertheless, in the middle of the 20th century, Koga initiated a study into the principles of quantum mechanics to eradicate quantum physics' riddles. Until then, he was not satisfied with the foundations of quantum mechanics. As a result of his efforts to elucidate the principles, he developed a new hypothesis regarding electrons: the deterministic theory of the electron. He explored the internal gravitational fields of electrons in detail. See the book [8] or the papers [2–7] for an account of Koga's work.

In 1926, Schrödinger published an equation which described an electron as a wave. This was based on de Broglie's idea that any particle has an associated wave. The equation is non-relativistic. Its solution is a complex-valued map from $\mathbb{R}^3 \times \mathbb{R}$ (Galilean spacetime).

K. V. Didimos (✉)
Department of Mathematics, Sacred Heart College, Thevara, Kochi 682013, KL, India
e-mail: didimos@shcollege.ac.in

A. A. Ambily and V. B. Kiran Kumar (eds.), *Semigroups, Algebras and Operator Theory*, Springer Proceedings in Mathematics & Statistics 436,
https://doi.org/10.1007/978-981-99-6349-2_11

Earlier, he tried out a relativistic equation but it didn't give correct predictions. This equation was found independently later by Klein and Gordon and is called the Klein-Gordon equation.

The electron was considered a sizeless particle and the solution of the Schrödinger equation was interpreted as giving probabilities of finding it at different locations.

The following year, Pauli published a paper accounting for the magnetic nature of the electron by taking two Schrödinger equations and adding a term to represent interaction with an external magnetic field.

Then, in 1928, Dirac presented the equation named after him. Its solution is a map from Minkowski space to 4-dimensional complex space. The coefficients of terms in the equation are 4×4 complex matrices. He interpreted the solution as a spinor which is Lorentz-covariant.

This work of Schrödinger and Dirac is described in Dirac's book [1].

In 1972, Koga exhibited a solution of the Schrödinger equation for a free electron which represented a localised field in 3-space, surrounding a singularity (the singularity arises because the equation is linear whereas a real electron is nonlinear). He completed this work in 1974 by considering the effect of an external electromagnetic field on an electron. He showed that a de Broglie wave could be obtained by averaging these solutions over an ensemble, thus confirming a conjecture of Einstein.

He then extended this work to the Dirac equation. In 1983, he showed in Chapter V of his book [9] how to get a solution of the Dirac equation along these lines. But he somehow faltered at this point and was unable to understand the solution and its mathematical properties.

Koga extended this work to incorporate the internal gravitational field of an electron but this is beyond the scope of this sketch.

His explanation that the de Broglie wave is obtained through an averaging process is presented in [2] and [8]. He did not deal much with Pauli's equations, probably because he emphasised studying the internal electromagnetism of electrons.

Although most academics study Dirac equation in its original matrix form, it is evident from the research of S. K. Panday and R. S. Chakravarti that geometric algebra may be used to examine the solution.

Initiated by David Hestenes in the 1960s, geometric algebra is the study of Clifford algebras from a geometrical perspective. In the 1860s, W. K. Clifford researched a class of associative algebras that brought together prior work on quaternions by W. R. Hamilton and exterior algebras by H. Grassmann. Hestenes developed the field of applied Clifford algebras, often known as geometric algebra, with the idea that even while it does not represent novel physics, it should provide fresh insights into the underlying physics and advance our understanding. Studying Toyoki Koga's presentation of the Schrödinger equation and the Dirac equation using geometric algebra is relevant since it offers a fresh perspective.

The importance of geometric algebra and how to effectively use the special 4-dimensional gradient is clearly stated in [12]. The 4-dimensional gradient is very effective when dealing with Maxwell's equations. It is also noted how geometry can be used to illustrate that in some special cases, the Dirac equation and Maxwell's equation are almost the same. As a result, the Bohr magneton got the correct value.

Numerous scientists, including Doran and Lasenby, have translated the standard version of quantum mechanics using geometric algebra. It is uncommon to examine the deterministic theory of electrons using geometric algebra. The first such treatment was given by S. K. Panday through his study of Koga's solution using geometric algebra. Pandey and Chakravarti [12] showed how to study the Dirac eqn using Geometric Algebra [10, 11] and get a solution. But they did not get the general solution.

Didimos and Chakravarti (not yet published) completed this work of Koga by getting his solution in final, general form and explaining its geometrical meaning. Earlier, they described how to solve the Pauli equation along Koga's lines [13].

2 Fundamentals of Geometric Algebra

2.1 Tensor Product

Vectors are well-known as vector space members and physical entities with magnitude and direction. Several operations on vectors can be defined by considering vector space members, such as dot product, cross product, triple scalar product and so on.

By defining the tensor product of their members, we define another relevant product of two vector spaces.

Let V and W be two real vectorspaces with basis $\{\alpha_i\}$ and $\{\beta_j\}$ respectively. Then the tensor product of V and W, denoted by $V \otimes W$, is the set of all formal sums $\sum\limits_{i,j} p_{ij}\alpha_i \otimes \beta_j$.

For instance, consider the vector spaces $\mathbb{R}^2$ and $\mathbb{R}^3$. Let $\{\alpha_1 = (1, 0), \alpha_2 = (0, 1)\}$ and $\{\beta_1 = (1, 0, 0), \beta_2 = (0, 1, 0), \beta_3 = (0, 0, 1)\}$ be the bases. Then

$$V \otimes W = span\ \{\alpha_1 \otimes \beta_1, \alpha_1 \otimes \beta_2, \alpha_1 \otimes \beta_3, \alpha_2 \otimes \beta_1, \alpha_2 \otimes \beta_2, \alpha_2 \otimes \beta_3\}$$

Definition 1 Let $\mathcal{B}_1$ and $\mathcal{B}_2$ be bases for V and W respectively. Then the dimension of $V \otimes W$ is the cardinality of its basis $\mathcal{B} = \{\alpha_i \otimes \beta_j : \alpha_i \in \mathcal{B}_1, \beta_j \in \mathcal{B}_2\}$

Example 1 Let $X = \begin{pmatrix} x_1 \ x_2 \ \ldots \ x_n \end{pmatrix} \in \mathbb{R}^n$ and $Y = \begin{pmatrix} y_1 \ y_2 \ \ldots \ y_m \end{pmatrix} \in \mathbb{R}^m$. Then $X \otimes Y = (z_{ij})$ where $z_{ij} = x_i y_j$.

For instance, $\begin{pmatrix} 3 \ 2 \end{pmatrix} \otimes \begin{pmatrix} 9 \ 8 \ 7 \end{pmatrix} = \begin{pmatrix} 27 & 24 & 21 \\ 18 & 16 & 14 \end{pmatrix}$

For a real, finite-dimensional vector space V with quadratic or symmetric bilinear form. The Clifford algebra $Cl(V)$ of V is defined as $T(V)/J$ where $T(V)$ is the tensor algebra of V, the ideal J generated by all elements $v \otimes v - B(v, v)$ and v is a vector in V. This tensor product can be used to define the geometric product. A geometric algebra can be embedded into the corresponding tensor algebra. The members of the geometric algebra are called the multivectors.

Let $\mathcal{B} = \{\alpha_1, \ldots, \alpha_n\}$ be an orthogonal ordered basis for V. Then a basis for the corresponding Clifford algebra of V will be

$$\{1\} \cup \{\alpha_i\} \cup \{\alpha_i\alpha_j | i < j\} \cup \{\alpha_i\alpha_j\alpha_k | i < j < k\} \cup \cdots \cup \{\alpha_1\alpha_2 \ldots \alpha_n\}$$

Therefore $Cl(V)$ has the dimension 2^n.

3 The Four-Dimensional Gradient

Now we discuss the specific geometric algebras(STA) that enable us to define a four-dimensional gradient. This algebra is generated by the unit vectors $\gamma_0, \gamma_1, \gamma_2$ and γ_3 such that $\gamma_0^2 = 1$ and $\gamma_1^2 = \gamma_2^2 = \gamma_3^2 = -1$. with the bilinear form due to Minkowski metric η associated with the matrix

$$\begin{pmatrix} 1 & 0 & 0 & 0 \\ 0 & -1 & 0 & 0 \\ 0 & 0 & -1 & 0 \\ 0 & 0 & 0 & -1 \end{pmatrix}$$

and we have

$$\gamma_i\gamma_j + \gamma_j\gamma_i = 2\eta_{ij}$$

A basis for the spacetime algebra is

$$\{1\} \cup \{\gamma_i\} \cup \{\gamma_i\gamma_j | i < j\} \cup \{\gamma_i\gamma_j\gamma_k | i < j < k\} \cup \{\gamma_0\gamma_1\gamma_2\gamma_3\}$$

The elements of the above sets are called scalars, vectors, bivectors, trivectors and pseudoscalar, respectively.

The subalgebra of STA generated by the even multivectors $\sigma_1 = \gamma_1\gamma_0, \sigma_2 = \gamma_2\gamma_0, \sigma_3 = \gamma_3\gamma_0$ is called even subalgebra and the basis consists of the multivectors $\{1\} \cup \{\sigma_i\} \cup \{\sigma_i\sigma_j | i < j\} \cup \{\sigma_1\sigma_2\sigma_3\}$. Obviously, the dimension of this subalgebra is 8. This is also called the geometric algebra of the Euclidean space of dimension three.

One of the merits of geometric algebra is the existence of the 4 dimensional gradient. For a function ψ defined on Minkowski space into STA, the gradient is defined as

$$\nabla\psi = \gamma^\mu \frac{\partial\psi}{\partial x^\mu}.$$

The corresponding Laplacian is $\nabla^2\psi = \nabla \cdot \nabla\psi$. That is

$$\nabla^2\psi = \frac{\partial\psi}{(\partial x^0)^2} - \frac{\partial\psi}{(\partial x^1)^2} - \frac{\partial\psi}{(\partial x^2)^2} - \frac{\partial\psi}{(\partial x^3)^2}.$$

4 The Dirac-Hestenes Equation

We now discuss the Dirac equation in the context of geometric algebra. Hestenes first gave it (sometimes called the Dirac-Hestenes equation [10, 11]) and it can be related to the Dirac equation, which is written in terms of matrices and column vectors of complex functions. The Dirac equation can be expressed in terms of a spinor called the Dirac spinor field ψ. The Dirac spinor field is defined on the spacetime called Minkowski space $\mathbb{R} \times \mathbb{R}^3$ and assuming values on the four-dimensional complex vector space.

An isomorphism (real vector space) ϕ can be defined on 2-dimensional complex space to the subspace of space time algebra(STA) spanned by $1, I\sigma_1, I\sigma_2, I\sigma_3$.

$$\phi\left[\begin{pmatrix} a^0 + ia^3 \\ -a^2 + ia^1 \end{pmatrix}\right] = a^0 + a^1 I\sigma_1 + a^2 I\sigma_2 + a^3 I\sigma_3.$$

This would help us to define an isomorphism between a 4-dimensional complex space and the even subalgebra of STA. That is, if $\psi = \begin{pmatrix} \psi_1 \\ \psi_2 \end{pmatrix}$ where $\psi_i \in \mathbb{C}^2$, define a map that sends

$$\psi = \begin{pmatrix} \psi_1 \\ \psi_2 \end{pmatrix} \mapsto \phi(\psi_1) + \phi(\psi_2)\,\sigma_3$$

Fix $k \in \{1, 2, 3\}$. Consider the linear operator $\psi \mapsto \sigma_k\psi\sigma_3$, where $\psi \in$ span $(1, I\sigma_1, I\sigma_2, I\sigma_3)$. It turns out that this corresponds to the operator $|\psi\rangle \mapsto \hat{\sigma}_k|\psi\rangle$ on complex 2-space where $\hat{\sigma}_k$ is a Pauli matrix:

$$\hat{\sigma}_1 = \begin{pmatrix} 0 & 1 \\ 1 & 0 \end{pmatrix}, \quad \hat{\sigma}_2 = \begin{pmatrix} 0 & -i \\ i & 0 \end{pmatrix}, \quad \hat{\sigma}_3 = \begin{pmatrix} 1 & 0 \\ 0 & -1 \end{pmatrix}.$$

Similarly, for $\mu = 0, 1, 2, 3$ consider the following operator:

$$\psi \mapsto \gamma_\mu \psi \gamma_0$$

on the even subalgebra of STA. In complex 4-space, this corresponds to

$$|\psi\rangle \mapsto \hat{\gamma}_\mu|\psi\rangle$$

where $\hat{\gamma}_0 = \begin{pmatrix} I_2 & 0 \\ 0 & -I_2 \end{pmatrix}$ ($I_2 = 2 \times 2$ identity matrix)

$$\hat{\gamma}_k = \begin{pmatrix} 0 & -\hat{\sigma}_k \\ \hat{\sigma}_k & 0 \end{pmatrix}$$

for $k = 1, 2, 3$. Similarly the operator

$$\psi \mapsto \psi I \sigma_3$$

corresponds to

$$|\psi\rangle \mapsto i|\psi\rangle.$$

Using these correspondences, the Dirac equation (expressed in terms of spinors)

$$i\hat{\gamma}^{\mu}\frac{\partial|\psi\rangle}{\partial x^{\mu}} = m|\psi\rangle$$

can be expressed in geometric algebra as

$$\nabla\psi I\sigma_3 = m\psi\gamma_0.$$

The Dirac-Hestenes equation and an explicit map for column spinors are expressed in the so-called Dirac-Pauli form. There are many depictions in the literature. Since they are equal via unitary transformations, similar formulations can be used for different representations.

This is one of the advantages of geometric algebra to use a 4-dimensional gradient vector so that the Dirac equation and other equations like Maxwell's equations can be expressed in the most straightforward representation.

5 A Solution to the Dirac-Hestenes Equation

Using geometric algebra, we now examine a solution to the Dirac-Hestenes equation. For convenience we take $c = 1$ and $\hbar = 1$.

Even though the empirical utility is limited, we begin with the Klein-Gordon equation to get a solution to the Dirac equation.

Let φ be defined from Minkowski's spacetime taking values in spacetime algebra, and $m > 0$ is the rest mass electron. Then Klein-Gordon equation is

$$\nabla^2\varphi + m^2\varphi = 0$$

Koga took as a solution (See [8]) to the original Klein-Gordon equation,

$$\varphi = ae^{iS}$$

where a and S are real scalar fields in spacetime. The fields a and S are defined in the following manner;

$$\begin{aligned} S &= -Et + \mathbf{p} \cdot \mathbf{r}, \\ a &= \exp(-\kappa r')/r' \end{aligned}$$

where

$$\mathbf{r} = x\mathbf{i} + y\mathbf{j} + z\mathbf{k} = x^1\gamma_1 + x^2\gamma_2 + x^3\gamma_3$$

$$\mathbf{r}' = (\mathbf{r} - \mathbf{u}t)/\sqrt{1-u^2}$$

where $\mathbf{u}$ is the electron's velocity, κ is a positive constant, $\mathbf{p} = \mathbf{u}E$ the momentum and $E^2 = (m^2 - \kappa^2)/(1-u^2)$ where E is the energy of the electron.

Consider now a frame of inertia in which the electron is at rest. Then $\mathbf{u} = 0$, $\mathbf{p} = 0$, $\mathbf{r}' = r$.

Following Koga's assumption that the electron's origin is, in some way, its "centre", a solution to the Klein-Gordon equation can be written as $\varphi = ae^{SI\sigma_3}$ (See [12]). This is because $(I\sigma_3)^2 = -1$ in STA.

By substituting $I\sigma_3$ for i the x^3-axis receives a unique role. Hestenes had the notion to substitute $I\sigma_3$ for $i = \sqrt{-1}$. Therefore φ takes the form

$$\varphi = a(\cos S + (\sin S)I\sigma_3)$$

where $a \cos S$ and $a \sin S$ are scalars, however $I\sigma_3 = \gamma_2\gamma_1$ is a bivector, hence its value is even.

These ideas can be utilised to solve the Dirac-Hestenes equation. If φ is a solution to the Klein-Gordon equation and $\psi = \nabla\varphi I\sigma_3 + m\varphi\gamma_0$, then ψ is also a solution to the Dirac-Hestenes equation. When φ is odd, ψ is even and vice versa.

We want ψ to be even because of the correspondence mentioned earlier between complex 4-space and the even subalgebra of STA.

In order to make ψ to be even, an odd solution φ of the Klein-Gordon equation can be taken as

$$\varphi = ae^{SI\sigma_3}\gamma_0$$

A solution ψ to the Dirac-Hestenes equation can be derived using this solution. Using 4-dimensional gradient, we write

$$\begin{aligned}\nabla\varphi &= \nabla\left(ae^{SI\sigma_3}\gamma_0\right)\\ &= (\nabla a)\, e^{SI\sigma_3}\gamma_0 + a\ \nabla(e^{SI\sigma_3})\gamma_0\\ &= a\mathbf{r}\left(\frac{1}{r^2}+\frac{\kappa}{r}\right)e^{SI\sigma_3}\gamma_0 + a(\nabla S)e^{SI\sigma_3}I\sigma_3\ \gamma_0,\end{aligned}$$

since $a = e^{-\kappa r}/r$ where $r = \sqrt{x^2+y^2+z^2}$ and $S = -Et$

Then

$$\begin{aligned}\nabla\psi = \psi &= \nabla\varphi I\sigma_3 + m\varphi\gamma_0\\ &= \left(a\mathbf{r}\left(\frac{1}{r^2}+\frac{\kappa}{r}\right)e^{SI\sigma_3}\gamma_0 + a(\nabla S)e^{SI\sigma_3}I\sigma_3\ \gamma_0\right)I\sigma_3 + m\left(ae^{SI\sigma_3}\gamma_0\right)\\ &= \mathbf{R}\varphi I\sigma_3 + (E+m)\varphi\gamma_0\end{aligned}$$

where $\mathbf{R} = \mathbf{r}(\frac{1}{r^2}+\frac{\kappa}{r})$

The reader can make use of [11, 12] for the correspondence used above and the consequences of the solution as well.

Koga investigated the properties of his solution in [4–6, 9] and claimed that there is a rotational field in the solution. Panday and Chakravarti studied his claim using geometric algebra in [12].

References

1. Dirac, P.A.M.: Principles of Quantum Mechanics, 4th edn. Oxford University Press, Oxford (1958)
2. Koga, T.: The motion of wavelets—an interpretation of the Schrödinger equation. Int. J. Theor. Phys. **2**, 49–78 (1972)
3. Koga, T.: Tunnel effect in the motion of deformable wavelets. Found. Phys. **4**, 261–266 (1974)
4. Koga, T.: Representation of the spin in the Dirac equation for the electron. Int. J. Theor. Phys. **12**, 205–215 (1975)
5. A rational interpretation of the Dirac equation for the electron. Int. J. Theor. Phys. **13**, 271–278 (1975)
6. Koga, T.: A relation between the Dirac field of the electron and electromagnetic fields. Int. J. Theor. Phys. **13**, 377–385 (1975)
7. Koga, T.: A relativistic field theory of the electron. Int. J. Theor. Phys. **15**, 99–119 (1976)
8. Koga, T.: Foundation of Quantum Physics. Wood & Jones, Pasadena (1980)
9. Koga, T.: Inquiries into Foundations of Quantum Physics. Wood & Jones, Pasadena (1983)
10. Hestenes, D.: Space-Time Algebra. Gordon and Breach, New York (1966)
11. Doran, C.J.L., Lasenby, A.N.: Geometric Algebra for Physicists. Cambridge University Press (2003)
12. Panday, S.K., Chakravarti, R.S.: The Dirac equation: an approach through geometric algebra. Annales de la Fondation Louis de Broglie **34**, 223–228 (2009)
13. Didimos, K.V., Chakravarti, R.S.: Deterministic electron spin. Annales de la Fondation Louis de Broglie **42**, 299–306 (2017)

Resolvent Algebra in Fock-Bargmann Representation

Wolfram Bauer and Robert Fulsche

Abstract The resolvent algebra $\mathfrak{R}(X, \sigma)$ associated to a symplectic space (X, σ) was introduced by D. Buchholz and H. Grundling as a convenient model of the canonical commutation relation (CCR) in quantum mechanics. We first study a representation of $\mathfrak{R}(\mathbb{C}^n, \sigma)$ with the standard symplectic form σ inside the full Toeplitz algebra over the Fock-Bargmann space. We prove that $\mathfrak{R}(\mathbb{C}^n, \sigma)$ itself is a Toeplitz algebra. In the sense of R. Werner's *correspondence theory*, we determine its corresponding shift-invariant and closed space of symbols. Finally, we discuss a representation of the resolvent algebra $\mathfrak{R}(\mathfrak{H}, \tilde{\sigma})$ for an infinite dimensional symplectic separable Hilbert space $(\mathfrak{H}, \tilde{\sigma})$. More precisely, we find a representation of $\mathfrak{R}(\mathfrak{H}, \tilde{\sigma})$ inside the full Toeplitz algebra over the Fock-Bargmann space in infinitely many variables.

Keywords Correspondence theory · Toeplitz algebra · Infinite dimensional phase space

Mathematical Subject Classification MSC 2020: 47L15 · 47L80 · 81S30

1 Introduction

The classical CCR algebra (see [5]) provides a standard C^* algebraic model for the *canonical commutation relation* (CCR) in quantum mechanics. In typical applications, however, these algebras have significant disadvantages. In particular, CCR algebras in general do not contain bounded functions of the Hamiltonian (physical *observables*). Only in rare and physically less relevant cases they are stable under

W. Bauer (✉) · R. Fulsche
Institut für Analysis, Welfengarten 1, 30167 Hannover, Germany
e-mail: bauer@math.uni-hannover.de

R. Fulsche
e-mail: fulsche@math.uni-hannover.de

A. A. Ambily and V. B. Kiran Kumar (eds.), *Semigroups, Algebras and Operator Theory*, Springer Proceedings in Mathematics & Statistics 436,
https://doi.org/10.1007/978-981-99-6349-2_12

dynamics. More precisely, it was noticed in [11] that time evolution of the Hamiltonian $\boldsymbol{H} := -\Delta + V$ on $L^2(\mathbb{R}^n)$ in the standard Schrödinger representation gives rise to $*$-automorphisms of the CCR algebra over the standard symplectic space $\mathbb{R}^{2n} \cong T^*\mathbb{R}^n$ only in the trivial case of an identically vanishing potential function V. To resolve such problems, D. Buchholz and H. Grundling have proposed a new approach to CCR by introducing the *resolvent algebra* $\mathfrak{R}(X, \sigma)$ associated to a symplectic space (X, σ) in [7]. This algebra abstractly is defined through certain algebraic relations that mimic properties of the resolvents of the (i.g. unbounded) *canonical operators* (see (1)).

In the present paper, we first consider complex n-space $\mathbb{C}^n \cong \mathbb{R}^{2n}$ equipped with the standard symplectic structure. As is well-known, in this setting the CCR algebra is identified with the C^* algebra generated by Toeplitz operators with symbols in the space TP of trigonometric polynomials. It is an interesting observation that this Toeplitz C^* algebra coincides with the linear closure of Toeplitz operators with symbols in TP (see [8] for a precise statement and further extensions of these results). In a similar manner, our goal is to study the resolvent algebra in the Fock-Bargmann representation and to describe it as a concrete algebra generated by Toeplitz operators with shift-invariant symbol space. Being a shift-invariant C^* algebra, it has turned out that a convenient mathematical framework for the analysis of the resolvent algebra is R. Werner's *quantum harmonic analysis* [20] and its extension by R. Fulsche in [13] to Toeplitz operator theory on the Fock-Bargmann space. A particularly useful tool is the *correspondence theorem* (Theorem 2) in [13, 20] which ensures the existence and uniqueness of a closed and shift-invariant space of bounded uniformly continuous functions on $\mathbb{C}^n$, which are the symbols of Toeplitz operators linearly generating the resolvent algebra.

In dimension $n = 1$, we show that the space corresponding to the resolvent algebra $\mathfrak{R}(\mathbb{C}, \sigma)$ is the uniform closure of the classical resolvent functions. However, in higher dimensions $n > 1$ we only can prove a weaker result (see Proposition 6). Nevertheless, we show that $\mathfrak{R}(\mathbb{C}^n, \sigma)$ coincides with a C^* algebra generated by a concrete set of Toeplitz operators with bounded symbols. It remains an open problem whether or not the closure of classical resolvent functions forms the corresponding space to the resolvent algebra in every complex dimension $n \in \mathbb{N}$ and in the sense of Theorem 2.

In the second part of the paper, we replace $\mathbb{C}^n$ by a separable infinite dimensional complex symplectic Hilbert spaces $(\mathfrak{H}, \tilde{\sigma})$. In [15, 16] the authors have proposed two definitions of Toeplitz operators on the infinitely many variable Fock-Bargmann space. Here, we follow the measure theoretical approach and consider C^* algebras generated by Toeplitz operators over $\mathfrak{H}$. Due to the non-nuclearity of $\mathfrak{H}$ (as well as of the space of entire functions over $\mathfrak{H}$ with the compact-open topology) as topological vector spaces and caused by features of the measure theory on infinite dimensional spaces, a variety of new effects can be observed in the theory of Toeplitz operators. In particular, in this setting a *correspondence theorem* so far is unknown, even though it may exist in a suitable formulation. Our main result of the last section (Theorem 9) states that there is a representation of the resolvent algebra corresponding to a symplectic Hilbert space $\mathfrak{H}_{1/2}$ with Hilbert-Schmidt embedding $\mathfrak{H}_{1/2} \hookrightarrow \mathfrak{H}$ inside the full Toeplitz algebra over $\mathfrak{H}$.

The structure of the paper is as follows: In Sect. 2 we recall the definition of the CCR and resolvent algebra [7]. We restate a well-known representation of the CCR algebra associated to $\mathbb{C}^n$ with the standard symplectic structure as a C^* algebra generated by Toeplitz operators.

Section 3 provides some basic facts on Fock-Bargmann spaces $F_{\mathfrak{t}}^2$ over $\mathbb{C}^n$ and Toeplitz operators acting on $F_{\mathfrak{t}}^2$. We define the Berezin transform and then express Weyl operators as Toeplitz operators with bounded symbols. Most of the material is standard and further details on the role of Toeplitz operators in quantum mechanics can be found in [3, 8].

Section 4 describes the Fock-Bargmann representation of the resolvent algebra. We present the *correspondence theorem* [20] in this setting (see [13]). In passing, we mention some applications to the analysis of Toeplitz operators on Fock-Bargmann spaces. We discuss the unique shift-invariant closed symbol space corresponding to the resolvent algebra. An essential ingredient to the analysis is an integral form of the Berezin transform on products of resolvents (Proposition 3). Moreover, in our proofs we essentially use a specific compactification of the *affine Grassmannian* which was introduced in [19].

Finally, in Sect. 5, we discuss the resolvent algebra associated to an infinite dimensional symplectic separable Hilbert space in the framework of Toeplitz operators on the Fock-Bargmann space of Gaussian square integrable entire functions in infinitely many variables.

Acknowledgement: We thank Prof. R. F. Werner for many enlightening discussions on *quantum harmonic analysis*, the *correspondence theorem* and the structure of the *resolvent algebra*. We thank the anonymous reviewer for helpful comments and corrections.

2 CCR and Resolvent Algebra

Let (X, σ) be a symplectic vector space. Consider a Hilbert space $\mathfrak{H}$ and a real linear map ϕ from (X, σ) into the space of (unbounded) self-adjoint operators on $\mathfrak{H}$. Let us assume that all operators $\phi(f)$ are essentially self-adjoint on a common domain $\mathfrak{D}$ which forms a core for each $\phi(f)$. Recall that the *canonical commutation relation* (CCR) have the form:

$$\left[\phi(f), \phi(g)\right] = i\sigma(f, g), \quad f, g \in X. \tag{1}$$

We call $\phi(f)$ and $\phi(g)$ *canonical operators*. Especially in an algebraic setup, the analysis of unbounded operators involves some difficulties (compare e.g. [17, Chapters VIII.5 and VIII.6]). Therefore, one may look for C^* algebraic models which suitably encode (CCR). One standard idea consists in replacing $\phi(f)$ above by suitable bounded functions of $\phi(f)$. A classical approach due to H. Weyl amounts in considering the *CCR algebra* generated by the unitary operators $\exp(i\phi(f))$, $f \in X$. The modified CCR are:

$$\exp\big(i\phi(f)\big)\exp\big(i\phi(g)\big) = e^{-i\sigma(f,g)}\exp\big(i\phi(f+g)\big)$$

and self-adjointness of $\phi(f)$ implies that $\exp(i\phi(f))^* = \exp(i\phi(-f))$.

More abstractly, one is interested in the C^* algebra CCR (X, σ) generated by the relations

$$\begin{aligned} W(f)\,W(g) &= e^{-i\sigma(f,g)}W(f+g), \qquad f, g \in X \\ W(f)^* &= W(-f). \end{aligned}$$

Such C^* algebras are usually called *CCR algebras*, see [5].

With $z, w \in \mathbb{C}^n$ put $\langle w, z\rangle := w_1\overline{z}_1 + \ldots + w_n\overline{z}_n$ and consider the standard symplectic space $(\mathbb{C}^n, \sigma)$ with symplectic form

$$\sigma(w, z) := \operatorname{Im}\langle w, z\rangle.$$

One obtains a representation of CCR $(\mathbb{C}^n, \sigma)$ as operators acting on the Fock-Bargmann space F_1^2 (see [8] and Sect. 3 for precise definitions). In fact, with our previous notation, we put:

$$\phi(z) := T_{2\sigma(\cdot,z)},$$

where T_f is a Toeplitz operator with complex valued symbol f on $\mathbb{C}^n$. The corresponding CCR algebra is generated by the unitary *Weyl operators* (6) below which satisfy

$$W_z = \exp\big(iT_{2\sigma(\cdot,z)}\big).$$

In other words, $iT_{2\sigma(\cdot,z)}$ is the generator of the unitary one-parameter group $(W_{tz})_{t\in\mathbb{R}}$. Hence,

$$\text{CCR}\big(\mathbb{C}^n, \sigma\big) \cong C^*\big(W_z \ : \ z \in \mathbb{C}^n\big). \tag{2}$$

The algebra CCR $(\mathbb{C}^n, \sigma)$ is a classical object, which in a more general setup is discussed in [5]. In particular, the Fock-Bargmann representation of CCR $(\mathbb{C}^n, \sigma)$ maps into the *full Toeplitz algebra* $\mathfrak{T}$, the C^* algebra generated by Toeplitz operators with arbitrary bounded measurable symbols.

Theorem 1 (L. A. Coburn [8]) *It holds*

$$\text{CCR}\big(\mathbb{C}^n, \sigma\big) \cong \overline{\{T_\phi \ : \ \phi \in TP\}},$$

where TP *is the space of trigonometric polynomials:*

$$TP := \operatorname{span}\big\{w \mapsto \exp\big(i\sigma(w, z)\big) \ : \ z \in \mathbb{C}^n\big\} \subset L^\infty\big(\mathbb{C}^n\big).$$

Here we write $\overline{\mathfrak{M}}$ *for the operator norm closure of a given set* $\mathfrak{M} \subset \mathfrak{L}\left(F_1^2\right)$.

As was mentioned in the introduction, there are certain drawbacks of using CCR algebras as quantum mechanical models. A new algebraic framework was suggested in [6, 7] which partly overcomes the above mentioned obstructions. Therein, the authors consider the *resolvent algebra*, which is the C^* algebra generated by the resolvents of the (in general unbounded) operators $\phi(f)$.

Definition 1 (*D. Buchholz, H. Grundling* [7]) Given a symplectic space (X, σ) we define $\mathfrak{R}_0(X, \sigma)$ as the universal unital $*$-algebra generated by the set $\{R(\lambda, f) : \lambda \in \mathbb{R} \setminus \{0\},\ f \in X\}$ together with the relations:

(1) $R(\lambda, 0) = -\frac{i}{\lambda}1$,
(2) $R(\lambda, f)^* = R(-\lambda, f)$,
(3) $\nu R(\nu\lambda, \nu f) = R(\lambda, f)$,
(4) $R(\lambda, f) - R(\mu, f) = i(\mu - \lambda) R(\lambda, f) R(\mu, f)$,
(5) $[R(\lambda, f), R(\mu, g)] = i\sigma(f, g) R(\lambda, f) R(\mu, g)^2 R(\lambda, f)$,
(6) $R(\lambda, f) R(\mu, g) = R(\lambda + \mu, f + g)\left[R(\lambda, f) + R(\mu, g) + i\sigma(f, g) R(\lambda, f)^2 R(\mu, g)\right]$,

for $\lambda, \mu, \nu \in \mathbb{R} \setminus \{0\}$ and $f, g \in X$.

The *resolvent algebra* $\mathfrak{R}(X, \sigma)$ is defined as the closure of $\mathfrak{R}_0(X, \sigma)$ with respect to a certain semi-norm obtained through the GNS construction (cf. [7]).

Remark 1 1. Note that Eqs. (1) and (2) encode $\phi(0) = 1$ and the self-adjointness of $\phi(f)$, respectively. The third and sixth equation reflect the $\mathbb{R}$-linearity of ϕ, whereas Equation (4) is the resolvent identity. Finally, the fifth equation is the substitute for the (CCR).
2. When one considers a representation of $\mathfrak{R}_0(X, \sigma)$ as a concrete $*$-algebra generated by resolvents of self-adjoint operators on a Hilbert space, then $\mathfrak{R}(X, \sigma)$ coincides with the operator norm closure of $\mathfrak{R}_0(X, \sigma)$.

In the present paper we study the *Fock-Bargmann representation* of the resolvent algebra over the standard symplectic space $\mathbb{C}^n \cong \mathbb{R}^{2n}$ as well as on an infinite dimensional Hilbert space with a symplectic structure. In the finite dimensional setting of $\mathbb{C}^n$ it is well-known (cf. [3, 8] and Sect. 3) that the generators are resolvents of self-adjoint Toeplitz operators $T_{2\sigma(\cdot,z)}$, $z \in \mathbb{C}^n$ defined below.

Therefore, the resolvent algebra $\mathfrak{R}(\mathbb{C}^n, \sigma)$ is the C^* algebra generated by the resolvents of Toeplitz operators:

$$\mathfrak{R}\left(\mathbb{C}^n, \sigma\right) \cong C^*\Big(\left(i\lambda - T_{2\sigma(\cdot,z)}\right)^{-1};\ \lambda \in \mathbb{R} \setminus \{0\}, z \in \mathbb{C}^n\Big). \tag{3}$$

In fact, this will be the starting point of our analysis.

3 Fock-Bargmann Space and Toeplitz Operators

Let $n \in \mathbb{N}$ and consider positive real numbers $t_j > 0$ where $j = 1, \ldots, n$. We write $\mathbf{t} := (t_1, \ldots, t_n)$ and on $\mathbb{C}^n$ we define the probability measure $\mu_{\mathbf{t}}$ by

$$d\mu_{\mathbf{t}}(z) = \frac{1}{\pi^n t_1 \cdot \ldots \cdot t_n} e^{-\left(\frac{|z_1|^2}{t_1} + \ldots + \frac{|z_n|^2}{t_n}\right)} dV(z), \qquad z = (z_1, \ldots, z_n) \in \mathbb{C}^n.$$

Here, V denotes the Lebesgue measure on $\mathbb{C}^n \cong \mathbb{R}^{2n}$. The Fock-Bargmann space $F_{\mathbf{t}}^2 = F_{\mathbf{t}}^2(\mathbb{C}^n)$ is defined as

$$F_{\mathbf{t}}^2 := L^2\left(\mathbb{C}^n, \mu_{\mathbf{t}}\right) \cap \operatorname{Hol}\left(\mathbb{C}^n\right), \tag{4}$$

where $\operatorname{Hol}(\mathbb{C}^n)$ denotes the space of entire functions on $\mathbb{C}^n$. The reader experienced with the analysis on such spaces may wonder why we introduce the parameters $t_1, \ldots, t_n$ in the above definition, as the standard approach corresponds to the choice $t_1 = t_2 = \cdots = t_n =: t > 0$. We will use the notation $F_t^2 = F_{(t,\ldots,t)}^2$ when we are in this standard situation. Conceptually, the more general setup does not cause any problems (apart from longer notations), but will be convenient when we pass to infinite dimensional symplectic spaces.

Throughout the paper we denote the standard inner product of $L^2(\mathbb{C}^n, \mu_{\mathbf{t}})$ and of $F_{\mathbf{t}}^2$ by

$$\langle f, g \rangle := \int_{\mathbb{C}^n} f(z) \overline{g(z)} d\mu_{\mathbf{t}}(z)$$

and we write $\|f\| := \langle f, f \rangle^{1/2}$ for the induced norm. As is well-known, $F_{\mathbf{t}}^2$ is a reproducing kernel Hilbert space with kernel function

$$K_w^{\mathbf{t}}(z) := K^{\mathbf{t}}(z, w) := e^{\frac{z_1 \cdot \overline{w_1}}{t_1} + \ldots + \frac{z_n \cdot \overline{w_n}}{t_n}},$$

where $w = (w_1, \ldots, w_n) \in \mathbb{C}^n$. In order to simplify notations, we write

$$\langle z, w \rangle_{\mathbf{t}} := \frac{z_1 \cdot \overline{w_1}}{t_1} + \cdots + \frac{z_n \cdot \overline{w_n}}{t_n} \quad \textit{and} \quad \|z\|_{\mathbf{t}}^2 := \langle z, z \rangle_{\mathbf{t}} \tag{5}$$

such that $K_w^{\mathbf{t}}(z) = e^{\langle z, w \rangle_{\mathbf{t}}}$. We express the *normalized reproducing kernels* $k_w^{\mathbf{t}} \in F_{\mathbf{t}}^2$ as

$$k_w^{\mathbf{t}}(z) := \frac{K_w^{\mathbf{t}}(z)}{\|K_w^{\mathbf{t}}\|} = e^{\langle z, w \rangle_{\mathbf{t}} - \frac{1}{2}\|w\|_{\mathbf{t}}^2}.$$

Let $A \in \mathfrak{L}\left(F_{\mathbf{t}}^2\right)$ be a bounded linear operator on $F_{\mathbf{t}}^2$. We define the *Berezin transform* $\widetilde{A}$ of A by

$$\widetilde{A}(z) := \langle A k_z^{\mathbf{t}}, k_z^{\mathbf{t}} \rangle, \qquad z \in \mathbb{C}^n.$$

Note that $\widetilde{A}(z)$ is a bounded real-analytic function. Moreover, the map $A \mapsto \widetilde{A}$ is known to be injective, [12].

As $F_{\mathbf{t}}^2$ is a closed subspace of $L^2(\mathbb{C}^n, \mu_{\mathbf{t}})$, there is an orthogonal projection $P^{\mathbf{t}} : L^2(\mathbb{C}^n, \mu_{\mathbf{t}}) \to F_{\mathbf{t}}^2$ acting on $f \in L^2(\mathbb{C}^n, \mu_{\mathbf{t}})$ as

$$P^{\mathbf{t}}(f)(w) = \langle f, K_w^{\mathbf{t}} \rangle, \quad w \in \mathbb{C}^n.$$

For a measurable function $\varphi : \mathbb{C}^n \to \mathbb{C}$, we let $T_\varphi^{\mathbf{t}}$ denote the *Toeplitz operator* with symbol φ, which is defined by

$$T_\varphi^{\mathbf{t}}(g) := P^{\mathbf{t}}(\varphi g)$$

on the natural domain $D\left(T_\varphi^{\mathbf{t}}\right) = \left\{g \in F_{\mathbf{t}}^2 \, : \, \varphi g \in L^2(\mathbb{C}^n, \mu_{\mathbf{t}})\right\}$.

If $\varphi \in L^\infty(\mathbb{C}^n)$, then $T_\varphi^{\mathbf{t}}$ is a bounded operator. For a given set $S \subset L^\infty(\mathbb{C}^n)$, we will denote by $\mathfrak{T}^{\mathbf{t}}(S)$ the C^* algebra generated by all Toeplitz operators with symbols in S and set

$$\mathfrak{T}^{\mathbf{t}} := \mathfrak{T}^{\mathbf{t}}\left(L^\infty(\mathbb{C}^n)\right)$$

for the *full Toeplitz algebra*. We also define $\mathfrak{T}_{\text{lin}}^{\mathbf{t}}(S)$ to be the closed linear span of Toeplitz operators with symbols in S.

Let us introduce *Weyl operators* on the Fock-Bargmann space $F_{\mathbf{t}}^2$: if $z \in \mathbb{C}^n$ then we define $W_z^{\mathbf{t}} \in \mathfrak{L}\left(F_{\mathbf{t}}^2\right)$ by

$$W_z^{\mathbf{t}}(g)(w) = k_z^{\mathbf{t}}(w)\, g(w - z). \tag{6}$$

Weyl operators are well-known to be unitary. Moreover, they fulfill the relation

$$\left(W_z^{\mathbf{t}}\right)^* = \left(W_z^{\mathbf{t}}\right)^{-1} = W_{-z}^{\mathbf{t}} \quad \textit{and} \quad W_z^{\mathbf{t}} W_w^{\mathbf{t}} = e^{-i\sigma_{\mathbf{t}}(z,w)} W_{z+w}^{\mathbf{t}}.$$

Here, the symplectic form $\sigma_{\mathbf{t}}$ on $\mathbb{C}^n$ with parameter $\mathbf{t}$ is defined by

$$\sigma_{\mathbf{t}}(w, z) := \frac{\operatorname{Im}(w_1 \cdot \overline{z_1})}{t_1} + \cdots + \frac{\operatorname{Im}(w_n \cdot \overline{z_n})}{t_n} = \operatorname{Im}\langle w, z\rangle_{\mathbf{t}}.$$

As a matter of fact, Weyl operators are Toeplitz operators themselves. If we define the family $\left(g_z^{\mathbf{t}}\right)_{z \in \mathbb{C}^n}$ of bounded functions on $\mathbb{C}^n$ by

$$g_z^{\mathbf{t}}(w) := e^{\frac{1}{2}\|z\|_{\mathbf{t}}^2 + 2i\sigma_{\mathbf{t}}(w,z)},$$

then it holds

$$W_z^{\mathbf{t}} = T_{g_z^{\mathbf{t}}}^{\mathbf{t}}. \tag{7}$$

In fact, (7) follows by showing that the Berezin transform of both sides coincide (see [3, 8] and the references therein).

4 Resolvent Algebra in Fock-Bargmann Representation

In this section, we study the resolvent algebra $\mathfrak{R}(\mathbb{C}^n, \sigma_{\mathbf{t}})$ in its Fock-Bargmann representation. Of course, it is not hard to reduce the analysis of $\mathfrak{R}(\mathbb{C}^n, \sigma_{\mathbf{t}})$ to that of $\mathfrak{R}(\mathbb{C}^n, \sigma)$. Again, we emphazise that the extra flexibility coming from the parameter set $\mathbf{t} = (t_1, \ldots, t_n)$ will be useful when passing to the infinite dimensional limit $n \to \infty$. Therefore, we take the (mostly notational) burden upon us to carry $\mathbf{t}$ all the way through this section. For readability, we will denote the Fock-Bargmann representation of the resolvent algebra also by $\mathfrak{R}(\mathbb{C}^n, \sigma_{\mathbf{t}})$. Since the Toeplitz operators $T^{\mathbf{t}}_{2\sigma_{\mathbf{t}}(\cdot,z)}$ fulfill the canonical commutation relation (equivalently, the Weyl operators $W^{\mathbf{t}}_z$ fulfill the relation of the CCR algebra), the resolvent algebra is the C^* algebra generated by resolvents of Toeplitz operators. The following integral representations allow us to study $\mathfrak{R}(\mathbb{C}^n, \sigma_{\mathbf{t}})$ more in detail:

Lemma 1 *For $z \in \mathbb{C}^n \setminus \{0\}$ the map $\mathbb{R} \ni s \mapsto W^t_{sz}$ defines a strongly continuous unitary one-parameter group with generator $iT^t_{2\sigma_t(\cdot,z)}$. In particular, for $\lambda > 0$ the following integral representations of the resolvents hold in the strong sense:*

$$\left(T^t_{2\sigma_t(\cdot,z)} + i\lambda\right)^{-1} = \left(T^t_{2\sigma_t(\cdot,z)+i\lambda}\right)^{-1} = -i\int_0^\infty e^{-\lambda s} W^t_{sz} ds \tag{8}$$

and

$$\left(T^t_{2\sigma_t(\cdot,z)} - i\lambda\right)^{-1} = \left(T^t_{2\sigma_t(\cdot,z)-i\lambda}\right)^{-1} = i\int_0^\infty e^{-\lambda s} W^t_{-sz} ds. \tag{9}$$

Proof It is easy to check that $\mathbb{R} \ni s \mapsto W^{\mathbf{t}}_{sz}$ is indeed a strongly continuous unitary one-parameter group. To determine its generator we compute

$$\lim_{s\to 0} \frac{W^{\mathbf{t}}_{sz} f - f}{s} = \lim_{s\to 0} \frac{T^{\mathbf{t}}_{g^{\mathbf{t}}_{sz}} f - f}{s} = \lim_{s\to 0} T^{\mathbf{t}}_{(g^{\mathbf{t}}_{sz}-1)/s} f$$

for all $f \in F^2_{\mathbf{t}}$ such that the limit exists. According to Stone's Theorem there is a unique self-adjoint operator A with domain $D(A)$ such that iA generates the group, i.e. for $f \in D(A)$ it holds

$$\lim_{s\to 0} T^{\mathbf{t}}_{(g^{\mathbf{t}}_{sz}-1)/s} f = iAf \tag{10}$$

in the norm sense. Since norm convergence in the reproducing kernel Hilbert space $F^2_{\mathbf{t}}$ implies pointwise convergence, it suffices to determine the pointwise limit of the right hand side of (10). Let $u \in \mathbb{C}^n$ and note that:

$$T^{\mathbf{t}}_{(g^{\mathbf{t}}_{sz}-1)/s} f(u) = \int_{\mathbb{C}^n} \frac{1}{s}\left(e^{\frac{s^2}{2}\|z\|^2_{\mathbf{t}}+2is\sigma_{\mathbf{t}}(w,z)} - 1\right) e^{\langle u,w\rangle_{\mathbf{t}}} f(w)\, d\mu_{\mathbf{t}}(w).$$

The integrand converges as $s \to 0$:

$$\frac{1}{s}\left(e^{\frac{s^2}{2}\|z\|^2_{\mathbf{t}}+2is\sigma_{\mathbf{t}}(w,z)} - 1\right) \xrightarrow{s\to 0} 2i\sigma_{\mathbf{t}}(w,z).$$

An easy application of the dominated convergence theorem yields

$$T^{\mathbf{t}}_{(g^{\mathbf{t}}_{sz}-1)/s} f(u) \to iT^{\mathbf{t}}_{2\sigma_{\mathbf{t}}(\cdot,z)} f(u)$$

for those $f \in F^2_{\mathbf{t}}$ such that $w \mapsto \sigma_{\mathbf{t}}(w,z) f(w) \in L^2(\mathbb{C}^n, \mu_{\mathbf{t}})$. Hence,

$$T^{\mathbf{t}}_{(g^{\mathbf{t}}_{sz}-1)/s} f \to iT^{\mathbf{t}}_{2\sigma_{\mathbf{t}}(\cdot,z)} f = iAf \quad \textit{for} \quad f \in D(A).$$

Therefore, $iT^{\mathbf{t}}_{2\sigma_{\mathbf{t}}(\cdot,z)}$ is the generator of the one-parameter group $s \mapsto W^{\mathbf{t}}_{sz}$. Since all operators $W^{\mathbf{t}}_{sz}$ are unitary, the group $\left(W^{\mathbf{t}}_{sz}\right)_s$ has growth bound $\omega_0 = 0$. For $\lambda > 0$ it follows (cf. [10, Theorem I.1.10])

$$\left(\lambda - iT^{\mathbf{t}}_{2\sigma_{\mathbf{t}}(\cdot,z)}\right)^{-1} = \int_0^\infty e^{-\lambda s} W^{\mathbf{t}}_{sz} ds$$

strongly, i.e.

$$\left(T^{\mathbf{t}}_{2\sigma_{\mathbf{t}}(\cdot,z)} + i\lambda\right)^{-1} = -i\int_0^\infty e^{-\lambda s} W^{\mathbf{t}}_{sz} ds.$$

The integral representation (9) follows from:

$$\left(T^{\mathbf{t}}_{2\sigma_{\mathbf{t}}(\cdot,z)} - i\lambda\right)^{-1} = \left(\left(T^{\mathbf{t}}_{2\sigma_{\mathbf{t}}(\cdot,z)} + i\lambda\right)^{-1}\right)^* = i\int_0^\infty e^{-\lambda s} W^{\mathbf{t}}_{-sz} ds,$$

which completes the proof. □

Remark 2 An inspection of the above arguments shows that the limit in Equation (10) exists for all $f = K^{\mathbf{t}}_w,\ w \in \mathbb{C}^n$. Moreover, one easily verifies that the space $\mathfrak{X} := \operatorname{span}\{K^{\mathbf{t}}_w : w \in \mathbb{C}^n\}$ is invariant under the action of $W^{\mathbf{t}}_{sz}$. Then, [17, Theorem VIII.10] implies that $T^{\mathbf{t}}_{2\sigma_{\mathbf{t}}(\cdot,z)}$ is essentially self-adjoint on $\mathfrak{X}$.

Define for $z \in \mathbb{C}^n,\ \lambda \in \mathbb{C} \setminus i\mathbb{R}$:

$$R_{\mathbf{t}}(\lambda, z) := \left(T^{\mathbf{t}}_{2\sigma_{\mathbf{t}}(\cdot,z)} - i\lambda\right)^{-1}.$$

For simplicity, we will occasionally suppress $\mathbf{t}$ in the notation and shortly write $R(\lambda, z) = R_{\mathbf{t}}(\lambda, z)$. Before we continue with our investigation, we need to recall the following well-known expansion of the resolvent.

Lemma 2 *Let $\lambda_0, \lambda \in \mathbb{C} \setminus i\mathbb{R}$ such that $|\lambda_0 - \lambda| < |\lambda_0|$. Then, it holds*

$$R(\lambda, z) = \sum_{k=0}^{\infty} (\lambda - \lambda_0)^k i^k R(\lambda_0, z)^{k+1}, \tag{11}$$

where the series converges in operator norm. In particular:

$$R(\lambda_0, z)^k = \frac{i^{k-1}}{(k-1)!} \frac{\mathrm{d}^{k-1}}{\mathrm{d}\lambda^{k-1}} |_{\lambda=\lambda_0} R(\lambda, z). \tag{12}$$

The previous lemma has the following important consequence:

Corollary 1 $\mathfrak{R}(\mathbb{C}^n, \sigma_t) = C^*(R_t(\lambda, z) : \lambda \in \mathbb{C} \setminus i\mathbb{R}, z \in \mathbb{C}^n)$.

We further note that Eqs. (8) and (9) extend to $\lambda \in \mathbb{C} \setminus i\mathbb{R}$ with the same proof as in the case $\lambda \in \mathbb{R} \setminus \{0\}$:

$$R(\lambda, z) = i \int_0^{\infty} e^{-\lambda s} W^{\mathbf{t}}_{-sz} ds, \quad \mathrm{Re}(\lambda) > 0 \tag{13}$$

and

$$R(\lambda, z) = -i \int_0^{\infty} e^{\lambda s} W^{\mathbf{t}}_{sz} ds, \quad \mathrm{Re}(\lambda) < 0. \tag{14}$$

We mention two ways of analyzing the connection between the algebra $\mathfrak{R}(\mathbb{C}^n, \sigma_{\mathbf{t}})$ in its Fock-Bargmann space representation and the theory of Toeplitz operators and, more precisely, its realization as a Toeplitz algebra. The computational-heavy method studies the resolvents through their Laplace transform representation, using that the Weyl operators themselves are Toeplitz operators. Alternatively, one can use "soft analysis" arguments from the theory of Toeplitz operators, most prominently those arising in the theory of *quantum harmonic analysis* [13, 20]. Both approaches have their advantages and so we will try to shed some light on either of them.

4.1 Correspondence Theory in the Fock-Bargmann Space

We recall some aspects of *quantum harmonic analysis* in the setting of the Fock-Bargmann space and explain a particular part of it: the *correspondence theory*. We refer to [13, 20] for further details. Roughly speaking, *correspondence theory* relates certain subspaces of $\mathrm{BUC}(\mathbb{C}^n)$, the C^* algebra of bounded, uniformly continuous functions on $\mathbb{C}^n$, with corresponding subspaces of $\mathfrak{L}(F^2_{\mathbf{t}})$. We start with some notation. Let $f : \mathbb{C}^n \to \mathbb{C}$ be a function and $z \in \mathbb{C}^n$, then put:

$$\alpha_z(f)(w) := f(w - z).$$

Given an operator $A \in \mathfrak{L}(F_{\mathbf{t}}^2)$ we define its shift by $z \in \mathbb{C}^n$ through

$$\alpha_z(A) = W_z^{\mathbf{t}} A W_{-z}^{\mathbf{t}} = W_z^{\mathbf{t}} A \left(W_z^{\mathbf{t}}\right)^*,$$

where $W_z^{\mathbf{t}}$ is a Weyl operator. Clearly, BUC $(\mathbb{C}^n)$ is the subalgebra of $L^\infty(\mathbb{C}^n)$ consisting of functions f for which $z \mapsto \alpha_z(f)$ is continuous with respect to the L^∞-norm. The analogous space on the operator side is

$$\mathfrak{C}_1^{\mathbf{t}} = \left\{A \in \mathfrak{L}(F_{\mathbf{t}}^2) : z \mapsto \alpha_z(A) \text{ is } \|\cdot\|_{op}\text{-continuous}\right\}.$$

The Fock-Bargmann space formulation of the correspondence theorem due to R. Werner in [20] specifically tailored for Toeplitz operators can be found in [13] within the standard situation $t_1 = t_2 = \cdots = t_n = t > 0$. The case of positive weight parameters $t_1, \ldots, t_n$ in the definition of the Gaussian measure $\mu_{\mathbf{t}}$ follows by identical proofs and obvious modifications:

Theorem 2 (Correspondence Theorem, [13, 20]) *Let $\mathfrak{D}_1 \subset \mathfrak{C}_1^t$ be a closed, α-invariant subspace (meaning $\alpha_z(A) \in \mathfrak{D}_1$ for every $z \in \mathbb{C}^n$, $A \in \mathfrak{D}_1$). Then, there is a unique α-invariant closed subspace $\mathfrak{D}_0$ of* BUC $(\mathbb{C}^n)$ *such that*

$$\mathfrak{D}_1 = \mathfrak{T}_{lin}^t(\mathfrak{D}_0).$$

Further, for an operator $A \in \mathfrak{C}_1^t$ the following statements are equivalent:

$$A \in \mathfrak{D}_1 \Longleftrightarrow \widetilde{A} \in \mathfrak{D}_0.$$

Finally, $\mathfrak{D}_0$ can be computed as follows: $\mathfrak{D}_0 = \overline{\{\widetilde{A} : A \in \mathfrak{D}_1\}}$. In the following we call $\mathfrak{D}_0$ and $\mathfrak{D}_1$ corresponding spaces.

Here, we used the notation

$$\mathfrak{T}_{lin}^{\mathbf{t}}(\mathfrak{D}_0) = \overline{\left\{T_f^{\mathbf{t}} : f \in \mathfrak{D}_0\right\}}.$$

The previous result is a useful tool in the study of Toeplitz operators and Toeplitz algebras when it is combined with the following:

Theorem 3 ([13]) *The following C^* algebras coincide: $\mathfrak{C}_1^{\mathbf{t}} = \mathfrak{T}^t$.*

We mention that the previous two results extend to the case of the p-Fock space $F_{\mathbf{t}}^p$ where $p \neq 2$, cf. [13, 14]. They have immediate applications such as simple proofs of a compactness characterization for operators acting on the Fock-Bargmann space. We only present the case $p = 2$.

Let $\mathfrak{K}(\mathfrak{H})$ denote the ideal of compact operators on a Hilbert space $\mathfrak{H}$.

Theorem 4 ([1]) *Let $A \in \mathfrak{L}(F_t^2)$. Then, the following are equivalent:*

$$A \in \mathfrak{K}(F_t^2) \Longleftrightarrow A \in \mathfrak{T}^t \text{ and } \widetilde{A} \in C_0(\mathbb{C}^n).$$

Here, $C_0(\mathbb{C}^n)$ denotes the space of continuous complex valued functions vanishing at infinity.

The short proofs of the last theorems based on correspondence theory can be found in [13]. We now want to demonstrate how Theorem 2 can be applied in order to gain a better understanding of the resolvent algebra. First, we note:

Lemma 3 *Let $z, w \in \mathbb{C}^n$ and $\lambda \in \mathbb{C} \setminus i\mathbb{R}$. Then:*

$$\alpha_w\big(R(\lambda, z)\big) = R\big(\lambda + 2i\sigma_t(z, w), z\big).$$

Proof Applying the CCR of Weyl operators, we get for $\mathrm{Re}(\lambda) > 0$:

$$\begin{aligned}\alpha_w\big(R(\lambda, z)\big) &= W_w^{\mathbf{t}} i \int_0^\infty e^{-\lambda s} W_{-sz}^{\mathbf{t}}\, ds\, W_{-w}^{\mathbf{t}}\\ &= i\int_0^\infty e^{-\lambda s} W_w^{\mathbf{t}} W_{-sz}^{\mathbf{t}} W_{-w}^{\mathbf{t}}\, ds\\ &= i\int_0^\infty e^{-\lambda t} e^{-2is\sigma_{\mathbf{t}}(z,w)} W_{-sz}^{\mathbf{t}}\, ds\\ &= R\big(\lambda + 2i\sigma_{\mathbf{t}}(z, w), z\big).\end{aligned}$$

The case $\mathrm{Re}(\lambda) < 0$ follows analogously. □

Proposition 1 *$\mathfrak{R}(\mathbb{C}^n, \sigma_t)$ is a closed, α-invariant subspace of $\mathfrak{C}_1^t$.*

Proof The α-invariance immediately follows from the previous lemma. Further, the resolvent algebra is closed by definition. We only need to prove that it is contained in $\mathfrak{C}_1^{\mathbf{t}}$. It suffices to show that $R_{\mathbf{t}}(\lambda, z) \in \mathfrak{C}_1^{\mathbf{t}}$ for $\lambda \in \mathbb{C} \setminus i\mathbb{R}$ and $z \in \mathbb{C}^n$. This is an easy consequence of Lemma 2 and Lemma 3. In fact, for $|w|$ sufficiently small such that simultaneously

$$2|\sigma_{\mathbf{t}}(z, w)| < |\lambda| \quad \textit{and} \quad |2\sigma_{\mathbf{t}}(z, w)|\|R(\lambda, z)\| < 1$$

it follows:

$$\begin{aligned}\|R(\lambda, z) - \alpha_w(R(\lambda, z))\| &= \|R(\lambda, z) - R(\lambda + 2i\sigma_{\mathbf{t}}(z, w), z)\|\\ &\le \sum_{k=1}^{\infty} |2\sigma_{\mathbf{t}}(z, w)|^k \|R(\lambda, z)\|^{k+1}\\ &= \frac{2|\sigma_{\mathbf{t}}(z, w)|\|R(\lambda, z)\|^2}{1 - 2|\sigma_{\mathbf{t}}(z, w)|\|R(\lambda, z)\|}.\end{aligned}$$

This last expression tends to 0 as $|w| \to 0$. □

Proposition 1 and the correspondence theorem (Theorem 2) imply that there is a closed, α-invariant subspace $\mathfrak{D}_0^{\mathbf{t}}$ of $\mathrm{BUC}(\mathbb{C}^n)$ (possibly depending on the parameter tuple $\mathbf{t}$) such that

$$\mathfrak{R}\left(\mathbb{C}^n, \sigma_{\mathbf{t}}\right) = \mathfrak{T}^{\mathbf{t}}_{\text{lin}}\left(\mathfrak{D}^{\mathbf{t}}_0\right).$$

Since $\mathfrak{R}\left(\mathbb{C}^n, \sigma_{\mathbf{t}}\right)$ itself is a C^* algebra, we get:

Theorem 5 *$\mathfrak{R}\left(\mathbb{C}^n, \sigma_t\right)$ is a Toeplitz algebra. More precisely,*

$$\mathfrak{R}\left(\mathbb{C}^n, \sigma_t\right) = \mathfrak{T}^t\left(\mathfrak{D}^t_0\right) := C^*\left(T^t_f : f \in \mathfrak{D}^t_0\right)$$

for a suitable α-invariant subspace $\mathfrak{D}^t_0 \subset \mathrm{BUC}\left(\mathbb{C}^n\right)$.

Our next task is to determine the space $\mathfrak{D}^{\mathbf{t}}_0$ explicitly.

4.2 Computing $\mathfrak{D}^t_0$

Correspondence theory has the nice flavour that in several examples the corresponding spaces (in the sense of Theorem 2) are what one might naively expect; for example, $C_0\left(\mathbb{C}^n\right)$ corresponds to the compact operators $\mathfrak{K}\left(F^2_{\mathbf{t}}\right)$ and the almost periodic functions correspond to the CCR algebra. In this respect, one might hope that the space $\mathfrak{D}^{\mathbf{t}}_0$ corresponding to $\mathfrak{R}\left(\mathbb{C}^n, \sigma_{\mathbf{t}}\right)$ is:

$$\mathfrak{R}_{cl} = C^*\left(\{(\lambda - 2i\sigma_{\mathbf{t}}(\cdot, z))^{-1} \ : \ \lambda \in \mathbb{C} \setminus i\mathbb{R},\ z \in \mathbb{C}^n\}\right).$$

Note that $\mathfrak{R}_{cl}$ does not depend on the choice of the parameter set $\mathbf{t} = (t_1, \ldots, t_n)$ and we can replace $\sigma_{\mathbf{t}}$ by $\sigma = \sigma_1$ in its definition (the "contribution" of the parameters t_k can be absorbed into z).

Since $\mathfrak{R}_{cl}$ is indeed a subalgebra of $\mathrm{BUC}\left(\mathbb{C}^n\right)$, which is further α-invariant as well as invariant under the parity operation $f \mapsto f(-\cdot)$, Theorem 3.13 in [13] implies that:

$$\mathfrak{T}^{\mathbf{t}}_{lin}\left(\mathfrak{R}_{\text{cl}}\right) = \mathfrak{T}^{\mathbf{t}}\left(\mathfrak{R}_{cl}\right).$$

Therefore, $\mathfrak{D}^{\mathbf{t}}_0 = \mathfrak{R}_{cl}$ if and only if $\mathfrak{R}\left(\mathbb{C}^n, \sigma_{\mathbf{t}}\right) = \mathfrak{T}^{\mathbf{t}}\left(\mathfrak{R}_{cl}\right)$. Next, we collect some useful facts:

Proposition 2 *The following inclusions hold:*

1. *$C_0\left(\mathbb{C}^n\right) \subset \mathfrak{R}_{cl}$.*
2. *$\mathfrak{K}\left(F^2_t\right) \subset \mathfrak{R}\left(\mathbb{C}^n, \sigma_t\right)$.*

Proof The first statement is an easy application of the Stone-Weierstrass Theorem; the second statement is [7, Theorem 5.4]. □

In what follows some computations can no longer be avoided. The Berezin transform of a product of resolvents can be computed explicitly, which turns out to be quite useful. We will not need the formula in full generality, but still provide the complete expression here.

Proposition 3 *Let $m \in \mathbb{N}$ and $z_1, \ldots, z_m \in \mathbb{C}^n$. Given a multi-index $\boldsymbol{k} = (k_1, \ldots, k_m) \in \mathbb{N}^m$ and $\lambda_1, \ldots, \lambda_m \in \mathbb{C} \setminus i\mathbb{R}$ we have:*

$$\Big(R(\lambda_1, z_1)^{k_1} \ldots R(\lambda_m, z_m)^{k_m}\Big)^{\sim}(w) =$$
$$= \frac{C i^{|\boldsymbol{k}|}}{(\mathbf{k}-\mathbf{1})!} \int_{(0,\infty)^m} \boldsymbol{s}^{\mathbf{k}-\mathbf{1}} e^{-\Lambda \cdot s - 2i\sigma_t(w, s\cdot z) - i\sum_{j<\ell} s_j s_\ell \sigma_t(z_j, z_\ell) - \frac{1}{2}\|s\cdot z\|_t^2}\, d\boldsymbol{s},$$

where

$$C = C_{k,\lambda,m} := (-1)^{|\boldsymbol{k}|-m} \cdot \prod_{j=1}^{m} \operatorname{sign}\big(\operatorname{Re}(\lambda_j)\big)^{k_j} \in \{-1, 1\}.$$

Here, we have used the standard multi-index notation together with:

$$\begin{aligned}
\mathbf{1} &:= (1, \ldots, 1) \in \mathbb{N}^m, \\
\Lambda &:= \big(\operatorname{sign}(\operatorname{Re}(\lambda_1))\,\lambda_1, \ldots, \operatorname{sign}(\operatorname{Re}(\lambda_m))\,\lambda_m\big), \\
\boldsymbol{s} \cdot \boldsymbol{z} &= s_1 z_1 + \cdots + s_m z_m \in \mathbb{C}^n.
\end{aligned}$$

Proof Using the relation $R(-\lambda, z) = -R(\lambda, -z)$, we can reduce the proof to the case where $\operatorname{Re}(\lambda_j) > 0$ for $j = 1, \ldots, m$. From Lemma 2 we obtain that

$$\Big(R(\lambda_1, z_1)^{k_1} \ldots R(\lambda_m, z_m)^{k_m}\Big)^{\sim}(w) =$$
$$= \frac{i^{|\mathbf{k}|-m}}{(\mathbf{k}-\mathbf{1})!} \left\langle \frac{\partial^{\mathbf{k}-\mathbf{1}}}{\partial \mu^{\mathbf{k}-\mathbf{1}}}\Big|_{\mu=\Lambda} R(\mu_1, z_1) \ldots R(\mu_m, z_m)\, k_w^{\mathbf{t}}, k_w^{\mathbf{t}} \right\rangle,$$

where we use the short notation:

$$\frac{\partial^{\mathbf{k}-\mathbf{1}}}{\partial \mu^{\mathbf{k}-\mathbf{1}}}\Big|_{\mu=\Lambda} := \frac{\partial^{k_1-1}}{\partial \mu_1^{k_1-1}} \cdots \frac{\partial^{k_m-1}}{\partial \mu_m^{k_m-1}}\Big|_{\mu_1=\lambda_1, \ldots, \mu_m=\lambda_m}.$$

According to Lemma 3 we have:

$$\begin{aligned}
\Big(R(\mu_1, z_1) \ldots R(\mu_m, z_m)\Big)^{\sim}(w) &= \alpha_{-w}\Big(R(\mu_1, z_1) \ldots R(\mu_m, z_m)\Big)^{\sim}(0) \\
&= \Big(\alpha_{-w}\big[R(\mu_1, z_1)\big] \ldots \alpha_{-w}\big[R(\mu_m, z_m)\big]\Big)^{\sim}(0) \\
&= \Big(R\big(\mu_1 - 2i\sigma_{\mathbf{t}}(z_1, w), z_1\big) \ldots R\big(\mu_m - 2i\sigma_{\mathbf{t}}(z_m, w), z_m\big)\Big)^{\sim}(0).
\end{aligned}$$

By using analyticity of the resolvent maps $\mu_j \mapsto R(\mu_j, z_j) \in \mathfrak{L}(F_{\mathbf{t}}^2)$ it follows that the difference quotients converge in operator norm. Hence, differentiation can be interchanged with the inner product, which yields:

$$\begin{aligned}
&\frac{(\mathbf{k}-\mathbf{1})!}{i^{|\mathbf{k}|-m}}\Big(R(\lambda_1,z_1)^{k_1}\dots R(\lambda_m,z_m)^{k_m}\Big)^{\sim}(w)\\
&=\frac{\partial^{\mathbf{k}-\mathbf{1}}}{\partial\mu^{\mathbf{k}-\mathbf{1}}}|_{\mu=\Lambda}\Big(R(\mu_1,z_1)\dots R(\mu_m,z_m)\Big)^{\sim}(w)\\
&=\frac{\partial^{\mathbf{k}-\mathbf{1}}}{\partial\mu^{\mathbf{k}-\mathbf{1}}}|_{\mu=\Lambda}\underbrace{\Big(R\big(\mu_1-2i\sigma_{\mathbf{t}}(z_1,w),z_1\big)\dots R\big(\mu_m-2i\sigma_{\mathbf{t}}(z_m,w)\big)\Big)^{\sim}(0)}_{=(*)}.
\end{aligned}$$

Now, we insert the integral expression of the resolvent in (13):

$$(*)=\int_{(0,\infty)^m} e^{-\sum_j[\mu_j+2i\sigma_{\mathbf{t}}(w,z_j))]s_j-i\sum_{j<\ell}s_js_\ell\sigma_{\mathbf{t}}(z_j,z_\ell)}\langle W^{\mathbf{t}}_{\mathbf{s}\cdot z}1,1\rangle\,ds.$$

The inner product in the integrand can be calculated explicitly:

$$\langle W^{\mathbf{t}}_{\mathbf{s}\cdot z}1,1\rangle=e^{-\frac{1}{2}\|\mathbf{s}\cdot z\|^2_{\mathbf{t}}}.$$

Finally, the assertion follows by inserting this expression into the last integral and performing the μ-derivatives. □

Similarly, one computes the Berezin transform of the classical resolvent functions:

Lemma 4 *Let $m\in\mathbb{N}$, $z_1,\dots,z_m\in\mathbb{C}^n$ and $\mathbf{k}:=(k_1,\dots,k_m)\in\mathbb{N}^m$. For any set of complex numbers $\lambda_1,\dots,\lambda_m\in\mathbb{C}\setminus i\mathbb{R}$ consider the function*

$$g:=\big(\lambda_1-2i\sigma_t(\cdot,z_1)\big)^{-k_1}\dots\big(\lambda_m-2i\sigma_t(\cdot,z_m)\big)^{-k_m}.$$

With the notation in Proposition 3 the Berezin transform of g is given by:

$$\begin{aligned}
\widetilde{g}^{(t)}(w):&=\big(T^t_g\big)^{\sim}(w)\\
&=\frac{1}{(\mathbf{k}-\mathbf{1})!}\int_{(0,\infty)^m}s^{\mathbf{k}-\mathbf{1}}e^{-\Lambda\cdot s+2i\sigma_t(w,s\cdot z)-\|s\cdot z\|^2_t}\,ds.
\end{aligned}$$

Proof The lemma follows by a direct calculation. First, note that

$$\widetilde{g}^{(\mathbf{t})}(w)=\big\langle\alpha_{-w}(g)1,1\big\rangle=\int_{\mathbb{C}^n}g(v+w)\,d\mu_{\mathbf{t}}(v). \tag{15}$$

Without loss of generality we may assume that $\mathrm{Re}\,(\lambda_j)>0$ for $j=1,\dots,m$ such that g has an integral representation (Laplace transform):

$$g(w)=\frac{1}{(\mathbf{k}-\mathbf{1})!}\int_{(0,\infty)^m}\mathbf{s}^{\mathbf{k}-\mathbf{1}}e^{-\Lambda\cdot\mathbf{s}+2i\sigma_{\mathbf{t}}(w,\mathbf{s}\cdot\mathbf{z})}ds.$$

Inserting the last expression into (15) and interchanging the order of integrations shows:

$$\widetilde{g}^{(\mathbf{t})}(w) = \frac{1}{(\mathbf{k}-\mathbf{1})!}\int_{(0,\infty)^m} \mathbf{s}^{\mathbf{k}-\mathbf{1}} e^{-\Lambda\cdot\mathbf{s}+2i\sigma_{\mathbf{t}}(w,\mathbf{s}\cdot\mathbf{z})} \int_{\mathbb{C}^n} e^{2i\sigma_{\mathbf{t}}(v,\mathbf{s}\cdot\mathbf{z})}\, d\mu_{\mathbf{t}}(v)\, d\mathbf{s}.$$

The inner integration can be evaluated explicitly:

$$\int_{\mathbb{C}^n} e^{2i\sigma_{\mathbf{t}}(v,\mathbf{s}\cdot\mathbf{z})}\, d\mu_{\mathbf{t}}(v) = \left\langle K^{\mathbf{t}}_{\mathbf{s}\cdot\mathbf{z}}, K^{\mathbf{t}}_{-\mathbf{s}\cdot\mathbf{z}}\right\rangle = e^{-\|\mathbf{s}\cdot z\|_{\mathbf{t}}^2},$$

which implies the statement of the lemma. □

We now prove a first inclusion of algebras:

Lemma 5 *It holds $R(\lambda, z) \in \mathfrak{T}^t(\mathfrak{R}_{cl})$ for every $\lambda \in \mathbb{C}\setminus i\mathbb{R}$ and $z \in \mathbb{C}^n$. In particular,*

$$\mathfrak{R}\left(\mathbb{C}^n, \sigma_t\right) \subset \mathfrak{T}^t(\mathfrak{R}_{cl}).$$

Proof Since we already know that $\mathfrak{R}(\mathbb{C}^n, \sigma) \subset \mathfrak{C}_1^{\mathbf{t}}$, it suffices to show that $\widetilde{R(\lambda, z)} \in \mathfrak{R}_{cl}$ by the correspondence theorem. Using $R(\lambda, z) = -R(-\lambda, -z)$ we may assume without loss of generality that $\mathrm{Re}(\lambda) > 0$. By transformation of the integral and according to Lemma 4:

$$\begin{aligned}\widetilde{R(\lambda, z)}(w) &= -i\int_0^\infty e^{-\lambda s-2is\sigma_{\mathbf{t}}(w,z)-\frac{s^2}{2}\|z\|_{\mathbf{t}}^2}\, ds \\ &= i\sqrt{2}\int_0^\infty e^{-\lambda\sqrt{2}s-2is\sigma_{\mathbf{t}}(\sqrt{2}w,z)-s^2\|z\|_{\mathbf{t}}^2} ds \\ &= i\sqrt{2}\,\widetilde{g}^{(\mathbf{t})}(\sqrt{2}w),\end{aligned}$$

where g is a classical resolvent, namely:

$$g(w) = \left(\sqrt{2}\lambda - 2i\sigma_{\mathbf{t}}(w, -z)\right)^{-1}.$$

Note that the Berezin transform $\sim^{(\mathbf{t})}$ behaves under dilations as follows:

$$\widetilde{f}^{(\mathbf{t})}(\rho w) = \rho^2 \cdot \widetilde{f_\rho}^{(\mathbf{t}/\rho^2)}(w), \qquad \rho > 0,$$

where $f \in L^\infty(\mathbb{C}^n)$, $f_\rho(w) := f(\rho w)$ and $\mathbf{t}/\rho^2 = \left(t_1/\rho^2, \ldots, t_n/\rho^2\right)$. Hence, we obtain:

$$\widetilde{R(\lambda, z)}(w) = i2\sqrt{2}\cdot \widetilde{g_{\sqrt{2}}}^{(\mathbf{t}/2)}(w).$$

Since $g_{\sqrt{2}}$ again is a resolvent function and $\mathfrak{R}_{cl}$ is invariant under the Berezin transform $\sim^{(\mathfrak{t}/2)}$ (which is simply the convolution by an appropriate Gaussian function), the inclusion $\widetilde{R(\lambda, z)} \in \mathfrak{R}_{cl}$ follows. □

Recall that an *isotropic subspace* $V \subset \mathbb{C}^n$ is a (real) subspace such that $\sigma_{\mathfrak{t}}(z, w) = 0$ for all $z, w \in V$. Every isotropic subspace V is of real dimension $\leq n$, and if $\dim_{\mathbb{R}}(V) = n$, then V is called *Lagrangian*. To every Lagrangian subspace $V \subset \mathbb{C}^n$ there exists a complementary Lagrangian subspace $V' \subset \mathbb{C}^n$, i.e. $\mathbb{C}^n = V \oplus V'$. Indeed, one can choose $V' := \{iz : z \in V\}$ and we will make this choice in the following for convenience.

If we now fix a Lagrangian subspace V, then Proposition 3 shows that the unital C^* algebra

$$\mathfrak{R}_V := C^*\big(R(\lambda, z) : \lambda \in \mathbb{C} \setminus i\mathbb{R},\ z \in V\big)$$

is commutative. Note that $\mathfrak{R}_V$ is also α-invariant according to Lemma 3.

It is our next aim to show that $\mathfrak{R}_V$, in the sense of Theorem 2, corresponds to the space

$$\mathfrak{R}_{cl,V} := C^*\big((\lambda - 2i\sigma_{\mathfrak{t}}(\cdot, z))^{-1} : z \in V,\ \lambda \in \mathbb{C} \setminus i\mathbb{R}\big),$$

i.e.

$$\mathfrak{R}_V = \mathfrak{T}^{\mathfrak{t}}_{lin}\left(\mathfrak{R}_{cl,V}\right).$$

Before we approach this goal, note the following facts:

First, by Theorem 3.13 of [14], we have $\mathfrak{T}^{\mathfrak{t}}_{lin}\left(\mathfrak{R}_{cl,V}\right) = \mathfrak{T}^{\mathfrak{t}}\left(\mathfrak{R}_{cl,V}\right)$, i.e. it suffices to prove that $\mathfrak{R}_V = \mathfrak{T}^{\mathfrak{t}}\left(\mathfrak{R}_{cl,V}\right)$. Secondly, since $\mathfrak{R}_{cl,V}$ is also invariant under dilations $f \mapsto f(\lambda\, \cdot)$ where $\lambda > 0$, we obtain $\mathfrak{R}_V \subseteq \mathfrak{T}^{\mathfrak{t}}\left(\mathfrak{R}_{cl,V}\right)$ as in the proof of Lemma 5. Therefore, we only need to prove that $T^{\mathfrak{t}}_g \in \mathfrak{R}_V$ where g is a product of classical resolvent functions $(\lambda - 2i\sigma_{\mathfrak{t}}(\cdot, z))^{-1}$ with $z \in V$. Since the operator algebra $\mathfrak{R}_V$ is commutative, we have more techniques at hand for obtaining this goal. In particular, Gelfand theory turns out to be useful here. The first conceptual goal is therefore describing the Gelfand spectrum of $\mathfrak{R}_V$.

Using the explicit formulas for the Berezin transform, one observes that point evaluations of the Berezin transforms are multiplicative linear functionals on $\mathfrak{R}_V$. Since $\widetilde{A}\left(v + v'\right) = \widetilde{A}\left(v'\right)$ for every $v \in V$ and $v' \in V'$ and $A \in \mathfrak{R}_V$, there is no loss of generality in considering only point evaluations of the Berezin transform at V'. Since the Berezin transform is injective, we can expect the Gelfand spectrum of $\mathfrak{R}_V$ to be a suitable compactification of V'. We will describe this compactification now and start by recalling a compactification of a real inner product space first constructed in [19][1]. Therein the author described the maximal ideal space of $\mathfrak{R}_{cl}$ as such a compactification of $\mathbb{R}^{2n}$.

Let X be a finite-dimensional real inner product space. We denote by P_Y the orthogonal projection onto a given subspace $Y \subset X$. By Graff (X) we denote the *affine Grassmannian* of X, i.e. the set of all affine subspaces of X. As a set, this can be written as

$$\text{Graff}\,(X) = \left\{x + Y : \; Y \textit{ a linear subspace of } X \textit{ and } x \perp Y\right\}.$$

The precise topology with which Graff (X) is endowed can be found in [19] and will not be described here. We will denote by γX the set Graff (X) endowed with this particular topology[2]. We collect some facts about γX in the following lemma:

Lemma 6 ([19]) *Let X be a finite-dimensional real inner product space.*

1. *γX is a compact Hausdorff space.*
2. *Together with the embedding $X \ni x \mapsto x + \{0\} \in \gamma X$, γX is a compactification of X.*
3. *A net $(x_\iota + Y_\iota)_{\iota \in I} \subset \gamma X$ converges to $x + Y \in \gamma X$ if and only if the following hold:*
   - $P_{Y^\perp} x_\iota \overset{\iota \in I}{\longrightarrow} x$
   - *eventually $Y_\iota \subseteq Y$*
   - *There is no affine subspace $x' + Y' \subsetneq x + Y$ such that there exists a subnet of $(x_\iota + Y_\iota)$ with $P_{(Y')^\perp} x_\iota \to x'$ and $Y_\iota \subseteq Y'$ eventually (along the subnet).*

One of the main theorems of [19] is the following:

Theorem 6 ([19]) *The Gelfand spectrum $\mathfrak{M}\,(\mathfrak{R}_{cl})$ of $\mathfrak{R}_{cl}$ can be identified with the above compactification of $\mathbb{R}^{2n}$, i.e. $\mathfrak{M}\,(\mathfrak{R}_{cl}) \cong \gamma \mathbb{R}^{2n}$.*

Let us briefly describe the identification of $\mathfrak{M}\,(\mathfrak{R}_{cl})$ with $\gamma \mathbb{R}^{2n}$ in more detail: Given a function $f \in \mathfrak{R}_{cl}$ and an affine line $x + \text{span}\{y\} \subset \mathbb{R}^{2n}$, $y \in \mathbb{R}^{2n}$, it is not hard to verify that $\lim_{\alpha \to \infty} f\,(x + \alpha y)$ exists. For an affine subspace $x + Y \subset \mathbb{R}^{2n}$ and any $y \in Y$ with $\|y\| = 1$, the value of this limit is almost everywhere, with respect to the surface measure on $\{y \in Y : \; \|y\| = 1\}$, the same. If we denote this value by $\varphi_{x+Y}\,(f)$, then this defines a multiplicative linear functional and every element of $\mathfrak{M}\,(\mathfrak{R}_{cl})$ can be obtained in this way.

Indeed, the same can be done for $\mathfrak{R}_{cl,V}$, and the following holds true:

Proposition 4 *The Gelfand spectrum $\mathfrak{M}\left(\mathfrak{R}_{cl,V}\right)$ of $\mathfrak{R}_{cl,V}$ can be identified with the above compactification of V', i.e. $\mathfrak{M}\left(\mathfrak{R}_{cl,V}\right) \cong \gamma V'$.*

[1] To be more precise, therein it was only described for $\mathbb{R}^n$. It is straightforward to generalize the procedure to any finite dimensional real inner product space.

[2] In [19], the compactification is denoted by Ω. We chose to name it differently, as the symbol Ω is somewhat ambiguous in a symplectic context

Here, we consider V' as a real inner product space with the $\mathbf{t}$-weighted inner product induced from $\mathbb{R}^{2n} \cong \mathbb{C}^n$, i.e. $\langle z, w\rangle = \sum_{j=1}^{n} \frac{\mathrm{Re}(z\cdot\overline{w})}{t_j}$. We will not need the result in its full strength and we leave it as an exercise to adapt the arguments from [19]. It is sufficient and elementary to verify that via the embedding $V' \ni v \mapsto v + \{0\} \in \gamma V'$, every classical resolvent $(\lambda - 2i\sigma_{\mathbf{t}}(\cdot, z))^{-1}$ with $z \in V$ extends to a function in $C\left(\gamma V'\right)$.

As we will see next, the Gelfand spectrum of $\mathfrak{R}_V$ is indeed the same as the one of $\mathfrak{R}_{cl,V}$:

Proposition 5 *The Gelfand spectrum* $\mathfrak{M}(\mathfrak{R}_V)$ *can be identified with* $\mathfrak{M}(\mathfrak{R}_V) \cong \mathfrak{M}\left(\mathfrak{R}_{cl,V}\right) \cong \gamma V'$*, where all multiplicative linear functionals on* $\mathfrak{R}_V$ *have the form:*

$$\psi_{x+Y}(A) = \varphi_{x+Y}\left(\widetilde{A}|_{V'}\right).$$

Here, we denote by φ_{x+Y} *the multiplicative functional of* $\mathfrak{R}_{cl,V}$ *introduced above.*

Before proving Proposition 5, let us derive our intended result from this:

Theorem 7 $\mathfrak{R}_V = \mathfrak{T}^t_{lin}\left(\mathfrak{R}_{cl,V}\right)$.

Proof The inclusion $\mathfrak{R}_V \subset \mathfrak{T}^{\mathbf{t}}_{lin}\left(\mathfrak{R}_{cl,V}\right)$ was already stated above. Let $g \in \mathfrak{R}_{cl,V}$. Since $\mathfrak{R}_{cl,V}$ is translation invariant and closed, we conclude that $\widetilde{g}^{(\mathbf{t})} = \widetilde{T^{\mathbf{t}}_g} \in \mathfrak{R}_{cl,V} = C\left(\mathfrak{M}(\mathfrak{R}_V)\right)$. Therefore, $\widetilde{g}^{(\mathbf{t})}$ is the Gelfand transform of an operator in $\mathfrak{R}_V$. This operator must be $T^{\mathbf{t}}_g$, i.e. $T^{\mathbf{t}}_g \in \mathfrak{R}_V$. □

We will present a sequence of lemmas that lead to a proof of Proposition 5.

Lemma 7 *For any resolvent* $R(\lambda, z) \in \mathfrak{R}_V$*, where* $z \in V$*, and* $x, y \in V'$ *it holds:*

$$\widetilde{R(\lambda, z)}(x + y) = \widetilde{R(\lambda, z)}\left(x + P_{\mathrm{span}\{iz\}} y\right).$$

Proof Let $z \in V$ and first observe that $\sigma_{\mathbf{t}}(z, y) = 0$ iff $y \perp iz$ (note that $iz \in V'$ by our choice $V' = \{iz : z \in V\}$). Hence, we have

$$\begin{aligned}\widetilde{R(\lambda, z)}(x + y) &= \widetilde{R(\lambda, z)}\left(x + \left(I - P_{\mathrm{span}\{iz\}}\right) y + P_{\mathrm{span}\{iz\}} y\right)\\ &= \widetilde{R(\lambda, z)}\left(x + P_{\mathrm{span}\{iz\}} y\right),\end{aligned}$$

where the last equality follows from the formula for the Berezin transform of a resolvent in Proposition 3. □

Lemma 8 *For any resolvent* $R(\lambda, z) \in \mathfrak{R}_V$ *where* $z \in V$ *and any affine line* $x + \mathrm{span}\{y\} \in \gamma V'$ *the limit* $\lim_{\alpha\to\infty} \widetilde{R(\lambda, z)}(x + \alpha y)$ *exists and is given by*

$$\lim_{\alpha\to\infty} \widetilde{R(\lambda, z)}(x + \alpha y) = \begin{cases} \widetilde{R(\lambda, z)}(x), & \sigma_t(z, y) = 0,\\ 0, & \sigma_t(z, y) \neq 0.\end{cases} \tag{16}$$

Proof If $\sigma_{\mathfrak{t}}(z, y) = 0$, then $y \perp iz$ such that $P_{\text{span}\{iz\}}y = 0$ and the equality follows from the previous lemma. Assume that $\sigma_{\mathfrak{t}}(z, y) \neq 0$ and let $\text{Re}(\lambda) > 0$ (the case $\text{Re}(\lambda) > 0$ follows similarly). According to Proposition 3 the Berezin transform of the resolvent at $x + \alpha y$ has the value:

$$\widetilde{R(\lambda, z)}(x + \alpha y) = \int_0^\infty e^{-\lambda s - 2is\alpha\sigma_{\mathfrak{t}}(y,z) - 2is\sigma_{\mathfrak{t}}(x,z) - \frac{s^2}{2}\|z\|_{\mathfrak{t}}^2} ds.$$

We use integration by parts in order to show that the right hand side converges to 0 as $\alpha \to \infty$. With the short notation $\sigma := \sigma_{\mathfrak{t}}(y, z) \neq 0$ we have:

$$\begin{aligned}
&\widetilde{R(\lambda, z)}(x + \alpha y) \\
&= -\frac{1}{2i\alpha\sigma}\int_0^\infty \frac{\mathrm{d}}{\mathrm{d}s}\left[e^{-2is\alpha\sigma}\right] e^{-\lambda s - 2is\sigma_{\mathfrak{t}}(x,z) - \frac{s^2}{2}\|z\|_{\mathfrak{t}}^2} ds \\
&= -\frac{1}{2i\alpha\sigma}\Bigg(\underbrace{\left[e^{-2is\alpha\sigma - \lambda s - 2is\sigma_{\mathfrak{t}}(x,z) - \frac{s^2}{2}\|z\|_{\mathfrak{t}}^2}\right]_{s=0}^\infty}_{=-1} \\
&\qquad - \int_0^\infty e^{-2is\alpha\sigma}\left(-\lambda - 2i\sigma_{\mathfrak{t}}(x, z) - s\|z\|_{\mathfrak{t}}^2\right)e^{-\lambda s - 2is\sigma_{\mathfrak{t}}(x,z) - \frac{s^2}{2}\|z\|_{\mathfrak{t}}^2} ds\Bigg).
\end{aligned}$$

Since the integral

$$\int_0^\infty \left|\lambda + 2i\sigma_{\mathfrak{t}}(x, z) + s\|z\|_{\mathfrak{t}}^2\right| e^{-\lambda s - \frac{s^2}{2}\|z\|_{\mathfrak{t}}^2} ds$$

over the absolute value is finite, we obtain $\widetilde{R(\lambda, z)}(x + \alpha y) \to 0$ as $\alpha \to \infty$. □

Lemma 9 *Let $R(\lambda, z)$ with $z \in V$ be a resolvent in $\mathfrak{R}_V$. Then, its restriction $\widetilde{R(\lambda, z)}|_{V'}$ to V' extends to a continuous function on $\gamma V'$.*

Proof We first describe the values that (the extension of) $\widetilde{R(\lambda, z)}|_{V'}$ takes on $\gamma V'$ and discuss continuity in a second step.

The value of the Berezin transform at $x \cong x + \{0\}$ is given by $\widetilde{R(\lambda, z)}(x)$. At each affine line $x + \text{span}\{y\}$ the value of the extension of the Berezin transform to $\gamma V'$ is defined by the right hand side of (16). For a given affine spaces $x + Y$ with $\dim_{\mathbb{R}}(Y) \geq 2$, we distinguish two cases: If $Y \subseteq \{w \in V' : \sigma_{\mathfrak{t}}(w, z) = 0\}$, then the value of $\widetilde{R(\lambda, z)}(x + Y)$ coincides with $\widetilde{R(\lambda, z)}(x)$. Otherwise, if $Y \nsubseteq \{w \in V' : \sigma_{\mathfrak{t}}(w, z) = 0\}$, then we set it to be zero.

Note that these choices are in accordance with the definition of the multiplicative linear functional φ_{x+Y}. We explain what we mean by this only in the last example: If $Y \nsubseteq \{w \in V' : \sigma_{\mathfrak{t}}(w, z) = 0\}$, then $\{w \in Y : \sigma_{\mathfrak{t}}(w, z) = 0\}$ is a subspace of Y of dimension strictly smaller than the dimension of Y. Hence, the unit sphere of $\{w \in Y : \sigma_{\mathfrak{t}}(w, z) = 0\}$ is a zero set with respect to the surface measure of the unit sphere of Y. Hence, $\varphi_{x+Y}\left(\widetilde{R(\lambda, z)}|_{V'}\right) = 0$.

It remains to prove that the above extension of $\widetilde{R(\lambda, z)}|_{V'}$ from V' to the compactification $\gamma V'$ is in fact continuous. According to *Bourbaki's Extension Theorem*, [4, Theorem 1, p. 82], it suffices to show that

$$\widetilde{R(\lambda, z)}(x_\iota) \to \varphi_{x+Y}\big(\widetilde{R(\lambda, z)}\big) \tag{17}$$

for any net $(x_\iota)_{\iota \in I} \subset V'$ such that $x_\iota + \{0\} \to x + Y$. We distinguish several cases:

1. If $x_\iota + \{0\} \to x + \{0\}$, then (17) follows from the continuity of the Berezin transform $\widetilde{R(\lambda, z)}$ on V'.
2. If $x_\iota + \{0\} \to x + Y$ with $Y \subseteq \{w \in V' : \sigma_{\mathfrak{t}}(w, z) = 0\}$, then:

$$\begin{aligned}\widetilde{R(\lambda, z)}(x_\iota) = \widetilde{R(\lambda, z)}\big(x_\iota - P_Y x_\iota\big) = \widetilde{R(\lambda, z)}\big(P_{Y^\perp} x_\iota\big) \xrightarrow{\iota} \widetilde{R(\lambda, z)}(x) \\ = \varphi_{x+Y}\big(\widetilde{R(\lambda, z)}\big).\end{aligned}$$

3. If $x_\iota + \{0\} \to x + Y$ with $Y \nsubseteq \{w \in V' : \sigma_{\mathfrak{t}}(w, z) = 0\}$, then we conclude that

$$\widetilde{R(\lambda, z)}(x_\iota) \xrightarrow{\iota} 0 = \varphi_{x+Y}\big(\widetilde{R(\lambda, z)}\big).$$

In fact, assume that there exists a subnet, also denoted by (x_ι), such that $\widetilde{R(\lambda, z)}(x_\iota)$ is bounded away from zero. Take $w_0 \in Y$ such that $\sigma_{\mathfrak{t}}(w_0, z) \neq 0$ and put $Z := \operatorname{span}\{w_0\} \subset Y$. By $Y' \subset Y$ denote the orthogonal complement of Z in Y. We obtain a decomposition of V':

$$V' = Y^\perp \oplus Y = Y^\perp \oplus Y' \oplus Z. \tag{18}$$

(a) If $P_Z x_\iota$ was unbounded, we could pass to a subnet such that $\|P_Z x_\iota\| \to \infty$. For this subnet, one could show by using the method of stationary phase (as in the proof of Lemma 8) that $\widetilde{R(\lambda, z)}(x_\iota) \to 0$. However, this is impossible as we assumed that $\widetilde{R(\lambda, z)}(x_\iota)$ is bounded away from zero.

(b) If $P_Z x_\iota$ is bounded, we can pass to a subnet such that $P_Z x_\iota$ converges (say, to some $x_0 \in Z$). From the orthogonal decomposition (18) we see that:

$$P_{(Y')^\perp} x_\iota = P_{Y^\perp} x_\iota + P_Z x_\iota \xrightarrow{\iota} x + x_0 \in Y^\perp \oplus Z.$$

From $x_0 \in Z \subset Y$ it follows: $x + x_0 + Y' \subset x + Y$. However this contradicts the assumption $x_\iota + \{0\} \xrightarrow{\iota} x + Y$. □

Since the Berezin transform is multiplicative on V' and $A_k \to A$ in $\mathfrak{R}_V$ implies $\widetilde{A_k} \to \widetilde{A}$ uniformly, we conclude that $\widetilde{A}$ for every $A \in \mathfrak{R}_V$ extends to a continuous function on $\gamma V'$. Now we can give the proof of Proposition 5:

Proof *(Proof of Proposition 5)* We claim that the map

$$\Phi : \mathfrak{R}_V \to C\left(\gamma V'\right) : \Phi(A) = \left[(x+Y) \mapsto \varphi_{x+Y}\left(\widetilde{A}_{|V'}\right)\right]$$

is a bijective and unital homomorphism of C^* algebras. By what has been said before, Φ defines a continuous and injective unital $*$-homomorphism. We are left with verifying surjectivity. Since the range of Φ is a $*$-algebra, we only need to show that it separates points of $\gamma V'$ in order to apply the Stone-Weierstrass Theorem. Showing that the points are separated by $\Phi(\mathfrak{R}_V)$ is easy, as we can always choose the Berezin transform of a resolvent for separating two given sets. Careful inspection of the proof of the previous lemma indeed gives a guideline on how to choose the resolvent. We briefly explain this here. In what follows, we always assume that $\mathrm{Re}(\lambda) > 0$.

1. $0 + V'$ is separated from any other affine subspace in the following way: If $Y \subsetneq V'$ is a proper subspace, then let $0 \neq v \perp Y$ and consider $R(\lambda, i\alpha v)$ for $\alpha \in \mathbb{R}$. Then, α can be chosen such that

$$\Phi\big(R(\lambda, i\alpha v)\big)(v+Y) = \psi_{v+Y}\Big(R(\lambda, i\alpha v)\Big) = \widetilde{R(\lambda, i\alpha v)}(v) \neq 0.$$

However, $\psi_{0+V'}(R(\lambda, i\alpha v)) = 0$ as long as $\alpha \neq 0$. Hence the affine spaces $v + Y$ and $0 + V'$ are separated. Separating $0 + V'$ from $0 + Y$ can be done by considering $R(\lambda, iv)$.

2. Affine spaces $x_1 + Y \neq x_2 + Y$ are separated as follows: We necessarily have $Y \neq V'$ and $x_1 \neq x_2$. If $x_1 = \rho x_2$ for some $\rho \in \mathbb{R}$, then consider $R(\lambda, i\alpha x_2)$ for suitable $\alpha \in \mathbb{R}$. Otherwise, we can find $z_0 \perp Y$ such that $P_{\mathrm{span}\{z_0\}}x_1 \neq P_{\mathrm{span}\{z_0\}}x_2$. Let $z = iz_0 \in V$ such that $Y \subset \{w \in V' : \sigma_t(w, z) = 0\}$. Hence for $\alpha \in \mathbb{R} \setminus \{0\}$:

$$\psi_{x_1+Y}\big(R(\lambda, \alpha z)\big) = \widetilde{R(\lambda, \alpha z)}\big(P_{\mathrm{span}\{z_0\}}x_1\big),$$
$$\psi_{x_2+Y}\big(R(\lambda, \alpha z)\big) = \widetilde{R(\lambda, \alpha z)}\big(P_{\mathrm{span}\{z_0\}}x_2\big).$$

Now, one has to choose α accordingly such that these values are different (see the formula in Proposition 3).

3. If $Y_1 \neq Y_2$ are proper subspaces of V', then we consider two cases:

   (a) Recall that we always have $x \perp Y$ for an affine subspace $x + Y$. If $x_1 = \rho x_2$ with $\rho \in \mathbb{R}$ are real multiples of each other, then we obtain with $\alpha \in \mathbb{R}$:

$$\begin{aligned}\psi_{x_1+Y_1}\big(R(\lambda, i\alpha x_2)\big) &= \widetilde{R(\lambda, i\alpha x_2)}(\rho x_2)\\ &= \int_0^\infty e^{-\lambda s} e^{-2is\rho\alpha\|x_2\|_t^2 - \frac{s^2\alpha^2}{2}\|x_2\|_t^2}\, ds,\\ \psi_{x_2+Y_2}\big(R(\lambda, i\alpha x_2)\big) &= \widetilde{R(\lambda, i\alpha x_2)}(x_2)\\ &= \int_0^\infty e^{-\lambda s} e^{-2is\alpha\|x_2\|_t^2 - \frac{s^2\alpha^2}{2}\|x_2\|_t^2}\, ds.\end{aligned}$$

Now, arrange α such that these are not equal.

(b) If x_1 and x_2 are not real multiples of each other, then it must be either $P_{\mathrm{span}\{x_1\}}x_2 \neq x_1$ or $P_{\mathrm{span}\{x_2\}}x_1 \neq x_2$. Assume that $P_{\mathrm{span}\{x_1\}}x_2 \neq x_1$. For $\alpha \in \mathbb{R}$:

$$\psi_{x_1+Y_1}\big(R(\lambda, i\alpha x_1)\big) = \widetilde{R(\lambda, i\alpha x_1)}(x_1) = \int_0^\infty e^{-\lambda s} e^{-2is\alpha\|x_1\|_{\mathfrak{t}}^2 - \frac{s^2\alpha^2}{2}\|x_1\|_{\mathfrak{t}}^2}\,ds.$$

The value at $x_2 + Y_2$ now depends on whether x_1 is orthogonal to Y_2 or not: If $x_1 \perp Y_2$, then

$$\psi_{x_2+Y_2}\big(R(\lambda, i\alpha x_1)\big) = \widetilde{R(\lambda, i\alpha x_1)}(x_2) = \int_0^\infty e^{-\lambda s} e^{-2is\alpha\sigma_{\mathfrak{t}}\left(P_{\mathrm{span}\{ix_1\}}x_2, x_1\right) - \frac{s^2\alpha^2}{2}\|x_1\|_{\mathfrak{t}}^2}\,ds.$$

Since $\sigma_{\mathfrak{t}}\left(P_{\mathrm{span}\{ix_1\}}x_2, x_1\right) \neq \|x_1\|_{\mathfrak{t}}^2$, we can arrange α such that the values of $\psi_{x_j+Y_j}(R(\lambda, i\alpha x_1))$ for $j = 1, 2$ are different. If, on the other hand, $x_1 \not\perp Y_2$, then

$$\psi_{x_2+Y_2}\big(R(\lambda, i\alpha x_1)\big) = 0.$$

Hence, we have to arrange α such that $\psi_{x_1+Y_1}(R(\lambda, i\alpha x_1)) \neq 0$ (e.g. by letting $\alpha = 0$). □

Before continuing, we state a consequence of the previous considerations:

Corollary 2 *Let $m \in \mathbb{N}$ and $\lambda_1, \ldots, \lambda_m \in \mathbb{C} \setminus i\mathbb{R}$. Consider the function*

$$g = \big(\lambda_1 - 2i\sigma_t(\cdot, z_1)\big)^{-k_1} \cdot \ldots \cdot \big(\lambda_m - 2i\sigma_t(\cdot, z_m)\big)^{-k_m},$$

where $z_1, \ldots, z_m \in V$ and $V \subset \mathbb{C}^n$ is Lagrangian. Then, $T_g^t \in \mathfrak{R}(\mathbb{C}^n, \sigma_t)$.

Proof By definition we have $g \in \mathfrak{R}_{cl,V}$. Now Theorem 7 implies that $T_g^{\mathfrak{t}} \in \mathfrak{T}_{lin}^{\mathfrak{t}}\left(\mathfrak{R}_{cl,V}\right) = \mathfrak{R}_V \subset \mathfrak{R}(\mathbb{C}^n, \sigma_{\mathfrak{t}})$. □

Now we consider resolvent functions for which the vectors z_j are either taken from V or from V'.

Corollary 3 *Let V and V' be two Lagrangian subspaces of $\mathbb{C}^n$ such that $\mathbb{C}^n = V \oplus V'$. Let $m, \ell \geq n$ and $z_1, \ldots, z_m \in V$ and $w_1, \ldots, w_\ell \in V'$ such that $V = \mathrm{span}_{\mathbb{R}}\{z_1, \ldots, z_m\}$ and $V' = \mathrm{span}_{\mathbb{R}}\{w_1, \ldots, w_\ell\}$. Set*

$$g = \prod_{j=1}^{m} \left(\lambda_j - 2i\sigma_t\left(\cdot, z_j\right)\right)^{-k_j} \cdot \prod_{j=1}^{\ell} \left(\lambda'_j - 2i\sigma_t\left(\cdot, w_j\right)\right)^{-k'_j},$$

where $k_j, k'_j \in \mathbb{N}$ and $\lambda_j, \lambda'_j \in \mathbb{C} \setminus i\mathbb{R}$. Then, $T_g^t \in \mathfrak{R}(\mathbb{C}^n, \sigma_t)$.

Proof For $v = v_1 + v_2 \in V \oplus V' = \mathbb{C}^n$ we have

$$g(v) = \prod_{j=1}^{m} \left(\lambda_j - 2i\sigma_{\mathrm{t}}\left(v_2, z_j\right)\right)^{-k_j} \cdot \prod_{j=1}^{\ell} \left(\lambda'_j - 2i\sigma_{\mathrm{t}}\left(v_1, w_j\right)\right)^{-k'_j}.$$

As $v_1 \to \infty$, the first factor tends to 0. As $v_2 \to \infty$, the second factor tends to 0. In conclusion, $g \in C_0(\mathbb{C}^n)$. Hence, $T_g^{\mathrm{t}} \in \mathfrak{K}\left(F_{\mathrm{t}}^2\right)$. Since $\mathfrak{K}\left(F_{\mathrm{t}}^2\right) \subset \mathfrak{R}(\mathbb{C}^n, \sigma_{\mathrm{t}})$ according to Proposition 2 the statement follows. □

Theorem 8 *Let $n = 1$, then $\mathfrak{D}_0^t = \mathfrak{R}_{cl}$.*

Proof If $n = 1$, then any classical resolvent function g either fulfills the assumption of Corollary 2 or the assumptions of Corollary 3. In either case, we obtain $T_g^{\mathrm{t}} \in \mathfrak{R}(\mathbb{C}^n, \sigma_{\mathrm{t}})$. □

For $n > 1$, the situation is more complicated. We will show only a weaker result for this case.

Since every $z \in \mathbb{C}^n$ is contained in some Lagrangian subspace V, we conclude that each resolvent $R(\lambda, z)$ is contained in some $\mathfrak{R}_V$. Thus, we see that

$$\mathfrak{R}_0 := \sum_{V \text{ Lagrangian subspace}} \mathfrak{R}_V$$

contains every resolvent $R(\lambda, z)$. Further, since $\mathfrak{R}_V = \mathfrak{T}_{lin}^{\mathrm{t}}\left(\mathfrak{R}_{cl,V}\right)$, we obtain

$$\overline{\mathfrak{R}_0} = \mathfrak{T}_{lin}^{\mathrm{t}}\left(\overline{\mathfrak{R}_{cl,0}}\right),$$

where $\mathfrak{R}_{cl,0}$ is given by

$$\mathfrak{R}_{cl,0} := \sum_{V \text{ Lagrangian subspace}} \mathfrak{R}_{cl,V}.$$

As a weak substitute for Theorem 8 in higher dimensions, we get:

Proposition 6 $\mathfrak{R}(\mathbb{C}^n, \sigma_t) = \mathfrak{T}^t\left(\overline{\mathfrak{R}_{cl,0}}\right)$.

It might of course be true that $\overline{\mathfrak{R}_{cl,0}} = \mathfrak{R}_{cl}$. In this case, the previous result would imply $\mathfrak{D}_0^t = \mathfrak{R}_{cl}$. Nevertheless, so far this remains an open question.

5 Infinite Dimensional Symplectic Space

As is well-known there are significant differences between the resolvent algebras of finite and infinite dimensional symplectic spaces: in finite dimensions every two regular irreducible representations are unitarily equivalent. However, in the case of infinite dimensional symplectic spaces, this statement is false. Nevertheless, we can build particular representations of the resolvent algebra by using ideas from the previous sections.

Let $\mathbf{t} = (t_k)_{k=1}^{\infty} \in \ell^1(\mathbb{N})$ be a summable sequence of strictly positive real numbers. Then, we set $\mathfrak{H} := \ell^2(\mathbb{N})$ with the usual norm $\|\cdot\|_{\ell^2}$ and define:

$$\mathfrak{H}_{1/2}^{\mathbf{t}} := \left\{ z \in \ell^2(\mathbb{N}) : \sum_{n=1}^{\infty} \frac{|z_n|^2}{t_n} < \infty \right\},$$

$$\mathfrak{H}_1^{\mathbf{t}} := \left\{ z \in \ell^2(\mathbb{N}) : \sum_{n=1}^{\infty} \frac{|z_n|^2}{t_n^2} < \infty \right\}.$$

We may think of $\mathfrak{H}_1^{\mathbf{t}}$ and $\mathfrak{H}_{1/2}^{\mathbf{t}}$ as the range of B and $\sqrt{B}$, respectively, where B is the trace class operator obtained by linearly extending the map $e_n \mapsto t_n e_n$ with $\{e_n\}$ being the standard orthonormal basis of $\ell^2(\mathbb{N})$. We consider $\mathfrak{H}_{1/2}^{\mathbf{t}}$ as a symplectic Hilbert space equipped with the symplectic form:

$$\sigma_{\mathbf{t}}(z, w) := \sum_{n=1}^{\infty} \frac{\operatorname{Im}(z_n \cdot \overline{w_n})}{t_n}, \qquad z = (z_n), w = (w_n) \in \mathfrak{H}_{1/2}^{\mathbf{t}}.$$

Note that $\sigma_{\mathbf{t}}(z, w)$ is well-defined even for $z \in \mathfrak{H}_1^{\mathbf{t}}$ and $w \in \mathfrak{H}$. We can write $\sigma_{\mathbf{t}} = \operatorname{Im}\langle\cdot,\cdot\rangle_{1/2}$, where $\langle\cdot,\cdot\rangle_{1/2}$ is the canonical inner product of $\mathfrak{H}_{1/2}^{\mathbf{t}}$:

$$\langle z, w\rangle_{1/2} := \sum_{n=1}^{\infty} \frac{z_n \cdot \overline{w_n}}{t_n}.$$

Moreover, $\|\cdot\|_{1/2}$ denotes the corresponding norm: $\|z\|_{1/2}^2 = \langle z, z\rangle_{1/2}$.

We recall some basic facts about Gaussian measures on infinite dimensional Hilbert spaces, cf. [2, 9, 18]. The infinite product measure $\mu_{\mathbf{t}} := \prod_{k=1}^{\infty} \mu_{t_k}$ of the Gaussian measures

$$d\mu_{t_k}(z) := \frac{1}{\pi t_k} e^{-\frac{|z|^2}{t_k}} dV(z), \qquad z \in \mathbb{C}$$

gives a well-defined probability measure on $\mathbb{C}^{\infty} = \prod_{k=1}^{\infty} \mathbb{C}$. It is concentrated on $\mathfrak{H} = \ell^2(\mathbb{N}) \subset \mathbb{C}^{\infty}$. By $\mu_{\mathbf{t}}$, we also denote its restriction to the measurable space $(\mathfrak{H}, \mathfrak{B})$, where $\mathfrak{B}$ is the Borel σ-algebra of $\mathfrak{H}$. Note that $\mathfrak{B}$ agrees with the σ-algebra generated by cylindrical Borel sets. The measure $\mu_{\mathbf{t}}$ is called the (centered) *Gaussian measure* on $\mathfrak{H}$ with *covariance operator* B: This is due to the fact that B is naturally related to $\mu_{\mathbf{t}}$ by

$$\int_{\mathfrak{H}} e^{i\,\mathrm{Re}\langle x,y\rangle_{\ell^2}}\, d\mu\,(x) = e^{-\frac{1}{2}\langle By,y\rangle_{\ell^2}}, \qquad (y \in \mathfrak{H})\,. \tag{19}$$

The space $\mathfrak{H}^{\mathbf{t}}_{1/2}$ introduced above is also known as the *Cameron-Martin space* of $\mu_{\mathbf{t}}$. It is a measurable set of measure zero: $\mu_{\mathbf{t}}\left(\mathfrak{H}^{\mathbf{t}}_{1/2}\right) = 0$. Hence, so is $\mathfrak{H}_1$.

We will now introduce the *Fock-Bargmann space* of holomorphic functions in infinitely many variables and Berezin-Toeplitz quantization in this setting. Two different approaches have been proposed in [15, 16]: the first being of *measure theoretic* nature and the second based on an *inductive limit construction*. Both approaches yield different theories and we deal with the measure theoretic approach here.

We will write $\mathfrak{Z}_0$ for the set of all sequences $(\alpha_n)_{n=1}^{\infty}$ with values in $\mathbb{N}_0$ such that all but finitely many entries are zero. In the following we use the standard multi-index notation: let $z = (z_1, z_2, \ldots) \in \ell^2\,(\mathbb{N})$ and $\alpha \in \mathfrak{Z}_0$, then we put:

$$z^{\alpha} = z_1^{\alpha_1} z_2^{\alpha_2} \ldots, \quad \alpha! = \alpha_1!\alpha_2! \ldots, \quad \mathbf{t}^{\alpha} = t_1^{\alpha_1} t_2^{\alpha_2} \ldots.$$

The monomials $\mathfrak{M}$ below form an orthonormal system of complex analytic functions on $\mathfrak{H}$ inside $L^2\,(\mathfrak{H}, \mu_{\mathbf{t}})$:

$$\mathfrak{M} := \left[e_{\alpha}\,(z) := \frac{1}{\sqrt{\mathbf{t}^{\alpha}\alpha!}} z^{\alpha} \;:\; \alpha \in \mathfrak{Z}_0\right].$$

We write $\langle \cdot, \cdot \rangle$ and $\| \cdot \|$ for the standard inner product and norm in $L^2\,(\mathfrak{H}, \mu_{\mathbf{t}})$, respectively. Elements in the linear span of $\mathfrak{M}$ are referred to as *analytic polynomials*. In analogy to the finite dimensional setting of $\mathbb{C}^n$ we define the Bargmann-Fock space $F^2_{\mathbf{t}}\,(\mathfrak{H})$ to be the L^2-closure of analytic polynomials:

$$F^2_{\mathbf{t}}\,(\mathfrak{H}) := \overline{\mathrm{span}\,\mathfrak{M}}.$$

Let $z \in \mathfrak{H}_1$ and $p \in \mathrm{span}\,\mathfrak{M}$ be an analytic polynomial. Extending (6) we define the *Weyl operator* $W^{\mathbf{t}}_z$ on $F^2_{\mathbf{t}}\,(\mathfrak{H})$ by:

$$W^{\mathbf{t}}_z p\,(w) := k^{\mathbf{t}}_z\,(w)\, p\,(w - z)\,, \qquad w \in \mathfrak{H}.$$

Here, $k^{\mathbf{t}}_z$ denotes the *normalized reproducing kernel* defined by:

$$k^{\mathbf{t}}_z\,(w) = e^{\langle w,z\rangle_{1/2} - \frac{1}{2}\|z\|^2_{1/2}}.$$

Let $N \in \mathbb{N}$ and assume that the analytic polynomial p only depends on the variables $z_1, \ldots, z_N$. We write $\mu_{\mathbf{t}} = \mu_{\mathbf{t}'} \otimes \mu_{\mathbf{t}_{N+1}}$, where $\mathbf{t}' := (t_1, \ldots, t_N)$ and

$\mathbf{t}_{N+1} = (t_{N+1}, t_{N+2}, \ldots)$. Define the projections $\pi_N(z) := (z_1, \ldots, z_N)$ for a sequence $z = (z_j)_j \in \mathfrak{H}$. Since $p = p \circ \pi_N$ we obtain:

$$\int_{\mathfrak{H}} |W_z^{\mathbf{t}} p(w)|^2 d\mu_{\mathbf{t}}(w) =$$

$$= \int_{\mathbb{C}^\infty} |k_{\pi_N(z)}^{\mathbf{t}'} \circ \pi_N(w) \;\; p \circ \pi_N(w-z)|^2 |k_{(I-\pi_N)(z)}^{\mathbf{t}_{N+1}} \circ (I - \pi_N)(w)|^2 \, d\mu_{\mathbf{t}}(w)$$

$$= \int_{\mathbb{C}^N} |W_{\pi_N(z)}^{\mathbf{t}'} p(w)|^2 \, d\mu_{\mathbf{t}'}(w) \times \underbrace{\int_{\prod_{k=N+1}^{\infty} \mathbb{C}} |k_{(I-\pi_N)(z)}^{\mathbf{t}_{N+1}}(w)|^2 \, d\mu_{\mathbf{t}_{N+1}}(w)}_{=1}$$

$$= \int_{\mathfrak{H}} |p(w)|^2 \, d\mu_{\mathbf{t}}(w).$$

Therefore, $W_z^{\mathbf{t}} p \in L^2(\mathfrak{H}, \mu_{\mathbf{t}})$ and $W_z^{\mathbf{t}}$ acts isometrically (say, on analytic polynomials). In finite dimensions, it would now be clear that $W_z^t p \in F_{\mathbf{t}}^2$, since $W_z^t p$ defines a holomorphic function. In infinite dimensions, this statement is not entirely trivial, as $F_{\mathbf{t}}^2$ is defined as the closure of the analytic polynomials. Nevertheless, $W_z^t p \in F_{\mathbf{t}}^2$ remains true, cf. [15]. Hence, $W_z^{\mathbf{t}}$ extends to an isometric operator on $F_{\mathbf{t}}^2(\mathfrak{H})$ for every $z \in \mathfrak{H}_1$.

Further, note that for $(z_k) \subset \mathfrak{H}_1^{\mathbf{t}}$, $z_k \to z$ as $k \to \infty$ in $\mathfrak{H}_1^{\mathbf{t}}$ we clearly have

$$k_{z_k}^{\mathbf{t}}(w)\, p(w - z_k) \to k_z^{\mathbf{t}}(w)\, p(w - z)$$

pointwise almost everywhere on $\ell^2(\mathbb{N})$. Hence, *Scheffé's Lemma* shows that the assignment $z \mapsto W_z^{\mathbf{t}} p$ is continuous from $\mathfrak{H}_1$ to $F_{\mathbf{t}}^2(\mathfrak{H})$ for every analytic polynomial p. Since these are dense in $F_{\mathbf{t}}^2(\mathfrak{H})$, the Weyl operators $W_z^{\mathbf{t}}$ extend to isometric operators on $F_{\mathbf{t}}^2(\mathfrak{H})$ such $z \mapsto W_z^{\mathbf{t}}$ is continuous on $\mathfrak{H}_1$ with respect to the strong operator topology.

An even stronger statement holds, namely the map $z \mapsto W_z^{\mathbf{t}}$ is even strongly continuous with respect to the coarser topology of $\mathfrak{H}_{1/2}$, cf. [15]. Therefore, it extends to a map from $\mathfrak{H}_{1/2}$ to the unitary operators on $F_{\mathbf{t}}^2(\mathfrak{H})$.

By applying the Weyl operators with $z, w \in \mathfrak{H}_1$ to analytic polynomials, it is not hard to see that

$$W_z^{\mathbf{t}} W_w^{\mathbf{t}} = e^{-i\sigma_{\mathbf{t}}(z,w)} W_{z+w}^{\mathbf{t}} \quad \textit{and} \quad \left(W_z^{\mathbf{t}}\right)^* = \left(W_z^{\mathbf{t}}\right)^{-1} = W_{-z}^{\mathbf{t}}.$$

These relations extend to $z, w \in \mathfrak{H}_{1/2}$, as well:

Lemma 10 ([15, 16]) *The Weyl operators $W_z^t \in \mathfrak{L}\left(F_t^2(\mathfrak{H})\right)$ depend continuously on $z \in \left(\mathfrak{H}_{1/2}, \|\cdot\|_{1/2}\right)$ with respect to the strong operator topology and*

$$W_z^t W_w^t = e^{-i\sigma_t(z,w)} W_{z+w}^t$$

for every $z, w \in \mathfrak{H}_{1/2}$.

Fix a bounded operator $A \in \mathfrak{L}\left(F_{\mathbf{t}}^2(\mathfrak{H})\right)$ and $z \in \mathfrak{H}_1$. We define the *Berezin transform* of A at z by

$$\widetilde{A}(z) = \langle Ak_z^{\mathbf{t}}, k_z^{\mathbf{t}} \rangle = \langle W_{-z}^{\mathbf{t}} A W_z^{\mathbf{t}} 1, 1 \rangle.$$

Lemma 11 *The map* $A \mapsto \widetilde{A}$ *is injective on* $\mathfrak{L}\left(F_t^2(\mathfrak{H})\right)$.

Proof If $\widetilde{A} = 0$, then $\widetilde{A}(z) = 0$ for every $z = (z_1, \dots, z_N, 0, 0, \dots)$. We denote by F_N^2 the closed linear span in $F_{\mathbf{t}}^2(\mathfrak{H})$ of the set of all analytic polynomials in the finitely many variables $z_1, \dots, z_N$. Moreover, we write $P_N \in \mathfrak{L}\left(F_{\mathbf{t}}^2(\mathfrak{H})\right)$ for the orthogonal projection onto F_N^2. Then, $k_z^{\mathbf{t}} \in F_N^2$ for $z = (z_1, \dots, z_N, 0, 0, \dots)$ and hence:

$$\widetilde{A}(z) = \langle Ak_z^{\mathbf{t}}, k_z^{\mathbf{t}} \rangle = \langle P_N A k_z^{\mathbf{t}}, k_z^{\mathbf{t}} \rangle.$$

Note that F_N^2 can be naturally identified with the N-variable Fock-Bargmann space $F_{\mathbf{t}'}^2 = F_{\mathbf{t}'}^2\left(\mathbb{C}^N\right)$ with weight parameter $\mathbf{t}' = (t_1, \dots, t_N)$. Moreover,

$$P_N A|_{F_N^2} \in \mathfrak{L}\left(F_N^2\right) \cong \mathfrak{L}\left(F_{\mathbf{t}'}^2\right) \quad \text{and} \quad 0 = \widetilde{A}(z) = \left(P_N A|_{F_N^2}\right)^{\sim}(z_1, \dots, z_N).$$

On the right hand side of the last formula we have used the Berezin transform on $F_{\mathbf{t}'}^2$. Since the Berezin transform is injective on $\mathfrak{L}\left(F_{\mathbf{t}'}^2\right)$ (see [12]), we conclude that $P_N A|_{F_N^2} = 0$. Hence, $P_N A p = 0$ for any analytic polynomial p and $N \in \mathbb{N}$ sufficiently large. Finally, we have

$$P_N A p \to Ap, \quad N \to \infty,$$

in $F_{\mathbf{t}}^2(\mathfrak{H})$, i.e. $Ap = 0$ for analytic polynomials p showing that $A = 0$. □

We denote by $P^{\mathbf{t}}$ the orthogonal projection from $L^2(\mathfrak{H}, \mu_{\mathbf{t}})$ onto $F_{\mathbf{t}}^2(\mathfrak{H})$. Given $\varphi \in L^\infty(\mathfrak{H}, \mu_{\mathbf{t}})$ we define the Toeplitz operator $T_\varphi^{\mathbf{t}} \in \mathfrak{L}\left(F_{\mathbf{t}}^2(\mathfrak{H})\right)$ by:

$$T_\varphi^{\mathbf{t}}(g) = P^{\mathbf{t}}(\varphi g).$$

By applying the injectivity of the Berezin transform in Lemma 11, it is not hard to verify that the Weyl operators $W_z^{\mathbf{t}}$ for $z \in \mathfrak{H}_1$ have a representation as Toeplitz operators. More precisely:

$$W_z^{\mathbf{t}} = T_{g_z}^{\mathbf{t}} \quad \text{where} \quad g_z(w) = e^{2i\sigma_{\mathbf{t}}(w,z) + \frac{1}{2}\|z\|_{\mathbf{1}/2}^2}. \tag{20}$$

Due to the importance of this fact, we derive an explicit formula for the Berezin transform of both operators:

$$
\begin{aligned}
\widetilde{W_z^{\mathbf{t}}}(w) &= e^{-\|w\|_{1/2}^2} \langle W_z^{\mathbf{t}} K_w^{\mathbf{t}}, K_w^{\mathbf{t}} \rangle \\
&= e^{-\|w\|_{1/2}^2 - \frac{1}{2}\|z\|_{1/2}^2} \langle e^{\langle \cdot, z\rangle_{1/2}} e^{\langle \cdot - z, w\rangle_{1/2}}, e^{\langle \cdot, w\rangle_{1/2}} \rangle \\
&= e^{-\|w\|_{1/2}^2 - \frac{1}{2}\|z\|_{1/2}^2} e^{-\langle z, w\rangle_{1/2}} \langle K_{w+z}^{\mathbf{t}}, K_w^{\mathbf{t}} \rangle \\
&= e^{-\|w\|_{1/2}^2 - \frac{1}{2}\|z\|_{1/2}^2} e^{-\langle z, w\rangle_{1/2} + \langle w, w+z\rangle_{1/2}}, \\
\widetilde{g_z}^{(\mathbf{t})}(w) &= e^{\frac{1}{2}\|z\|_{1/2}^2 - \|w\|_{1/2}^2} \langle e^{2i\sigma_{\mathbf{t}}(\cdot, z)} K_w^{\mathbf{t}}, K_w^{\mathbf{t}} \rangle \\
&= e^{\frac{1}{2}\|z\|_{1/2}^2 - \|w\|_{1/2}^2} \langle e^{\langle \cdot, z\rangle_{1/2}} K_w^{\mathbf{t}}, e^{-\langle \cdot, z\rangle_{1/2}} K_w^{\mathbf{t}} \rangle \\
&= e^{\frac{1}{2}\|z\|_{1/2}^2 - \|w\|_{1/2}^2} \langle K_{w+z}^{\mathbf{t}}, K_{w-z}^{\mathbf{t}} \rangle \\
&= e^{\frac{1}{2}\|z\|_{1/2}^2 - \|w\|_{1/2}^2} e^{\langle w-z, w+z\rangle_{1/2}} = \widetilde{W_z^{\mathbf{t}}}(w)\,.
\end{aligned}
$$

It is an important but non-trivial fact that (20) extends to $z \in \mathfrak{H}_{1/2}$. First, we have to explain in what sense the map $w \mapsto \langle w, z\rangle_{1/2}$ defines a measurable function on $\mathfrak{H}$ for $z \in \mathfrak{H}_{1/2}$. Clearly, the expression $\langle \cdot, z\rangle$ is pointwise well-defined on $\mathfrak{H}_{1/2}$. However, since $\mathfrak{H}_{1/2}$ is a set of measure zero, this is of no big help. We need a preliminary result:

Lemma 12 $\|\langle \cdot, z\rangle_{\frac{1}{2}}\|^2 = 2\|z\|_{1/2}^2$ *for* $z \in \mathfrak{H}_1$.

Proof We introduce a complex parameter $\lambda \in \mathbb{C}$. Then, it is not hard to verify the following equalities through standard results on differentiation of parameter integrals, where we use that $\langle \cdot, z\rangle \in L^2(\mathfrak{H}, \mu_{\mathbf{t}})$ (this is easily established):

$$
\begin{aligned}
\frac{\partial^2}{\partial\overline{\lambda}\partial\lambda} \int_{\mathfrak{H}} e^{i2\,\mathrm{Re}\langle w, \lambda z\rangle_{1/2}}\, d\mu_{\mathbf{t}}(w) &= \frac{\partial^2}{\partial\overline{\lambda}\partial\lambda} \int_{\mathfrak{H}} e^{i\overline{\lambda}\langle w, z\rangle_{1/2} + i\lambda\langle z, w\rangle_{1/2}}\, d\mu_{\mathbf{t}}(w) \\
&= -\int_{\mathfrak{H}} |\langle z, w\rangle_{1/2}|^2 e^{i\overline{\lambda}\langle w, z\rangle_{1/2} + i\lambda\langle z, w\rangle_{1/2}}\, d\mu_{\mathbf{t}}(w)\,.
\end{aligned}
$$

On the other hand, using Equality (19):

$$
\begin{aligned}
\frac{\partial^2}{\partial\overline{\lambda}\partial\lambda} \int_{\mathfrak{H}} e^{i2\,\mathrm{Re}\langle w, \lambda z\rangle_{1/2}}\, d\mu_{\mathbf{t}}(w) &= \frac{\partial^2}{\partial\overline{\lambda}\partial\lambda} \int_{\mathfrak{H}} e^{i\,\mathrm{Re}\langle w, 2\lambda B^{-1} z\rangle_{\ell^2}}\, d\mu_{\mathbf{t}}(w) \\
&= \frac{\partial^2}{\partial\overline{\lambda}\partial\lambda} e^{-\frac{1}{2}\langle 2\lambda B B^{-1} z, 2\lambda B^{-1} z\rangle_{\ell^2}} \\
&= \frac{\partial^2}{\partial\overline{\lambda}\partial\lambda} e^{-2\lambda\overline{\lambda}\langle z, B^{-1} z\rangle_{\ell^2}} \\
&= \frac{\partial^2}{\partial\overline{\lambda}\partial\lambda} e^{-2\lambda\overline{\lambda}\langle z, z\rangle_{1/2}} \\
&= \Big(-2\|z\|_{1/2}^2 + 4|\lambda|^2\|z\|_{1/2}^4\Big) e^{-2\lambda\overline{\lambda}\|z\|_{1/2}^2}.
\end{aligned}
$$

We therefore obtain:

$$\int_{\mathfrak{H}} |\langle z, w\rangle_{1/2}|^2 e^{i2\,\mathrm{Re}\langle w,\lambda z\rangle_{1/2}}\, d\mu_{\mathfrak{t}}(w) = \left(2 - 4|\lambda|^2\|z\|_{1/2}^2\right)\|z\|_{1/2}^2 e^{-2|\lambda|^2\|z\|_{1/2}^2}.$$

Choosing $\lambda = 0$ yields the desired equality. □

Hence, if $(z_k) \subset \mathfrak{H}_1$ is a Cauchy sequence with respect to $\|\cdot\|_{1/2}$, then $\langle\cdot, z_k\rangle_{1/2}$ is Cauchy in $L^2(\mathfrak{H}, \mu_{\mathfrak{t}})$. In conclusion, the family of functions $\langle\cdot, z\rangle_{1/2} \in L^2(\mathfrak{H}, \mu_{\mathfrak{t}})$ continuously extends to $z \in \mathfrak{H}_{1/2}$. In particular, for each $z \in \mathfrak{H}_{1/2}$, $\langle w, z\rangle_{1/2}$ is a well-defined expression for almost every $w \in \mathfrak{H}$ and, as a function of w, measurable. In particular,

$$g_z(w) = e^{2i\sigma_{\mathfrak{t}}(w,z)+\frac{1}{2}\|z\|_{1/2}^2} = e^{\langle w,z\rangle_{1/2}} e^{-\langle z,w\rangle_{1/2}} e^{\frac{1}{2}\|z\|_{1/2}^2}$$

is an almost everywhere well-defined function with

$$\|g_z\|_\infty \le e^{\frac{1}{2}\|z\|_{1/2}^2}, \qquad z \in \mathfrak{H}_{1/2}.$$

We also note that, since $\langle\cdot, z\rangle_{1/2} \in F_{\mathfrak{t}}^2(\mathfrak{H})$ for every $z \in \mathfrak{H}_1$, the same is true for $z \in \mathfrak{H}_{1/2}$. Iterating the procedure from the proof of Lemma 12, one easily obtains:

$$\|\langle\cdot, z\rangle_{1/2}^k\|^2 = 2^k\|z\|_{1/2}^{2k} \quad \textit{for} \quad k \in \mathbb{N}.$$

Next, we calculate:

$$\sum_{k=0}^\infty \frac{\|\langle\cdot, z\rangle_{1/2}^k\|}{k!} = \sum_{k=0}^\infty \frac{2^{\frac{k}{2}}\|z\|_{1/2}^k}{k!} = e^{\sqrt{2}\|z\|_{1/2}}.$$

We obtain:

$$\sum_{k=0}^\infty \frac{\langle\cdot, z\rangle_{1/2}^k}{k!} = e^{\langle\cdot,z\rangle_{1/2}} = K_z^{\mathfrak{t}} \in F_{\mathfrak{t}}^2(\mathfrak{H})$$

for every $z \in \mathfrak{H}_{1/2}$. By a standard density argument, we have:

$$\langle K_z, K_w\rangle = e^{\langle w,z\rangle_{1/2}} \quad \textit{for} \quad z, w \in \mathfrak{H}_{1/2}.$$

Now, one can compute the Berezin transform of g_z as above. For every $z \in \mathfrak{H}_{1/2}$ and $w \in \mathfrak{H}_1$ one obtains:

$$\widetilde{T_{g_z}^{\mathfrak{t}}}(w) = e^{\frac{1}{2}\|z\|_{1/2}^2 - \|w\|_{1/2}^2} e^{\langle w-z, w+z\rangle_{1/2}}.$$

Further, since $W_z^{\mathbf{t}}$ continuously (in strong operator topology) depends on the parameter $z \in \mathfrak{H}_{1/2}$, we conclude that the Berezin transform $\widetilde{W_z^{\mathbf{t}}}$ continuously (in the topology of pointwise convergence) depends on $z \in \mathfrak{H}_{1/2}$, which gives

$$\widetilde{W_z^{\mathbf{t}}}(w) = e^{\frac{1}{2}\|z\|_{1/2}^2 - \|w\|_{1/2}^2} e^{\langle w-z, w+z\rangle_{1/2}}$$

for every $z \in \mathfrak{H}_{1/2}$ and $w \in \mathfrak{H}_1$. Comparing Berezin transforms, we have shown that $T_{g_z}^{\mathbf{t}} = W_z^{\mathbf{t}}$ extends to $z \in \mathfrak{H}_{1/2}$.

In an abuse of notation, we will write $\mathfrak{R}\left(\mathfrak{H}_{1/2}, \sigma_{\mathbf{t}}\right)$ for the representation of the resolvent algebra on $F_{\mathbf{t}}^2(\mathfrak{H})$ with respect to the symplectic space $\left(\mathfrak{H}_{1/2}, \sigma_{\mathbf{t}}\right)$. Since the Weyl operators $W_z^{\mathbf{t}}$, $z \in \mathfrak{H}_{1/2}$ satisfy the CCR in Lemma 10, we start again by expressing the resolvent $R(\lambda, z)$ in form of a Laplace transform of the one-parameter group $(W_{tz})_{t\in\mathbb{R}}$. For $\mathrm{Re}(\lambda) > 0$ we have:

$$R(\lambda, z) = i \int_0^\infty e^{-\lambda s} W_{-sz}^{\mathbf{t}}\, ds,$$

which exists as an integral in strong operator topology. There is an analogous formula in the case $\mathrm{Re}(\lambda) < 0$. Consider the space of (classical) symbols:

$$\mathfrak{R}_{cl}^{\mathbf{t}} = C^*\Big(\{(\lambda - 2i\sigma_{\mathbf{t}}(\cdot, z))^{-1} : \lambda \in \mathbb{C} \setminus i\mathbb{R},\ z \in \mathfrak{H}_{1/2}\}\Big).$$

Moreover, we write $\mathfrak{T}^{\mathbf{t}}\left(\mathfrak{R}_{cl}^{\mathbf{t}}\right)$ for the C* algebra generated by Toeplitz operators over $F_{\mathbf{t}}^2(\mathfrak{H})$ with symbols in $\mathfrak{R}_{cl}^{\mathbf{t}}$.

Theorem 9 *The following inclusion holds true:* $\mathfrak{R}\left(\mathfrak{H}_{1/2}, \sigma_t\right) \subset \mathfrak{T}^t\left(\mathfrak{R}_{cl}^t\right)$.

Proof Without loss of generality we assume $\mathrm{Re}(\lambda) > 0$. We will verify that the representation of the resolvent in form of a Laplace transform of Weyl operators defines an element in the Toeplitz algebra. The following integrals are to be understood as improper Riemann integrals in strong operator topology:

$$\begin{aligned}
\int_0^\infty e^{-\lambda s} W_{-sz}^{\mathbf{t}}\, ds &= \int_0^\infty e^{-\lambda s} T_{g_{-sz}}^{\mathbf{t}}\, ds \\
&= \int_0^\infty e^{-\lambda s} e^{\frac{s^2}{2}\|z\|_{1/2}^2} T_{\exp(-2is\sigma_{\mathbf{t}}(\cdot,z))}^{\mathbf{t}}\, ds \\
&= \int_0^\infty e^{-\lambda s} \sum_{k=0}^\infty \frac{s^{2k}\|z\|_{1/2}^{2k}}{2^k k!} T_{\exp(-2is\sigma_{\mathbf{t}}(\cdot,z))}^{\mathbf{t}}\, ds.
\end{aligned}$$

Since the Weyl operators are unitary, and hence satisfy $\|W_z^{\mathbf{t}}\| = 1$, it follows that:

$$\|T_{\exp(-2is\sigma_{\mathbf{t}}(\cdot,z))}^{\mathbf{t}}\| = e^{-\frac{s^2}{2}\|z\|_{1/2}^2}.$$

Therefore, the dominated convergence theorem gives as $m \to \infty$:

$$\Big\| iR(\lambda, z) - \sum_{k=0}^{m} \frac{\|z\|_{1/2}^{2k}}{2^k k!} \int_0^\infty s^{2k} e^{-\lambda s} T^{\mathbf{t}}_{\exp(-2is\sigma_{\mathbf{t}}(\cdot,z))}\, ds \Big\|$$

$$\leq \int_0^\infty e^{-\lambda s} \left| e^{\frac{s^2}{2}\|z\|_{1/2}^2} - \sum_{k=0}^{m} \frac{s^{2k}\|z\|_{1/2}^{2k}}{k! 2^k} \right| e^{-\frac{s^2}{2}\|z\|_{1/2}^2}\, ds \to 0, \quad m \to \infty.$$

We have therefore seen that

$$iR(\lambda, z) = \sum_{k=0}^{\infty} \frac{1}{k!} \frac{\|z\|_{1/2}^{2k}}{2^k} \int_0^\infty T^{\mathbf{t}}_{s^{2k}\exp(-\lambda s + 2is\sigma_{\mathbf{t}}(\cdot,z))}\, ds,$$

where the series converges in operator norm. Fix for the moment $N > 0$. Since the Riemann integral $\int_0^N T^{\mathbf{t}}_{s^{2k}\exp(-\lambda s+2is\sigma_{\mathbf{t}}(\cdot,z))}\, ds$ exists as a limit of Riemann sums in strong operator topology (this easily follows from the fact that the mapping $s \mapsto W^{\mathbf{t}}_{sz}$ is strongly continuous), we obtain for the Berezin transform:

$$\left(\int_0^N T^{\mathbf{t}}_{s^{2k}\exp(-\lambda s+2is\sigma_{\mathbf{t}}(\cdot,z))}\, ds \right)^{\sim}(u) = \left\langle T^{\mathbf{t}}_{\int_0^N s^{2k}\exp(-\lambda s+2is\sigma_{\mathbf{t}}(\cdot,z))ds}\, k_u^{\mathbf{t}}, k_u^{\mathbf{t}} \right\rangle.$$

Therefore, injectivity of the Berezin transform (Lemma 11) shows:

$$\int_0^N T^{\mathbf{t}}_{s^{2k}\exp(-\lambda s+2is\sigma_{\mathbf{t}}(\cdot,z))} ds = T^{\mathbf{t}}_{\int_0^N s^{2k}\exp(-\lambda s+2is\sigma_{\mathbf{t}}(\cdot,z))ds}.$$

Since

$$\int_0^N s^{2k} \exp(-\lambda s + 2is\sigma_{\mathbf{t}}(\cdot, z))\, ds \overset{N\to\infty}{\longrightarrow} \int_0^\infty s^{2k} \exp(-\lambda s + 2is\sigma_{\mathbf{t}}(\cdot, z))\, ds,$$

uniformly, this gives

$$\int_0^\infty T^{\mathbf{t}}_{s^{2k}\exp(-\lambda s+2is\sigma_{\mathbf{t}}(\cdot,z))}\, ds = T^{\mathbf{t}}_{\int_0^\infty s^{2k}\exp(-\lambda s+2is\sigma_{\mathbf{t}}(\cdot,z))\, ds}.$$

For the symbol, standard facts on the Laplace transform yield

$$\int_0^\infty s^{2k} e^{-\lambda s + 2is\sigma_{\mathbf{t}}(w,z)}\, ds = (2k)!\, (\lambda - 2i\sigma_{\mathbf{t}}(w, z))^{-(2k+1)}.$$

This is now, as a function of $w \in \mathfrak{H}$, bounded and measurable, which finishes the proof. □

Discussion: The careful reader may have noticed that the proofs of Proposition 3 and Lemma 4 generalize to the infinite dimensional phase space. Analogous formulas for the Berezin transforms of products of resolvents and the classical resolvent functions, respectively, are valid. Nevertheless, since a *correspondence theorem* in the infinite dimensional setup (similar to Theorem 2) is not available at the moment, this observation is of no further use for proving a refinement of Theorem 9. The biggest issue is the lack of a Haar measure on the infinite dimensional symplectic space. In the finite dimensional framework of $\mathbb{C}^n$, this measure coincides with the Lebesgue measure and is at the heart of *quantum harmonic analysis*. An important ingredient to the theory is a correspondence between the spaces $L^1(\mathbb{C}^n)$ and $\mathfrak{T}^1\left(F_t^2\right)$ (the latter being the space of trace class operators). It is not clear what an appropriate interpretation of "$L^1\left(\mathfrak{H}_{1/2}\right)$" should be in order to develop a *quantum harmonic analysis* and *correspondence theory* for infinite dimensional phase spaces.

Even in the case of a finite dimensional symplectic vector space various algebraic-analytic properties of the resolvent algebra seem to be only poorly understood. In particular, this concerns its automorphism group, a description of the set of affiliated self-adjoint operators or stability under taking certain spectral functions. As was mentioned above, the construction of the resolvent algebra in [7] starts from a symplectic vector space and an assignment of canonical operators. It is an interesting task to generalize this construction to the framework of symplectic manifolds or groups and to encode a geometric structure inside a C^* algebra. In a concrete model a quantization scheme is required and as a first example one may consider Toeplitz quantization over the complex unit disc with the standard Kähler structure. However, so far it is unclear whether the tools from *quantum harmonic analysis* can be suitably generalized to this setup. Such questions will be a key issue for our future work.

References

1. Bauer, W., Isralowitz, J.: Compactness characterization of operators in the Toeplitz algebra of the Fock space F_α^p. J. Funct. Anal. **263**, 1323–1355 (2012)
2. Bauer, W., Rodriguez Rodriguez, M.A.: Commutative Toeplitz algebras and their Gelfand theory: old and new results. Complex Anal. Oper. Theory **16**, 77 (2022)
3. Berger, C.L., Coburn, L.A.: Toeplitz operators on the Segal-Bargmann space. Trans. Amer. Math. Soc. **301**, 813–829 (1987)
4. Bourbaki, N.: Elements of Mathematics: General Topology Chapters 1–4. Springer (1989)
5. Bratteli, O., Robinson, D.W.: Operator Algebras and Quantum Statistical Mechanics, vol. 2, 2nd ed. Springer (1997)
6. Buchholz, D.: The resolvent algebra: ideals and dimension. J. Funct. Anal. **266**, 3286–3302 (2014)
7. Buchholz, D., Grundling, H.: The resolvent algebra: a new approach to canonical quantum systems. J. Funct. Anal. **254**, 2725–2779 (2008)
8. Coburn, L.A.: The measure algebra of the Heisenberg group. J. Funct. Anal. **161**, 509–525 (1999)
9. Da Prato, G.: An Introduction to Infinite-Dimensional Analysis. Universitext, Springer (2006)
10. Engel, K.-J., Nagel, R.: One-Parameter Semigroups for Linear Evolution Equations. Graduate Texts in Mathematics, vol. 194. Springer (2000)

11. Fannes, M., Verbeure, A.: On the time evolution automorphisms of the CCR-algebra for quantum mechanics. Comm. Math. Phys. **35**, 257–264 (1974)
12. Folland, G.B.: Harmonic Analysis in Phase Space. Princeton University Press (1989)
13. Fulsche, R.: Correspondence theory on p-Fock spaces with applications to Toeplitz algebras. J. Funct. Anal. **279**, 108661 (2020)
14. Fulsche, R.: Toeplitz operators on non-reflexive Fock spaces (2021), preprint available under arXiv:2202.11440
15. Janas, J., Rudol, K.: Toeplitz operators on the Segal-Bargmann space of infinitely many variables, linear operators in function spaces. In: Helson, H., Nagy, B.S., Vasilescu, F.-H. (eds.), 12th International Conference on Operator Theory Timişoara (Romania) June 6–16 1988. Operator Theory: Advances and Applications, vol. 43, pp. 217–228. Birkhäuser Verlag (1990)
16. Janas, J., Rudol, K.: Toeplitz operators in infinitely many variables, Topics in operator theory, operator algebras and applications. In: 15th International Conference on Operator Theory, Timişoara, Romania, June 6–10, 1994, pp. 147–160. IMAR Bucharest (1995)
17. Reed, M., Simon, B.: Methods of Modern Mathematical Physics 1: Functional Analysis. Academic (1972)
18. Skorohod, A.V.: Integration in Hilbert space. Springer, Ergebnisse der Mathematik und ihrer Grenzgebiete (1974)
19. van Nuland, T.D.H.: Quantization and the resolvent algebra. J. Funct. Anal. **277**, 2815–2838 (2019)
20. Werner, R.: Quantum harmonic analysis on phase space. J. Math. Phys. **25**, 1404–1411 (1984)
21. Wick, B.W., Wu, S.: Fock space on $\mathbb{C}^\infty$ and Bose-Fock space. J. Math. Anal. Appl. **505**, 125499 (2022)

(n, ε)-Condition Spectrum of Operator Pencils

G. Krishna Kumar and Judy Augustine

Abstract The goal of the study is to generalize the ε-condition spectrum of operators to (n, ε)-condition spectrum of operator pencils, which is a Ransford spectrum ($0 < \varepsilon < 1$ and $n \in \mathbb{Z}_+ \cup \{0\}$). Some salient attributes of the (n, ε)-condition spectrum of operator pencils are deduced. We provide an analogue of the Spectral Mapping Theorem for the (n, ε)-condition spectrum of operator pencils. Finally, we elucidate the (n, ε)-condition spectrum as an effective technique for the stability analysis of the operator equations.

Keywords Operator pencil · (n, ε)-condition spectrum · Spectral mapping theorem · Stability

1 Introduction

Let X denotes a complex Banach space, $BL(X)$ is the Banach algebra of all bounded linear operators on X, and $GL(X)$ is the set of all invertible operators in $BL(X)$. For $A, B \in BL(X)$, the spectrum of the linear operator pencil (A, B) is defined by

$$\sigma(A, B) = \{\lambda \in \mathbb{C} : \lambda B - A \notin GL(X)\}.$$

This article deals with linear operator pencils and is subsequently called operator pencils. The resolvent set of the operator pencil (A, B) is defined by $\rho(A, B) = \mathbb{C} \setminus \sigma(A, B)$. The generalized eigenvalue problem defined above becomes so significant that an unbounded operator's spectrum is equivalent to an operator pencil's spectrum [6, 8]. The spectrum of operator pencil appears in dynamical theory, mathematical modeling, perturbation theory, quantum mechanics, etc. For more information on various applications of operator pencils, one may refer to [3–5, 16]. The condition

G. Krishna Kumar (✉) · J. Augustine
Department of Mathematics, University of Kerala, Kariavattom Campus, Thiruvananthapuram 695581, Kerala, India
e-mail: krishna.math@keralauniversity.ac.in

A. A. Ambily and V. B. Kiran Kumar (eds.), *Semigroups, Algebras and Operator Theory*, Springer Proceedings in Mathematics & Statistics 436,
https://doi.org/10.1007/978-981-99-6349-2_13

spectrum of an element in a complex unital Banach algebra was introduced as a particular case of generalized Ransford spectrum, refer [13, 17].

Definition 1 Let $A \in BL(X)$ and $0 < \varepsilon < 1$. The ε-condition spectrum of A is defined by

$$\sigma_{\varepsilon}(A) = \sigma(A) \cup \left\{\lambda \in \mathbb{C} : \|(\lambda I - A)\| \left\|(\lambda I - A)^{-1}\right\| \geq \varepsilon^{-1}\right\}.$$

$\lambda \notin \sigma_{\varepsilon}(A)$ for some $0 < \varepsilon < 1$ guarantees a stable solution for the operator equation $Ax - \lambda x = b$, and the condition spectrum becomes useful for solving differential equations numerically. Refer to [1, 2, 11, 12, 19] for more properties on the condition spectrum. In this paper, we generalize the ε-condition spectrum of an operator to (n, ε)-condition spectrum of an operator pencil. The (n, ε)-condition spectrum is a better approximation to the spectrum than the usual condition spectrum and comes in handy in the stability analysis, as mentioned in Section 4. An analogous generalization of the usual pseudospectrum to the (n, ε)-pseudospectrum is available in the literature [7, 20]. The results developed here are generally true for elements of a complex unital Banach algebra. Here $\mathbb{Z}_{+}$ is the set of all positive integers.

Definition 2 Let $A, B \in BL(X)$, $n \in \mathbb{Z}_{+} \cup \{0\}$, and $0 < \varepsilon < 1$. The (n, ε)-condition spectrum of the operator pencil (A, B) is denoted as $\sigma_{n,\varepsilon}(A, B)$ and is defined by

$$\sigma_{n,\varepsilon}(A, B) = \sigma(A, B) \cup \left\{\lambda \in \mathbb{C} : \left\|(\lambda B - A)^{2^n}\right\|^{\frac{1}{2^n}} \left\|(\lambda B - A)^{-2^n}\right\|^{\frac{1}{2^n}} \geq \varepsilon^{-1}\right\}.$$

The generalized (n, ε)-condition resolvent of the operator pencil (A, B) is defined by

$$\rho_{n,\varepsilon}(A, B) = \rho(A, B) \cap \left\{\lambda \in \mathbb{C} : \left\|(\lambda B - A)^{2^n}\right\|^{\frac{1}{2^n}} \left\|(\lambda B - A)^{-2^n}\right\|^{\frac{1}{2^n}} < \varepsilon^{-1}\right\}.$$

We organize the article as follows. In Sect. 2, it is shown that the (n, ε)-condition spectrum of an operator pencil is a Ransford spectrum (Theorem 1). Certain properties of the (n, ε)-condition spectrum is discussed, and equivalent definitions of (n, ε)-condition spectrum of a matrix pencil (Theorem 4) are found. In section 3, we find an analogue of the Spectral Mapping Theorem for (n, ε)-condition spectrum of operator pencils (Theorems 7, 8). Finally, section 4 demonstrates the stability analysis of operator equations using (n, ε)-condition spectrum. To illustrate the conclusions drawn, some numerical findings are also given.

2 On Some Properties of (n, ε)-Condition Spectrum of Operator Pencils

$I, \tilde{0}$, respectively, denote the identity operator and zero operator in $BL(X)$.

Remark 1 Let $A, B \in BL(X)$, $n \in \mathbb{Z}_+ \cup \{0\}$, and $0 < \varepsilon < 1$. Then

1. $\sigma(A, B) \subseteq \sigma_{n,\varepsilon}(A, B)$.
2. $\sigma_{n,\varepsilon}(A, I) = \sigma_{n,\varepsilon}(A)$.
3. $\sigma_{0,\varepsilon}(A, B) = \sigma_{\varepsilon}(A, B)$.

Lemma 1 *For $n \in \mathbb{Z}_+ \cup \{0\}$ and a fixed $0 < \varepsilon < 1$, define*

$$\Omega_{n,\varepsilon} = \left\{A \in GL(X) : \left\|A^{2^n}\right\|^{\frac{1}{2^n}} \left\|A^{-2^n}\right\|^{\frac{1}{2^n}} < \varepsilon^{-1}\right\}.$$

Then $\Omega_{n,\varepsilon}$ is a Ransford set.

Proof 1. $\Omega_{n,\varepsilon}$ is an open subset of $GL(X)$.
2. $I \in \Omega_{n,\varepsilon}$ and $\Omega_{n,\varepsilon}$ is non empty.
3. $\tilde{0} \notin \Omega_{n,\varepsilon}$.
4. $z\,\Omega_{n,\varepsilon} \subseteq \Omega_{n,\varepsilon}$ for every $z \neq 0$.

□

Theorem 1 *Let $A, B \in BL(X)$, $n \in \mathbb{Z}_+ \cup \{0\}$, and $0 < \varepsilon < 1$. Then $\sigma_{n,\varepsilon}(A, B)$ is a Ransford spectrum for every $B \in \Omega_{n,\varepsilon}$.*

Proof Since $B \in \Omega_{n,\varepsilon}$, the Ransford spectrum of A corresponding to the Ransford set $\Omega_{n,\varepsilon}$ is defined by, [17],

$$\sigma_{\Omega_{n,\varepsilon}}(A) = \{\lambda \in \mathbb{C} : \lambda B - A \notin \Omega_{n,\varepsilon}\}.$$

It follows that $\sigma_{n,\varepsilon}(A, B) = \sigma_{\Omega_{n,\varepsilon}}(A)$.

Theorem 2 *Let $A, B \in BL(X)$, $n \in \mathbb{Z}_+ \cup \{0\}$, and $0 < \varepsilon < 1$. Then the following holds.*

(i) $\sigma(A, B) = \bigcap_{0<\varepsilon<1} \sigma_{n,\varepsilon}(A, B)$.
(ii) $\sigma_{n,\varepsilon_1}(A, B) \subseteq \sigma_{n,\varepsilon_2}(A, B)$ *for every* $0 < \varepsilon_1 \leq \varepsilon_2 < 1$.
(iii) $\lambda \mapsto \left\|(\lambda B - A)^{2^n}\right\|^{\frac{-1}{2^n}} \left\|(\lambda B - A)^{-2^n}\right\|^{\frac{-1}{2^n}}$ *is continuous for every* $B \in GL(X)$.
(iv) $\sigma_{n+1,\varepsilon}(A, B) \subseteq \sigma_{n,\varepsilon}(A, B)$.
(v) $\sigma_{n,\varepsilon}(\alpha A, \alpha B) = \sigma_{n,\varepsilon}(A, B)$ *for every* $\alpha \neq 0$.
(vi) $\sigma_{n,\varepsilon}(\beta A + \alpha B, B) = \alpha + \beta\sigma_{n,\varepsilon}(A, B)$ *for every* $\alpha, \beta \in \mathbb{C}$ *with* $\beta \neq 0$.
(vii) If X is a Hilbert space, then $\lambda \in \sigma_{n,\varepsilon}(A, B)$ if and only if $\overline{\lambda} \in \sigma_{n,\varepsilon}(A^*, B^*)$.

Proof (i) Since $\sigma(A, B) \subseteq \sigma_{n,\varepsilon}(A, B)$ for every $0 < \varepsilon < 1$,

$$\sigma(A, B) \subseteq \bigcap_{0<\varepsilon<1} \sigma_{n,\varepsilon}(A, B).$$

Now if $\lambda \in \bigcap_{0<\varepsilon<1} \sigma_{n,\varepsilon}(A, B)$, then $\lambda \in \sigma_{n,\varepsilon}(A, B)$ for each $0 < \varepsilon < 1$, and hence there are two possibilities.

(1) If $\lambda \in \sigma(A, B)$, then we get the desired result.

(2) If $\left\|(\lambda B - A)^{2^n}\right\|^{\frac{1}{2^n}} \left\|(\lambda B - A)^{-2^n}\right\|^{\frac{1}{2^n}} \geq \varepsilon^{-1}$ for every $0 < \varepsilon < 1$, then as $\varepsilon \to 0^+$

$$\left\|(\lambda B - A)^{2^n}\right\|^{\frac{1}{2^n}} \left\|(\lambda B - A)^{-2^n}\right\|^{\frac{1}{2^n}} \to \infty.$$

Hence $\lambda \in \sigma(A, B)$.

(ii) If $\lambda \in \sigma(A, B)$, the result is true. Let $\lambda \in \sigma_{n,\varepsilon_1}(A, B) \setminus \sigma(A, B)$ and $0 < \varepsilon_1 \leq \varepsilon_2 < 1$. Then

$$\left\|(\lambda B - A)^{2^n}\right\|^{\frac{1}{2^n}} \left\|(\lambda B - A)^{-2^n}\right\|^{\frac{1}{2^n}} \geq \frac{1}{\varepsilon_1} \geq \frac{1}{\varepsilon_2}.$$

(iii) Suppose $B \in GL(X)$, then $\sigma(A, B) = \sigma\left(AB^{-1}\right)$. Further $\lambda_m \in \mathbb{C} \setminus \sigma(A, B)$ and $\lambda_m \to \lambda$ for some $\lambda \notin \sigma(A, B)$. Then,

$$\left\|(\lambda_m B - A)^{2^n}\right\|^{\frac{-1}{2^n}} \left\|(\lambda_m B - A)^{-2^n}\right\|^{\frac{-1}{2^n}} \to \left\|(\lambda B - A)^{2^n}\right\|^{\frac{-1}{2^n}} \left\|(\lambda B - A)^{-2^n}\right\|^{\frac{-1}{2^n}}.$$

If $\lambda_m \in \mathbb{C} \setminus \sigma(A, B)$ and $\lambda_m \to \lambda$ for some $\lambda \in \sigma(A, B)$. From Lemma 10.17 of [18], $\left\|\left(\lambda_m I - AB^{-1}\right)^{-2^n}\right\| \to \infty$ and

$$\left\|(\lambda_m B - A)^{2^n}\right\|^{\frac{-1}{2^n}} \left\|(\lambda_m B - A)^{-2^n}\right\|^{\frac{-1}{2^n}} \to 0.$$

Hence $\lambda \mapsto \left\|(\lambda B - A)^{2^n}\right\|^{\frac{-1}{2^n}} \left\|(\lambda B - A)^{-2^n}\right\|^{\frac{-1}{2^n}}$ is continuous for every $B \in GL(X)$.

(iv) The result follows from the fact that

$$\frac{1}{\left\|(\lambda B - A)^{2^{n+1}}\right\|^{\frac{1}{2^{n+1}}}} \geq \frac{1}{\left\|(\lambda B - A)^{2^n}\right\|^{\frac{1}{2^{n+1}}} \left\|(\lambda B - A)^{2^n}\right\|^{\frac{1}{2^{n+1}}}}$$
$$= \frac{1}{\left\|(\lambda B - A)^{2^n}\right\|^{\frac{1}{2^n}}},$$

and

$$\frac{1}{\left\|(\lambda B - A)^{-2^{n+1}}\right\|^{\frac{1}{2^{n+1}}}} \geq \frac{1}{\left\|(\lambda B - A)^{-2^n}\right\|^{\frac{1}{2^{n+1}}} \left\|(\lambda B - A)^{-2^n}\right\|^{\frac{1}{2^{n+1}}}}$$
$$= \frac{1}{\left\|(\lambda B - A)^{-2^n}\right\|^{\frac{1}{2^n}}}.$$

(v) Let $\alpha \neq 0$, then $\sigma(\alpha A, \alpha B) = \sigma(A, B)$. Also $\lambda \in \sigma_{n,\varepsilon}(\alpha A, \alpha B) \setminus \sigma(\alpha A, \alpha B)$,

$$\iff \left\|(\lambda \alpha B - \alpha A)^{2^n}\right\|^{\frac{1}{2^n}} \left\|(\lambda \alpha B - \alpha A)^{-2^n}\right\|^{\frac{1}{2^n}} \geq \varepsilon^{-1}$$
$$\iff |\alpha| \left\|(\lambda B - A)^{2^n}\right\|^{\frac{1}{2^n}} \frac{1}{|\alpha|} \left\|(\lambda B - A)^{-2^n}\right\|^{\frac{1}{2^n}} \geq \varepsilon^{-1}$$
$$\iff \lambda \in \sigma_{n,\varepsilon}(A, B) \setminus \sigma(A, B).$$

(vi) Let $\alpha, \beta \in \mathbb{C}$ and $\beta \neq 0$, then

$$\sigma(\beta A + \alpha B, B) = \{\lambda \in \mathbb{C} : \lambda B - \beta A - \alpha B \notin GL(X)\}$$
$$= \left\{\lambda \in \mathbb{C} : \frac{\lambda - \alpha}{\beta} B - A \notin GL(X)\right\}.$$

i.e., $\lambda \in \sigma(\beta A + \alpha B, B) \iff \frac{\lambda-\alpha}{\beta} \in \sigma(A, B) \iff \lambda \in \alpha + \beta\sigma(A, B)$.
If $\lambda \in \sigma_{n,\varepsilon}(\beta A + \alpha B, B) \setminus \sigma(\beta A + \alpha B, B)$, then

$$\left\|(\lambda B - \beta A - \alpha B)^{2^n}\right\|^{\frac{1}{2^n}} \left\|(\lambda B - \beta A - \alpha B)^{-2^n}\right\|^{\frac{1}{2^n}} \geq \frac{1}{\varepsilon}$$
$$|\beta| \left\|\left(\frac{\lambda - \alpha}{\beta} B - A\right)^{2^n}\right\|^{\frac{1}{2^n}} \frac{1}{|\beta|} \left\|\left(\frac{\lambda - \alpha}{\beta} B - A\right)^{-2^n}\right\|^{\frac{1}{2^n}} \geq \frac{1}{\varepsilon}$$
$$\left\|\left(\frac{\lambda - \alpha}{\beta} B - A\right)^{2^n}\right\|^{\frac{1}{2^n}} \left\|\left(\frac{\lambda - \alpha}{\beta} B - A\right)^{-2^n}\right\|^{\frac{1}{2^n}} \geq \frac{1}{\varepsilon}$$

Hence $\lambda \in \alpha + \beta\sigma_{n,\varepsilon}(A, B) \setminus \sigma(A, B)$. The other inclusion also follows similarly.

(vii) We have $\lambda \in \sigma(A, B)$ if and only if $\bar{\lambda} \in \sigma(A^*, B^*)$. If $\lambda \in \sigma_{n,\varepsilon}(A, B) \setminus \sigma(A, B)$, then

$$\left\|(\lambda B - A)^{2^n}\right\|^{\frac{1}{2^n}} \left\|(\lambda B - A)^{-2^n}\right\|^{\frac{1}{2^n}} = \left\|\left(\bar{\lambda} B^* - A^*\right)^{2^n}\right\|^{\frac{1}{2^n}} \left\|\left(\bar{\lambda} B^* - A^*\right)^{-2^n}\right\|^{\frac{1}{2^n}}$$
$$\geq \frac{1}{\varepsilon}.$$

The other inclusion also follows similarly. □

Theorem 3 *Let $A, B \in BL(X)$, $n \in \mathbb{Z}_+ \cup \{0\}$ and $0 < \varepsilon < 1$. Further if $B \in \Omega_{n,\varepsilon}$, then the following are true.*

(i) If $\lambda \in \sigma_{n,\varepsilon}(A, B)$, then $|\lambda| \leq \frac{1+\varepsilon\|B\|\|B^{-1}\|}{1-\varepsilon\|B\|\|B^{-1}\|}\|AB^{-1}\|$.
(ii) $\sigma_{n,\varepsilon}(A, B)$ is compact.
(iii) If $AB = BA$ and $k := \|B^{2^n}\|^{\frac{1}{2^n}}\|B^{-2^n}\|^{\frac{1}{2^n}}$, then $0 < \varepsilon k < 1$ and

$$\sigma_{n,\frac{\varepsilon}{k}}\left(AB^{-1}\right) \subseteq \sigma_{n,\varepsilon}(A, B) \subseteq \sigma_{n,\varepsilon k}\left(AB^{-1}\right).$$

(iv) If $A \in GL(X)$, $AB = BA$, and $k := \|A^{2^n}\|^{\frac{1}{2^n}}\|A^{-2^n}\|^{\frac{1}{2^n}}\|B^{2^n}\|^{\frac{1}{2^n}}\|B^{-2^n}\|^{\frac{1}{2^n}}$. For $0 < \varepsilon k < 1$,

$$0 \neq \lambda \in \sigma_{n,\varepsilon}\left(A^{-1}, B\right) \Longrightarrow \frac{1}{\lambda} \in \sigma_{n,\varepsilon k}\left(A, B^{-1}\right),$$

and

$$0 \neq \lambda \in \sigma_{n,\varepsilon}\left(A, B^{-1}\right) \Longrightarrow \frac{1}{\lambda} \in \sigma_{n,\varepsilon k}\left(A^{-1}, B\right).$$

Proof (i) If $\lambda \in \sigma(A, B) = \sigma\left(AB^{-1}\right)$ then

$$|\lambda| \leq \|AB^{-1}\| \leq \frac{1 + \varepsilon\|B\|\|B^{-1}\|}{1 - \varepsilon\|B\|\|B^{-1}\|}\|AB^{-1}\|.$$

Suppose $|\lambda| > \|AB^{-1}\|$, then $\left(\lambda I - AB^{-1}\right)$ is invertible and

$$\|(\lambda B - A)^{-2^n}\|^{\frac{1}{2^n}} \leq \|B^{-1}\|\left\|\left(\lambda I - AB^{-1}\right)^{-1}\right\| \leq \|B^{-1}\|\frac{1}{|\lambda| - \|AB^{-1}\|}.$$

Thus

$$1 \leq \varepsilon\|(\lambda B - A)^{2^n}\|^{\frac{1}{2^n}}\|(\lambda B - A)^{-2^n}\|^{\frac{1}{2^n}} \leq \varepsilon\|B\|\|B^{-1}\|\frac{|\lambda| + \|AB^{-1}\|}{|\lambda| - \|AB^{-1}\|}.$$

On simplification,

$$|\lambda| \leq \frac{1 + \varepsilon\|B\|\|B^{-1}\|}{1 - \varepsilon\|B\|\|B^{-1}\|}\|AB^{-1}\|.$$

(ii) Since $\sigma_{n,\varepsilon}(A, B)$ is a Ransford spectrum (Theorem 1), $\sigma_{n,\varepsilon}(A, B)$ is compact ([17]).
(iii) We have $\sigma(A, B) = \sigma\left(AB^{-1}\right)$. Since $B \in \Omega_{n,\varepsilon}$, $\|B^{2^n}\|^{\frac{1}{2^n}}\|B^{-2^n}\|^{\frac{1}{2^n}} < \varepsilon^{-1}$ and hence $0 < \varepsilon k < 1$ where $k = \|B^{2^n}\|^{\frac{1}{2^n}}\|B^{-2^n}\|^{\frac{1}{2^n}}$.

Suppose $AB = BA$ and $\lambda \in \sigma_{n,\frac{\varepsilon}{k}}\left(AB^{-1}\right) \setminus \sigma\left(AB^{-1}\right)$ where $k = \left\|B^{2^n}\right\|^{\frac{1}{2^n}} \left\|B^{-2^n}\right\|^{\frac{1}{2^n}}$. Then $\frac{\left\|B^{2^n}\right\|^{\frac{1}{2^n}} \left\|B^{-2^n}\right\|^{\frac{1}{2^n}}}{\varepsilon}$ is

$$
\begin{aligned}
&\leq \left\|\left(\lambda I - AB^{-1}\right)^{2^n}\right\|^{\frac{1}{2^n}} \left\|\left(\lambda I - AB^{-1}\right)^{-2^n}\right\|^{\frac{1}{2^n}} \\
&= \left\|\left[(\lambda B - A)\, B^{-1}\right]^{2^n}\right\|^{\frac{1}{2^n}} \left\|\left[B\,(\lambda B - A)^{-1}\right]^{2^n}\right\|^{\frac{1}{2^n}} \\
&\leq \left\|(\lambda B - A)^{2^n}\right\|^{\frac{1}{2^n}} \left\|B^{-2^n}\right\|^{\frac{1}{2^n}} \left\|B^{2^n}\right\|^{\frac{1}{2^n}} \left\|(\lambda B - A)^{-2^n}\right\|^{\frac{1}{2^n}}.
\end{aligned}
$$

Hence $\lambda \in \sigma_{n,\varepsilon}(A, B) \setminus \sigma(A, B)$. Further assume that $\lambda \in \sigma_{n,\varepsilon}(A, B) \setminus \sigma(A, B)$, then

$$
\begin{aligned}
\frac{1}{\varepsilon} &\leq \left\|(\lambda B - A)^{2^n}\right\|^{\frac{1}{2^n}} \left\|(\lambda B - A)^{-2^n}\right\|^{\frac{1}{2^n}} \\
&= \left\|\left[\left(\lambda I - AB^{-1}\right) B\right]^{2^n}\right\|^{\frac{1}{2^n}} \left\|\left[\left(\lambda I - AB^{-1}\right) B\right]^{-2^n}\right\|^{\frac{1}{2^n}} \\
&\leq \left\|\left(\lambda I - AB^{-1}\right)^{2^n}\right\|^{\frac{1}{2^n}} \left\|B^{2^n}\right\|^{\frac{1}{2^n}} \left\|B^{-2^n}\right\|^{\frac{1}{2^n}} \left\|\left(\lambda I - AB^{-1}\right)^{-2^n}\right\|^{\frac{1}{2^n}}.
\end{aligned}
$$

Hence $\lambda \in \sigma_{n,\varepsilon k}\left(AB^{-1}\right) \setminus \sigma\left(AB^{-1}\right)$ where $k = \left\|B^{2^n}\right\|^{\frac{1}{2^n}} \left\|B^{-2^n}\right\|^{\frac{1}{2^n}}$.

(iv) Suppose $A \in GL(X)$, $AB = BA$, $k := \left\|A^{2^n}\right\|^{\frac{1}{2^n}} \left\|A^{-2^n}\right\|^{\frac{1}{2^n}} \left\|B^{2^n}\right\|^{\frac{1}{2^n}} \left\|B^{-2^n}\right\|^{\frac{1}{2^n}}$, and $0 < \varepsilon k < 1$. Let $\lambda \neq 0$, then

$$
\begin{aligned}
\lambda \in \sigma\left(A^{-1}, B\right) &\Longleftrightarrow \lambda B - A^{-1} \text{ is not invertible} \\
&\Longleftrightarrow -\lambda B\left(\frac{B^{-1}}{\lambda} - A\right) A^{-1} \text{ is not invertible} \\
&\Longleftrightarrow \frac{1}{\lambda} \in \sigma\left(A, B^{-1}\right).
\end{aligned}
$$

Further if $\lambda \in \sigma_{n,\varepsilon}\left(A^{-1}, B\right) \setminus \sigma\left(A^{-1}, B\right)$, then

$$
\begin{aligned}
\frac{1}{\varepsilon} &\leq \left\|\left(\lambda B - A^{-1}\right)^{2^n}\right\|^{\frac{1}{2^n}} \left\|\left(\lambda B - A^{-1}\right)^{-2^n}\right\|^{\frac{1}{2^n}} \\
&= \left\|\left[-\lambda B\left(\frac{B^{-1}}{\lambda} - A\right) A^{-1}\right]^{2^n}\right\|^{\frac{1}{2^n}} \left\|\left[-(\lambda)^{-1} A\left(\frac{B^{-1}}{\lambda} - A\right)^{-1} B^{-1}\right]^{2^n}\right\|^{\frac{1}{2^n}} \\
&\leq |\lambda| \left\|B^{2^n}\right\|^{\frac{1}{2^n}} \left\|\left(\frac{B^{-1}}{\lambda} - A\right)^{2^n}\right\|^{\frac{1}{2^n}} \left\|A^{-2^n}\right\|^{\frac{1}{2^n}} \left|\lambda^{-1}\right| \left\|A^{2^n}\right\|^{\frac{1}{2^n}} \\
&\quad \left\|\left(\frac{B^{-1}}{\lambda} - A\right)^{-2^n}\right\|^{\frac{1}{2^n}} \left\|B^{-2^n}\right\|^{\frac{1}{2^n}}.
\end{aligned}
$$

Hence $\frac{1}{\lambda} \in \sigma_{n,\varepsilon k}\left(A, B^{-1}\right) \setminus \sigma\left(A, B^{-1}\right)$. Next assume that $\lambda \in \sigma_{n,\varepsilon}\left(A, B^{-1}\right) \setminus \sigma\left(A, B^{-1}\right)$. Then,

$$\begin{aligned}
\frac{1}{\varepsilon} &\leq \left\|\left(\lambda B^{-1} - A\right)^{2^n}\right\|^{\frac{1}{2^n}} \left\|\left(\lambda B^{-1} - A\right)^{-2^n}\right\|^{\frac{1}{2^n}} \\
&= \left\|\left[-(\lambda) B^{-1}\left(\frac{B}{\lambda} - A^{-1}\right) A\right]^{2^n}\right\|^{\frac{1}{2^n}} \left\|\left[-(\lambda)^{-1} A^{-1}\left(\frac{B}{\lambda} - A^{-1}\right)^{-1} B\right]^{2^n}\right\|^{\frac{1}{2^n}} \\
&\leq |\lambda| \left\|B^{-2^n}\right\|^{\frac{1}{2^n}} \left\|\left(\frac{B}{\lambda} - A^{-1}\right)^{2^n}\right\|^{\frac{1}{2^n}} \left\|A^{2^n}\right\|^{\frac{1}{2^n}} \left|\lambda^{-1}\right| \left\|A^{-2^n}\right\|^{\frac{1}{2^n}} \left\|B^{2^n}\right\|^{\frac{1}{2^n}} \\
&\quad \left\|\left[\left(\frac{B}{\lambda} - A^{-1}\right)^{-1}\right]^{2^n}\right\|^{\frac{1}{2^n}}.
\end{aligned}$$

Hence $\frac{1}{\lambda} \in \sigma_{n,\varepsilon k}\left(A^{-1}, B\right) \setminus \sigma\left(A^{-1}, B\right)$. □

Proposition 1 *Let* $A, B \in BL(X)$, $V \in GL(X)$, $n \in \mathbb{Z}_+ \cup \{0\}$, *and* $0 < \varepsilon < 1$. *Define* $T = VAV^{-1}$, $S = VBV^{-1}$, *and* $k = \|V\|^{\frac{1}{2^n}} \left\|V^{-1}\right\|^{\frac{1}{2^n}}$. *Then*

$$\sigma_{n,\frac{\varepsilon}{k^2}}(T, S) \subseteq \sigma_{n,\varepsilon}(A, B) \subseteq \sigma_{n,\varepsilon k^2}(T, S),$$

for every $0 < \frac{\varepsilon}{k^2} \leq \varepsilon k^2 < 1$.

Proof Since $T = VAV^{-1}$ and $S = VBV^{-1}$,

$$\sigma(A, B) = \{\lambda \in \mathbb{C} : \lambda V^{-1} S V - V^{-1} T V \notin GL(X)\} = \sigma(T, S).$$

Define $k = \|V\|^{\frac{1}{2^n}} \left\|V^{-1}\right\|^{\frac{1}{2^n}}$, then $\left\|(\lambda S - T)^{2^n}\right\|^{\frac{1}{2^n}} \left\|[(\lambda S - T)^{-1}]^{2^n}\right\|^{\frac{1}{2^n}}$ is

$$\begin{aligned}
&= \left\|\left(\lambda VBV^{-1} - VAV^{-1}\right)^{2^n}\right\|^{\frac{1}{2^n}} \left\|\left[\left(\lambda VBV^{-1} - VAV^{-1}\right)^{-1}\right]^{2^n}\right\|^{\frac{1}{2^n}} \\
&= \left\|\left[V(\lambda B - A) V^{-1}\right]^{2^n}\right\|^{\frac{1}{2^n}} \left\|\left[V^{-1}(\lambda B - A)^{-1} V\right]^{2^n}\right\|^{\frac{1}{2^n}} \\
&\leq \left[\|V\| \left\|(\lambda B - A)^{2^n}\right\| \left\|V^{-1}\right\| \left\|V^{-1}\right\| \left\|(\lambda B - A)^{-2^n}\right\| \|V\|\right]^{\frac{1}{2^n}} \\
&\leq k^2 \left\|(\lambda B - A)^{2^n}\right\|^{\frac{1}{2^n}} \left\|(\lambda B - A)^{-2^n}\right\|^{\frac{1}{2^n}}.
\end{aligned}$$

Hence $\sigma_{n,\frac{\varepsilon}{k^2}}(T, S) \subseteq \sigma_{n,\varepsilon}(A, B)$. Similarly $\left\|(\lambda B - A)^{2^n}\right\|^{\frac{1}{2^n}} \left\|\left[(\lambda B - A)^{-1}\right]^{2^n}\right\|^{\frac{1}{2^n}}$ is

$$= \left\| \left[V^{-1} (\lambda S - T) V \right]^{2^n} \right\|^{\frac{1}{2^n}} \left\| \left[V (\lambda S - T)^{-1} V^{-1} \right]^{2^n} \right\|^{\frac{1}{2^n}}$$
$$\leq \left[\left\| V^{-1} \right\| \left\| (\lambda S - T)^{2^n} \right\| \|V\| \|V\| \left\| (\lambda S - T)^{-2^n} \right\| \left\| V^{-1} \right\| \right]^{\frac{1}{2^n}}$$
$$\leq k^2 \left\| (\lambda S - T)^{2^n} \right\|^{\frac{1}{2^n}} \left\| (\lambda S - T)^{-2^n} \right\|^{\frac{1}{2^n}}.$$

Hence $\sigma_{n,\varepsilon} (A, B) \subseteq \sigma_{n,\varepsilon k^2} (T, S)$. □

The following theorem presents some equivalent definitions for (n, ε)-condition spectrum of matrix pencils.

Theorem 4 *Let X be finite dimensional, $A, B \in BL(X)$, $n \in \mathbb{Z}_+ \cup \{0\}$, and $0 < \varepsilon < 1$. Then the following are equivalent.*

(i) $\lambda \in \sigma_{n,\varepsilon} (A, B)$.
(ii) $\left\| (\lambda B - A)^{2^n} x \right\| \leq \varepsilon^{2^n} \left\| (\lambda B - A)^{2^n} \right\|$ *for some* $x \in X$ *with* $\|x\| = 1$.
(iii) $\left[(\lambda B - A)^{2^n} - D \right] x = 0$ *for some* $D \in BL(X)$ *with* $\|D\| \leq \varepsilon^{2^n} \left\| (\lambda B - A)^{2^n} \right\|$ *and* $x \in X$ *with* $\|x\| = 1$.

Proof If $\lambda \in \sigma (A, B)$, then take x as the normalized eigenvector corresponding to the generalized eigenvalue λ and $D = 0$. Hence (i), (ii) and (iii) are equivalent.
(i) $\Longrightarrow$ (ii). If $\lambda \in \sigma_{n,\varepsilon} (A, B) \setminus \sigma (A, B)$, then $\left\| (\lambda B - A)^{-2^n} \right\| \geq \frac{1}{\varepsilon^{2^n} \left\| (\lambda B - A)^{2^n} \right\|}$.
Hence there exists $y \in X$ with $y \neq 0$ such that $\left\| (\lambda B - A)^{-2^n} y \right\| \geq \frac{\|y\|}{\varepsilon^{2^n} \left\| (\lambda B - A)^{2^n} \right\|}$.
Define $\tilde{x} = (\lambda B - A)^{-2^n} y$, then we get

$$\left\| (\lambda B - A)^{2^n} \frac{\tilde{x}}{\|\tilde{x}\|} \right\| \leq \varepsilon^{2^n} \left\| (\lambda B - A)^{2^n} \right\|.$$

Define $x = \frac{\tilde{x}}{\|\tilde{x}\|}$, then $\|x\| = 1$, and $\left\| (\lambda B - A)^{2^n} x \right\| \leq \varepsilon^{2^n} \left\| (\lambda B - A)^{2^n} \right\|$.
(ii) $\Longrightarrow$ (iii). Suppose there exists $x \in X$ with $\|x\| = 1$ and $\left\| (\lambda B - A)^{2^n} x \right\| \leq \varepsilon^{2^n} \left\| (\lambda B - A)^{2^n} \right\|$. Then there exists $\phi \in X'$ with $\|\phi\| = 1$ and $\phi (x) = 1$. Define the rank one operator $D : X \to X$ by

$$Dy = \phi (y) (\lambda B - A)^{2^n} x.$$

Then $||D|| \leq \varepsilon^{2^n} \left\| (\lambda B - A)^{2^n} \right\|$ and $\left[(\lambda B - A)^{2^n} - D \right] x = 0$.
(iii) $\Longrightarrow$ (i). Suppose there exists $D \in BL(X)$ with $\|D\| \leq \varepsilon^{2^n} \left\| (\lambda B - A)^{2^n} \right\|$ and a unit vector $x \in X$ such that $\left[(\lambda B - A)^{2^n} - D \right] x = 0$. For $\lambda \notin \sigma (A, B)$, define $x = (\lambda B - A)^{-2^n} Dx$. Then

$$1 = \|x\| = \left\| (\lambda B - A)^{-2^n} Dx \right\| \leq \left\| (\lambda B - A)^{-2^n} \right\| \varepsilon^{2^n} \left\| (\lambda B - A)^{2^n} \right\|.$$

Hence $\left\| (\lambda B - A)^{2^n} \right\| \left\| (\lambda B - A)^{-2^n} \right\| \geq \frac{1}{\varepsilon^{2^n}}$, and $\lambda \in \sigma_{n,\varepsilon} (A, B) \setminus \sigma (A, B)$. □

The following result shows that $\sigma_{n,\varepsilon}(A, B)$ contains a neighbourhood of $\sigma(A, B)$. We denote $D(z_0, r) = \{z \in \mathbb{C} : |z - z_0| \le r\}$.

Theorem 5 *Let $A \in BL(X)$, $B \in GL(X)$, $n \in \mathbb{Z}_+ \cup \{0\}$, and $0 < \varepsilon < 1$. Then for each $\lambda \in \sigma(A, B)$ there exists $r > 0$ such that $D(\lambda, r) \subseteq \sigma_{n,\varepsilon}(A, B)$.*

Proof Suppose there exists $D(\lambda, r) \cap \sigma_{n,\varepsilon}(A, B)^c \ne \emptyset$ for some $\lambda \in \sigma(A, B)$ and each $r > 0$. Then there exists a sequence $\lambda_m \in \sigma_{n,\varepsilon}(A, B)^c$ such that $\lambda_m \to \lambda$. From (iii) of Theorem 2 $\left\|(\lambda_m B - A)^{2^n}\right\|^{\frac{-1}{2^n}} \left\|(\lambda_m B - A)^{-2^n}\right\|^{\frac{-1}{2^n}} > \varepsilon$ for every m and $\left\|(\lambda B - A)^{2^n}\right\|^{\frac{-1}{2^n}} \left\|(\lambda B - A)^{-2^n}\right\|^{\frac{-1}{2^n}} = 0$. This is a contradiction. □

3 (n, ε)-Condition Spectral Mapping Theorem

Let $A \in BL(X)$ and f be an analytic function on an open set containing $\sigma(A)$. The Spectral Mapping Theorem gives $f(\sigma(A)) = \sigma(f(A))$. In this section, we give an analogue of the spectral mapping theorem for (n, ε)-condition spectrum of operator pencils. Similar studies have been done for pseudospectrum, (n, ε)-pseudospectrum, condition spectrum and determinant spectrum, refer to [1, 9, 10, 15]. The following is the spectral mapping theorem for operator pencils. For proof one may refer to [10].

Theorem 6 *Let $A \in BL(X)$, $B \in GL(X)$, f be analytic on Ω, an open set containing $\sigma(A, B)$, and Γ be the any closed contour enclosing Ω. Define*

$$f(A, B) = \frac{1}{2\pi i} \int_\Gamma f(z)\ (zB - A)^{-1}\ dz.$$

Then $B f(A, B) = f\left(AB^{-1}\right)$ and $f(\sigma(A, B)) = \sigma(B f(A, B))$.

The following example shows the Spectral Mapping Theorem of operator pencils is not true as it is for the (n, ε)-condition spectrum.

Example Consider $A, B \in \left(BL\left(\ell^1\right), \|\cdot\|_1\right)$ defined by $A(x_1, x_2, \dots) = (0, 2x_2, 0, 0, \dots)$ and $B(x_1, x_2, \dots) = (2x_1, -2x_2, 2x_3, 2x_4, \dots)$. Then $\sigma(A, B) = \{0, -1\}$ and for $\lambda \notin \sigma(A, B)$,

$$(\lambda B - A)^2 (x_1, x_2, \dots) = \left(4\lambda^2 x_1, 4(\lambda + 1)^2 x_2, 4\lambda^2 x_3, \dots\right),$$

$$(\lambda B - A)^{-2} (x_1, x_2, \dots) = \left(\frac{x_1}{4\lambda^2}, \frac{x_2}{4(\lambda + 1)^2}, \frac{x_3}{4\lambda^2}, \dots\right).$$

For $\lambda = x + iy \in \mathbb{C} \setminus \{0, -1\}$,

$$\left\|(\lambda B - A)^2\right\|_1 \left\|(\lambda B - A)^{-2}\right\|_1 = \begin{cases} \dfrac{4|\lambda + 1|^2}{4|\lambda|^2}, & \text{if } x \ge \frac{-1}{2} \\ \dfrac{4|\lambda|^2}{4|\lambda + 1|^2}, & \text{if } x < \frac{-1}{2}. \end{cases}$$

Now consider $f(z) = z^2 + z$, $\lambda = x + iy$, and for $0 < \varepsilon < 1$, then

$$f\left(\sigma_{1,\varepsilon}(A, B)\right) = \begin{cases} \lambda^2 + \lambda \in \mathbb{C} : \dfrac{|\lambda + 1|}{|\lambda|} \geq \frac{1}{\varepsilon} & \text{if } x \geq \frac{-1}{2} \\ \lambda^2 + \lambda \in \mathbb{C} : \dfrac{|\lambda + 1|}{|\lambda|} \leq \varepsilon & \text{if } x < \frac{-1}{2}. \end{cases}$$

From Theorem 6 for $f(z) = z^2 + z$, we have $B f(A, B) = f\left(AB^{-1}\right)$.

$$\left(AB^{-1}\right)(x_1, x_2, x_3, \ldots) = (0, -x_2, 0, 0, \ldots),$$

$$\left(AB^{-1}\right)^2(x_1, x_2, x_3, \ldots) = (0, x_2, 0, \ldots),$$

and

$$\left(AB^{-1}\right)^2 + \left(AB^{-1}\right)(x_1, x_2, x_3, \ldots) = (0, 0, \ldots).$$

Then for $0 < \varepsilon < 1$ and $f(z) = z^2 + z$

$$\sigma_{1,\varepsilon}(B f(A, B)) = \{0\}.$$

Hence for $0 < \varepsilon < 1$ and $f(z) = z^2 + z$,

$$f\left(\sigma_{1,\varepsilon}(A, B)\right) \neq \sigma_{1,\varepsilon}(B f(A, B)).$$

Theorem 7 *Let $A \in BL(X)$, $B \in GL(X)$, $n \in \mathbb{Z}_+ \cup \{0\}$, and f be an analytic function defined on Ω, an open set containing $\sigma(A, B)$. For $0 < \varepsilon < 1$, define*

$$\phi(\varepsilon) = \sup_{\lambda \in \sigma_{n,\varepsilon}(A,B)} \frac{1}{\left\| \left(f(\lambda) I - f\left(AB^{-1}\right)\right)^{2^n} \right\|^{\frac{1}{2^n}} \left\| \left(f(\lambda) I - f\left(AB^{-1}\right)\right)^{-2^n} \right\|^{\frac{1}{2^n}}}.$$

Then $\phi(\varepsilon)$ is well defined, $\lim_{\varepsilon \to 0} \phi(\varepsilon) = 0$ and $f\left(\sigma_{n,\varepsilon}(A, B)\right) \subseteq \sigma_{n,\phi(\varepsilon)}(B f(A, B))$.

Proof Define $g : \mathbb{C} \to \mathbb{R}^+$ by

$$g(\lambda) = \begin{cases} \dfrac{1}{\left\| \left(f(\lambda) I - f\left(AB^{-1}\right)\right)^{2^n} \right\|^{\frac{1}{2^n}} \left\| \left(f(\lambda) I - f\left(AB^{-1}\right)\right)^{-2^n} \right\|^{\frac{1}{2^n}}}, & \text{if } \lambda \notin \sigma(A, B) \\ 0, & \text{if } \lambda \in \sigma(A, B). \end{cases}$$

We claim that g is continuous. If $\lambda \notin \sigma(A, B)$, then $f(\lambda) \notin f(\sigma(A, B)) = \sigma\left(f\left(AB^{-1}\right)\right)$. Suppose $\lambda_m \in \mathbb{C} \setminus \sigma(A, B)$ such that $\lambda_m \to \lambda$. Since f is analytic, $g(\lambda_m) \to g(\lambda)$ and g is continuous on $\mathbb{C} \setminus \sigma(A, B)$. For $\lambda \in \sigma(A, B)$, by Spectral Mapping Theorem for operator pencils

$$f(\lambda) \in f(\sigma(A, B)) = \sigma(B f(A, B)) = \sigma\left(f\left(AB^{-1}\right)\right).$$

If $\lambda_m \in \mathbb{C} \setminus \sigma(A, B)$ be such that $\lambda_m \to \lambda$, then

$$\left\| \left(f(\lambda_m) I - f\left(AB^{-1}\right)\right)^{2^n} \right\|^{\frac{1}{2^n}} \to \left\| \left(f(\lambda) I - f\left(AB^{-1}\right)\right)^{2^n} \right\|^{\frac{1}{2^n}}.$$

Hence it is a bounded sequence, from Lemma 10.17 of [18],

$$\left\| \left(f(\lambda_m) I - f\left(AB^{-1}\right)\right)^{-2^n} \right\| \to \infty.$$

Thus

$$\left\| \left(f(\lambda_m) I - f\left(AB^{-1}\right)\right)^{2^n} \right\|^{\frac{1}{2^n}} \left\| \left(f(\lambda_m) I - f\left(AB^{-1}\right)\right)^{-2^n} \right\|^{\frac{1}{2^n}} \to \infty.$$

Hence $g(\lambda_m) \to 0 = g(\lambda)$. This proves the claim. Also

$$\phi(\varepsilon) = \sup\left\{g(\lambda) : \lambda \in \sigma_{n,\varepsilon}(A, B)\right\}.$$

Since $\sigma_{n,\varepsilon}(A, B)$ is compact $\phi(\varepsilon)$ is well defined. Next we claim that $\lim_{\varepsilon \to 0} \phi(\varepsilon) = 0$. As $\varepsilon \to 0^+$, $\lambda \in \sigma(A, B)$. We have

$$\begin{aligned}\lim_{\varepsilon \to 0} \phi(\varepsilon) &= \sup_{\lambda \in \sigma(A,B)} \frac{1}{\left\| \left(f(\lambda) I - f\left(AB^{-1}\right)\right)^{2^n} \right\|^{\frac{1}{2^n}} \left\| \left(f(\lambda) I - f\left(AB^{-1}\right)\right)^{-2^n} \right\|^{\frac{1}{2^n}}} \\ &= 0.\end{aligned}$$

If $\lambda \in \sigma_{n,\varepsilon}(A, B)$, then

$$\frac{1}{\left\| \left(f(\lambda) I - f\left(AB^{-1}\right)\right)^{2^n} \right\|^{\frac{1}{2^n}} \left\| \left(f(\lambda) I - f\left(AB^{-1}\right)\right)^{-2^n} \right\|^{\frac{1}{2^n}}} = g(\lambda) \le \phi(\varepsilon),$$

and

$$\left\| \left(f(\lambda) I - f\left(AB^{-1}\right)\right)^{2^n} \right\|^{\frac{1}{2^n}} \left\| \left(f(\lambda) I - f\left(AB^{-1}\right)\right)^{-2^n} \right\|^{\frac{1}{2^n}} = \frac{1}{g(\lambda)} \ge \frac{1}{\phi(\varepsilon)}.$$

Thus $f(\lambda) \in \sigma_{n,\phi(\varepsilon)}\left(f\left(AB^{-1}\right)\right) = \sigma_{n,\phi(\varepsilon)}(B\, f(A, B))$ (follows from Theorem 6). Hence

$$f\left(\sigma_{n,\varepsilon}(A, B)\right) \subseteq \sigma_{n,\phi(\varepsilon)}(B\, f(A, B)).$$

□

Theorem 8 *Let $A \in BL(X)$, $B \in GL(X)$, $AB = BA$, and $n \in \mathbb{Z}_+ \cup \{0\}$. Further, let f be an analytic injective function defined on Ω, an open set containing $\sigma(A, B)$ also there exists $0 < \varepsilon' < 1$ such that $\sigma_{n,\varepsilon'}(B f(A, B)) \subseteq f(\Omega)$. For $0 < \varepsilon < \varepsilon' < 1$, define*

$$\psi(\varepsilon) = \sup_{\omega \in f^{-1}(\sigma_{n,\varepsilon}(B f(A,B))) \bigcap \Omega} \frac{\left\|B^{2^n}\right\|^{\frac{1}{2^n}} \left\|B^{-2^n}\right\|^{\frac{1}{2^n}}}{\left\|\left(\omega I - AB^{-1}\right)^{2^n}\right\|^{\frac{1}{2^n}} \left\|\left(\omega I - AB^{-1}\right)^{-2^n}\right\|^{\frac{1}{2^n}}}.$$

Then $\psi(\varepsilon)$ is well defined, $\lim_{\varepsilon \to 0} \psi(\varepsilon) = 0$, and $\sigma_{n,\varepsilon}(B f(A, B)) \subseteq f\left(\sigma_{n,\psi(\varepsilon)}(A, B)\right)$.

Proof Define $h : \mathbb{C} \to \mathbb{R}^+$ by

$$h(\omega) = \begin{cases} \dfrac{\left\|B^{2^n}\right\|^{\frac{1}{2^n}} \left\|B^{-2^n}\right\|^{\frac{1}{2^n}}}{\left\|\left(\omega I - AB^{-1}\right)^{2^n}\right\|^{\frac{1}{2^n}} \left\|\left(\omega I - AB^{-1}\right)^{-2^n}\right\|^{\frac{1}{2^n}}}, & \text{if } \omega \notin \sigma(A, B) \\ 0, & \text{if } \omega \in \sigma(A, B). \end{cases}$$

We claim that h is continuous. Suppose $\omega \notin \sigma(A, B)$ and $\omega_m \in \mathbb{C} \setminus \sigma(A, B)$ such that $\omega_m \to \omega$. Then $\left\|\left(\omega_m I - AB^{-1}\right)^{2^n}\right\|^{\frac{1}{2^n}} \left\|\left(\omega_m I - AB^{-1}\right)^{-2^n}\right\|^{\frac{1}{2^n}}$ converges to $\left\|\left(\omega I - AB^{-1}\right)^{2^n}\right\|^{\frac{1}{2^n}} \left\|\left(\omega I - AB^{-1}\right)^{-2^n}\right\|^{\frac{1}{2^n}}$ and $h(\omega_m) \to h(\omega)$. Hence h is continuous on $\mathbb{C} \setminus \sigma(A, B)$. Suppose $\omega \in \sigma(A, B)$, then $\omega \in \sigma\left(AB^{-1}\right)$ and $h(\omega) = 0$. If $\omega_m \in \mathbb{C} \setminus \sigma(A, B)$ such that $\omega_m \to \omega$, then $\left\|\left(\omega_m I - AB^{-1}\right)^{2^n}\right\| \to \left\|\left(\omega I - AB^{-1}\right)^{2^n}\right\|$ and from Lemma 10.17 of [18],

$$\left\|\left(\omega_m I - AB^{-1}\right)^{-2^n}\right\| \to \infty.$$

Thus $h(\omega_m) \to 0 = h(\omega)$. Hence h is continuous on $\sigma(A, B)$. This proves the claim. Also

$$\psi(\varepsilon) = \sup\{h(\omega) : \omega \in f^{-1}\left(\sigma_{n,\varepsilon}(B f(A, B))\right)\}.$$

Since $f^{-1}\left(\sigma_{n,\varepsilon}(B f(A, B))\right)$ is compact $\psi(\varepsilon)$ is well defined. Next we claim that $\lim_{\varepsilon \to 0} \psi(\varepsilon) = 0$. As $\varepsilon \to 0^+$, $\omega \in f^{-1}(\sigma(B f(A, B)))$. If $\omega \in f^{-1}(\sigma(B f(A, B)))$, then $f(\omega) \in \sigma(B f(A, B)) = f(\sigma(A, B)) = f\left(\sigma\left(AB^{-1}\right)\right)$. Since f is injective, $\omega \in \sigma(A, B)$ and $\left\|\left(\omega I - AB^{-1}\right)^{-2^n}\right\|^{\frac{1}{2^n}} \to \infty$. Thus $h(w) = 0$.

$$\lim_{\varepsilon\to 0}\psi(\varepsilon) = \sup_{w\in f^{-1}(\sigma_{n,\varepsilon}(B\,f(A,B)))} \frac{\|B^{2^n}\|^{\frac{1}{2^n}}\|B^{-2^n}\|^{\frac{1}{2^n}}}{\left\|\left(\omega I - AB^{-1}\right)^{2^n}\right\|^{\frac{1}{2^n}}\left\|\left(\omega I - AB^{-1}\right)^{-2^n}\right\|^{\frac{1}{2^n}}} = 0.$$

If $0 < \varepsilon < \varepsilon' < 1$, then $\sigma_{n,\varepsilon}(B\,f(A,B)) \subseteq \sigma_{n,\varepsilon'}(B\,f(A,B)) \subseteq f(\Omega)$. For $z \in \sigma_{n,\varepsilon}(B\,f(A,B))\bigcap f(\Omega)$, consider $\omega \in \Omega$ such that $z = f(\omega)$, then $\omega \in f^{-1}\left(\sigma_{n,\varepsilon}(B\,f(A,B))\right)\bigcap \Omega$. Thus

$$\frac{\|B^{2^n}\|^{\frac{1}{2^n}}\|B^{-2^n}\|^{\frac{1}{2^n}}}{\left\|\left(\omega I - AB^{-1}\right)^{2^n}\right\|^{\frac{1}{2^n}}\left\|\left(\omega I - AB^{-1}\right)^{-2^n}\right\|^{\frac{1}{2^n}}} = h(\omega) \leq \psi(\varepsilon),$$

and

$$\begin{aligned}\left\|\left(\omega I - AB^{-1}\right)^{2^n}\right\|^{\frac{1}{2^n}}\left\|\left(\omega I - AB^{-1}\right)^{-2^n}\right\|^{\frac{1}{2^n}} &= \frac{\|B^{2^n}\|^{\frac{1}{2^n}}\|B^{-2^n}\|^{\frac{1}{2^n}}}{h(\omega)}\\ &\geq \frac{\|B^{2^n}\|^{\frac{1}{2^n}}\|B^{-2^n}\|^{\frac{1}{2^n}}}{\psi(\varepsilon)}.\end{aligned}$$

From (iii) of Theorem 3,

$$\omega \in \sigma_{n,\frac{\psi(\varepsilon)}{\|B^{2^n}\|^{\frac{1}{2^n}}\|B^{-2^n}\|^{\frac{1}{2^n}}}}\left(AB^{-1}\right) \subseteq \sigma_{n,\psi(\varepsilon)}(A,B),$$

and

$$z = f(\omega) \in f\left(\sigma_{n,\frac{\psi(\varepsilon)}{\|B^{2^n}\|^{\frac{1}{2^n}}\|B^{-2^n}\|^{\frac{1}{2^n}}}}\left(AB^{-1}\right)\right) \subseteq f\left(\sigma_{n,\psi(\varepsilon)}(A,B)\right).$$

Hence for $0 < \varepsilon < \varepsilon' < 1$,

$$\sigma_{n,\varepsilon}(B\,f(A,B)) \subseteq f\left(\sigma_{n,\psi(\varepsilon)}(A,B)\right).$$

□

Remark 2 1. Combining the inclusions of Theorem 7 and 8,

$$f\left(\sigma_{n,\varepsilon}(A,B)\right) \subseteq \sigma_{n,\phi(\varepsilon)}(B\,f(A,B)) \subseteq f\left(\sigma_{n,\psi(\phi(\varepsilon))}(A,B)\right),$$

and

$$\sigma_{n,\varepsilon}(B\,f(A,B)) \subseteq f\left(\sigma_{n,\psi(\varepsilon)}(A,B)\right) \subseteq \sigma_{n,\phi(\psi(\varepsilon))}(B\,f(A,B)).$$

2. If $n = 0$ and $B = I$, the above theorems becomes the analogue of the Spectral Mapping Theorem for condition spectrum of operators ([11]).

3. Since $\lim_{\varepsilon \to 0} \phi(\varepsilon) = 0 = \lim_{\varepsilon \to 0} \psi(\varepsilon)$, and $\sigma(A, B) = \bigcap_{0<\varepsilon<1} \sigma_{n,\varepsilon}(A, B)$ the usual Spectral Mapping Theorem for operator pencils can be deduced from the above theorems.
4. The set inclusions are sharp in the sense that the functions ϕ and ψ cannot be replaced by other functions.

Remark 3 If $AB = BA$, $f(z) = \alpha + \beta z$ where $\alpha, \beta \in \mathbb{C}$ and $\beta \neq 0$. Then

$$\begin{aligned}\phi(\varepsilon) &= \sup_{\lambda \in \sigma_{n,\varepsilon}(A,B)} \frac{1}{|\beta| \left\| \left(\lambda I - AB^{-1}\right)^{2^n} \right\|^{\frac{1}{2^n}} |\beta^{-1}| \left\| \left(\lambda I - AB^{-1}\right)^{-2^n} \right\|^{\frac{1}{2^n}}} \\ &\leq \left\| B^{2^n} \right\|^{\frac{1}{2^n}} \left\| B^{-2^n} \right\|^{\frac{1}{2^n}} \varepsilon.\end{aligned}$$

In the similar way,

$$\psi(\varepsilon) \leq \left\| B^{2^n} \right\|^{\frac{1}{2^n}} \left\| B^{-2^n} \right\|^{\frac{1}{2^n}} \varepsilon.$$

Whenever $\|B\|\|B^{-1}\| = 1$ we have $\psi(\phi(\varepsilon)) = \phi(\psi(\varepsilon)) = \varepsilon$. Thus for $AB = BA$ and $f(z) = \alpha + \beta z$ with $\beta \neq 0$,

$$\sigma_{n,\varepsilon}(B f(A, B)) = f\left(\sigma_{n, \left\| B^{2^n} \right\|^{\frac{1}{2^n}} \left\| B^{-2^n} \right\|^{\frac{1}{2^n}} \varepsilon}(A, B)\right).$$

Example Consider $A, B \in BL\left(\mathbb{C}^2, \|\cdot\|_1\right)$ defined by $A = \begin{bmatrix} 0 & -1 \\ 0 & 0 \end{bmatrix}$ and $B = \begin{bmatrix} 2 & 0 \\ 0 & 2 \end{bmatrix}$. Then $\sigma(A, B) = \{0\}$ and for $\lambda \neq 0$ and $0 < \varepsilon < 1$,

$$(\lambda B - A)^2 = \begin{bmatrix} 4\lambda^2 & 4\lambda \\ 0 & 4\lambda^2 \end{bmatrix} \quad \text{and} \quad \left\|(\lambda B - A)^2\right\|_1 = 4|\lambda|^2 + 4|\lambda|.$$

$$(\lambda B - A)^{-2} = \begin{bmatrix} \frac{1}{4\lambda^2} & \frac{-1}{4\lambda^3} \\ 0 & \frac{1}{4\lambda^2} \end{bmatrix} \quad \text{and} \quad \left\|(\lambda B - A)^{-2}\right\|_1 = \frac{1}{4|\lambda|^2} + \frac{1}{4|\lambda|^3}.$$

$$\begin{aligned}\sigma_{1,\varepsilon}(A, B) &= \left\{\lambda \in \mathbb{C} : \left\|(\lambda B - A)^2\right\|_1 \left\|(\lambda B - A)^{-2}\right\|_1 \geq \frac{1}{\varepsilon^2}\right\} \\ &= \left\{\lambda \in \mathbb{C} : \frac{(1 + |\lambda|)^2}{|\lambda|^2} \geq \frac{1}{\varepsilon^2}\right\} \\ &= \left\{\lambda \in \mathbb{C} : |\lambda| \leq \frac{\varepsilon}{1 - \varepsilon}\right\}.\end{aligned}$$

If $f(z) = e^z$, then

$$\phi(\varepsilon) = \sup_{\lambda \in \sigma_{1,\varepsilon}(A,B)} \frac{1}{\left\| \left(e^{\lambda} I - \exp\left(AB^{-1}\right)\right)^{2} \right\|_1^{\frac{1}{2}} \left\| \left(e^{\lambda} I - \exp\left(AB^{-1}\right)\right)^{-2} \right\|_1^{\frac{1}{2}}}$$

$$= \sup_{\lambda \in \sigma_{1,\varepsilon}(A,B)} \frac{1}{\left(|e^{\lambda} - 1|^2 + |e^{\lambda} - 1|\right)^{\frac{1}{2}} \left(\frac{1}{|e^{\lambda}-1|^2} + \frac{1}{|e^{\lambda}-1|^3}\right)^{\frac{1}{2}}}$$

$$= \sup_{\lambda \in \sigma_{1,\varepsilon}(A,B)} \left(\frac{|e^{\lambda} - 1|}{|e^{\lambda} - 1| + 1}\right) \leq \sum_{k=1}^{\infty} \frac{|\lambda|^k}{k!} \leq \left(e^{\frac{\varepsilon}{1-\varepsilon}} - 1\right).$$

$$\sigma_{1,\varepsilon}\left(B f(A,B)\right) = \sigma_{1,\varepsilon}\left(\exp\left(AB^{-1}\right)\right)$$

$$= \left\{\lambda \in \mathbb{C} : \left\| \left(\lambda I - \exp\left(AB^{-1}\right)\right)^{2} \right\|_1 \left\| \left(\lambda I - \exp\left(AB^{-1}\right)\right)^{-2} \right\|_1 \geq \frac{1}{\varepsilon^2}\right\}$$

$$= \left\{\lambda \in \mathbb{C} : \left(|\lambda - 1| + |\lambda - 1|^2\right) \left(\frac{1}{|\lambda - 1|^3} + \frac{1}{|\lambda - 1|^2}\right) \geq \frac{1}{\varepsilon^2}\right\}$$

$$= \left\{\lambda \in \mathbb{C} : \frac{(1 + |\lambda - 1|)^2}{|\lambda - 1|^2} \geq \frac{1}{\varepsilon^2}\right\}$$

$$= \left\{\lambda \in \mathbb{C} : |\lambda - 1| \leq \frac{\varepsilon}{1 - \varepsilon}\right\}.$$

$$\psi(\varepsilon) = \sup_{e^{\omega} \in \sigma_{1,\varepsilon}\left(\exp\left(AB^{-1}\right)\right)} \frac{\left\|B^2\right\|_1^{\frac{1}{2}} \left\|B^{-2}\right\|_1^{\frac{1}{2}}}{\left\| \left(\omega I - AB^{-1}\right)^{2} \right\|_1^{\frac{1}{2}} \left\| \left(\omega I - AB^{-1}\right)^{-2} \right\|_1^{\frac{1}{2}}}$$

$$= \sup_{e^{\omega} \in \sigma_{1,\varepsilon}\left(\exp\left(AB^{-1}\right)\right)} \frac{1}{\left(|\omega|^2 + |\omega|\right)^{\frac{1}{2}} \left(\frac{1}{|\omega|^2} + \frac{1}{|\omega|^3}\right)^{\frac{1}{2}}}$$

$$= \sup_{e^{\omega} \in \sigma_{1,\varepsilon}\left(\exp\left(AB^{-1}\right)\right)} \left(\frac{|\omega|}{|\omega| + 1}\right)$$

$$\leq \log\left(\frac{1}{1 - \varepsilon}\right).$$

Consequently,

$$e^{(\sigma_{1,\varepsilon}(A,B))} \subseteq \sigma_{1,\varepsilon_1}\left(\exp\left(AB^{-1}\right)\right) \quad \text{where} \quad \varepsilon_1 = \left(e^{\frac{\varepsilon}{1-\varepsilon}} - 1\right),$$
$$\sigma_{1,\varepsilon}\left(\exp\left(AB^{-1}\right)\right) \subseteq e^{(\sigma_{1,\varepsilon_2}(A,B))} \quad \text{where} \quad \varepsilon_2 = \log\left(\frac{1}{1-\varepsilon}\right).$$

4 Stability Analysis of Operator Equations Using (n, ε)-Condition Spectrum

In this section, we demonstrate how the (n, ε)-condition spectrum can be used to measure the stability of the system of operator equations. We calculate an upper bound and lower bound for the relative change in the solution of the operator equations with respect to a relative change in the initial data using the (n, ε)-condition spectrum. Consider the operator equation defined by

$$(A - \lambda B)^{2^n} x = b,$$

where $A, B \in BL(X)$, $b \in X$, and $n \in \mathbb{Z}_+ \cup \{0\}$. Suppose $\lambda \notin \sigma_{n,\varepsilon}(A, B)$ for some $0 < \varepsilon < 1$. Further let δb be small perturbation on b and δx be the corresponding change in the solution x. Then

$$(A - \lambda B)^{2^n} (x + \delta x) = b + \delta b.$$

Since $\lambda \notin \sigma(A, B)$,

$$\|x\| \leq \left\|(A - \lambda B)^{-2^n}\right\| \|b\| \quad , \quad \|b\| \leq \left\|(A - \lambda B)^{2^n}\right\| \|x\|, \tag{1}$$

and

$$\|\delta b\| \leq \left\|(A - \lambda B)^{2^n}\right\| \|\delta x\| \quad , \quad \|\delta x\| \leq \left\|(A - \lambda B)^{-2^n}\right\| \|\delta b\|. \tag{2}$$

From (1) and (2),

$$\frac{1}{\left\|(A - \lambda B)^{2^n}\right\| \left\|(A - \lambda B)^{-2^n}\right\|} \frac{\|\delta b\|}{\|b\|} \leq \frac{\|\delta x\|}{\|x\|} \leq \left\|(A - \lambda B)^{2^n}\right\| \left\|(A - \lambda B)^{-2^n}\right\| \frac{\|\delta b\|}{\|b\|}.$$

Define

$$\kappa(A, B, \lambda, n) := \sup\left\{0 < \varepsilon < 1 : \lambda \notin \sigma_{n,\varepsilon}(A, B)\right\}.$$

Then

$$\kappa(A, B, \lambda, n)^{2^n} \frac{\|\delta b\|}{\|b\|} < \frac{\|\delta x\|}{\|x\|} < \frac{1}{\kappa(A, B, \lambda, n)^{2^n}} \frac{\|\delta b\|}{\|b\|}.$$

Example Consider the operator equation $Ax - \lambda x = y$ defined by

$$A = \begin{bmatrix} 0 & 1 & & & \\ \frac{1}{4} & 0 & 1 & & \\ & \ddots & \ddots & \ddots & \\ & & \frac{1}{4} & 0 & 1 \\ & & & \frac{1}{4} & 0 \end{bmatrix}_{10\times 10} \quad \text{and} \quad y = \begin{bmatrix} 1 \\ 1 \\ \vdots \\ 1 \\ 1 \end{bmatrix}_{10\times 1}.$$

The following table gives the relative perturbation of the solution corresponding to $\delta y = \left[0.1\ 0\ 0\ 0\ 0\ 0\ 0\ 0\ 0\ 0\right]^T$ for various values of λ. The upper bound and lower bound found for the relative perturbation of the solution from the condition spectrum of A are also given. The $\|\cdot\|$ denotes the 2-norm.

λ	$\frac{\|\delta x\|}{\|x\|}$	$\kappa(A, I, \lambda, 0)$	$\kappa(A, I, \lambda, 0)\frac{\|\delta y\|}{\|y\|}$	$\frac{1}{\kappa(A,I,\lambda,0)}\frac{\|\delta y\|}{\|y\|}$
2	0.01451478	0.25562528	0.00808358	0.12370754
$2+i$	0.01916041	0.34480349	0.01090364	0.09171245
$3-i$	0.02157139	0.22394260	0.00708168	0.14120929
$1-i$	0.02192375	0.22394260	0.00708168	0.14120929
$2-i$	0.01916041	0.34480349	0.01090364	0.09171245
$3+3i$	0.02634769	0.62518073	0.01976995	0.05058181
$4i$	0.03238426	0.69642891	0.02202301	0.04540704

Example Consider the system of operator equation

$$(A - \lambda B)^{2^4} x = b,$$

where $A = \begin{bmatrix} 1 & 2 & 1 \\ 4 & 2 & 3 \\ 1 & 2 & 2 \end{bmatrix}$, $B = \begin{bmatrix} 4 & 2 & 1 \\ 0 & 4 & 3 \\ 0 & 0 & 5 \end{bmatrix}$ and $b = \begin{bmatrix} 1 \\ 2 \\ 1 \end{bmatrix}$.

The following table gives the relative perturbation of the solution corresponding to $\lambda = 2$ and for certain perturbations in b. The $\|\cdot\|$ denotes the 2-norm.

δb	δx	$\frac{\kappa(A,B,2,4)^{16}\|\delta b\|}{\|b\|}$
$(0.1, 0, 0)^T$	$\left(1.13\times 10^{-16}, 5.52\times 10^{-16}, 6.60\times 10^{-16}\right)^T$	0.0205
$(0.02, 0, 0)^T$	$\left(2.26\times 10^{-17}, 1.10\times 10^{-16}, 1.32\times 10^{-16}\right)^T$	0.0041
$(0, 0.2, 0)^T$	$\left(-1.09\times 10^{-15}, -1.54\times 10^{-16}, 1.65\times 10^{-16}\right)^T$	0.0411
$(0, 0, 0.4)^T$	$\left(1.69\times 10^{-15}, -2.66\times 10^{-15}, 9.06\times 10^{-16}\right)^T$	0.0822
$(0, 0.03, 0)^T$	$\left(-1.64\times 10^{-16}, -2.32\times 10^{-17}, 2.48\times 10^{-17}\right)^T$	0.0061

$\frac{\|\delta x\|}{\|x\|}$	$\frac{\|\delta b\|}{\kappa(A,B,2,4)^{16}\|b\|}$
0.0711	0.0810
0.0142	0.0162
0.0914	0.1621
0.2688	0.3243
0.0137	0.0243

The following table finds the relative perturbation of the solution corresponding to $\lambda = 5 - 0.5i$ and for certain perturbations in b.

δb	$\frac{\kappa(A,B,5-0.5i,4)^{16}\|\delta b\|}{\|b\|}$	$\frac{\|\delta x\|}{\|x\|}$	$\frac{\|\delta b\|}{\kappa(A,B,5-0.5i,4)^{16}\|b\|}$
$(0.1, 0, 0)^T$	0.0047	0.0267	0.3519
$(0.02, 0, 0)^T$	0.0009	0.0053	0.0703
$(0, 0.2, 0)^T$	0.0094	0.0355	0.7039
$(0, 0, 0.04)^T$	0.0018	0.0345	0.1407
$(0, 0.03, 0)^T$	0.0014	0.0053	0.1055

Acknowledgements The first author thanks Department of Higher Education, Government of Kerala, Kerala, India for the financial support through PLEASE scheme. The second author thanks Kerala State Council for Science, Technology and Environment (KSCSTE), Kerala, India (Ref No: KSCSTE/990/2021-FSHP-MS) for financial support.

References

1. Ammar, A., Jeribi, A., Mahfoudhi, K.: The condition pseudospectrum of a operator pencil, Asian-Eur. J. Math. **14**(4), Paper No. 2150057, 12 (2021). MR4224960
2. Ammar, A., Jeribi, A., Mahfoudhi, K.: The condition pseudospectrum subset and related results. J. Pseudo-Differ. Oper. Appl. **11**(1), 491–504 (2020). MR4064425
3. Barreiro, A.K., Bronski, J.C., Rapti, Z.: Applications of a class of Herglotz operator pencils. SIAM J. Math. Anal. **51**(1), 256–275 (2019). MR3907925
4. Ben-Artzi, A., Gohberg, I.: Inertia theorems for operator pencils and applications. Integral Equ. Oper. Theory **21**(3), 270–318 (1995). MR1316544
5. Engström, C., Richter, M.: On the spectrum of an operator pencil with applications to wave propagation in periodic and frequency dependent materials. SIAM J. Appl. Math. **70**(1), 231–247 (2009). MR2505087
6. Guebbai, H.: Generalized spectrum approximation and numerical computation of eigenvalues for Schrödinger's operators. Lobachevskii J. Math. **34**(1), 45–60 (2013). MR3031555
7. Hansen, A.C.: Infinite-dimensional numerical linear algebra: theory and applications. Proc. R. Soc. Lond. Ser. A Math. Phys. Eng. Sci. **466**(2124), 3539–3559 (2010). MR2734774
8. Kamouche et al. S.: Generalized quadratic spectrum approximation in bounded and unbounded cases. Probl. Anal. Issues Anal. **10**(28), 3, 53–70 (2021). MR4338649

9. Khellaf et al. A.: The pseudo-spectrum of operator pencils. Asian-Eur. J. Math. **13**(5), 2050100, 16 (2020). MR4134208
10. Krishna Kumar, G., Augustine, J.: A pseudospectral mapping theorem for operator pencils. Oper. Matrices **16**(3), 709–732 (2022). MR4502031
11. Krishna Kumar, G., Kulkarni, S.H.: An analogue of the spectral mapping theorem for condition spectrum. In: Concrete Operators, Spectral Theory, Operators in Harmonic Analysis and Approximation, pp. 299–316, Oper. Theory Adv. Appl., vol. 236. Birkhäuser/Springer, Basel. MR3203068
12. Kumar, G.K., Kulkarni, S.H.: Linear maps preserving pseudospectrum and condition spectrum. Banach J. Math. Anal. **6**(1), 45–60 (2012). MR2862542
13. Kulkarni, S.H., Sukumar, D.: The condition spectrum, Acta Sci. Math. (Szeged) **74**(3–4), 625–641 (2008). MR2487937
14. Lui, S.-H.: A pseudospectral mapping theorem. Math. Comp. **72**(244), 1841–1854 (2003). MR1986807
15. Mahfoudhi, K.: Generalized determinant pseudo-spectrum of matrix pencils. J. Anal. **29**(1), 163–176 (2021). MR4232811
16. Möller, M., Pivovarchik, V.: Spectral Theory of Operator Pencils, Hermite-Biehler Functions, and Their Applications. Operator Theory: Advances and Applications, vol. 246. Birkhäuser/Springer, Cham (2015). MR3379142
17. Ransford, T.J.: Generalised spectra and analytic multivalued functions. J. London Math. Soc. **29**(2), 306–322. MR0744102
18. Rudin, W.: Functional Analysis. International Series in Pure and Applied Mathematics, 2nd edn. McGraw-Hill, Inc., New York (1991). MR1157815
19. Sukumar, D., Veeramani, S.: Continuity of a condition spectrum and its level sets. J. Aust. Math. Soc. **108**(3), 412–430 (2020). MR4092874
20. Trefethen, L.N., Embree, M.: Spectra and Pseudospectra. Princeton University Press, Princeton (2005). MR2155029

Numerical Solution of One-Dimensional Hyperbolic Telegraph Equation Using Collocation of Cubic B-Splines

Athira Babu and Noufal Asharaf

Abstract In this article, we numerically solve the second-order one-dimensional hyperbolic telegraph equation using cubic B-spline collocation. By selecting newly modified cubic B-splines of accuracy order four, we could numerically approximate the problem's solution with fewer numerical errors than other related literary works. In the procedure, the equation is converted into a system of first-order ordinary differential equations using a transformation and a discretization in the spatial domain. The tridiagonal matrix property of the chosen modified cubic B-splines helps to convert it into a sparse matrix system easily and is then solved for the next time level using the SSP-RK54 scheme. The stability of the numerical technique has been analyzed. The method is stable because all the eigenvalues of the coefficient matrix are found to lie on the left half plane and possess a second-order convergence. Experimental results of some test problems are provided to illustrate the effectiveness of the proposed scheme, and the results show that the method outperforms some of the related numerical methods in the literature.

Keywords Telegraph equation · Cubic B-spines · SSP-RK54 · Collocation · Stability

1 Introduction

Hyperbolic partial differential equations naturally arise in many branches of science, engineering, and atomic physics for modeling vibrations of structures like beams, machines, buildings, etc. The telegraph equation is a critical hyperbolic partial differential equation more suitable than the ordinary diffusion equation in many applications. Oliver Heaviside proposed this model from a transmission line that describes the current or voltage against distance and time on an electrical transmission line.

A. Babu · N. Asharaf (✉)
Cochin University of Science and Technology, Kochi, India
e-mail: noufalasharaf@gmail.com

A. A. Ambily and V. B. Kiran Kumar (eds.), *Semigroups, Algebras and Operator Theory*, Springer Proceedings in Mathematics & Statistics 436,
https://doi.org/10.1007/978-981-99-6349-2_14

He demonstrated that this model could observe wave patterns along the line when there is a reflection of electromagnetic waves on the wire. Further gaining popularity among transmission engineers and researchers, they used the model to characterize telegraph wires. Currently, these equations arise in many real-life applications, where we use radio frequency communication channels, random walk theory, signal analysis, electrodynamics of superconducting materials, and the theory of wave propagations (for details, see [3, 10, 19, 20] and references therein).

The second-order one-dimensional telegraph equation with initial and boundary conditions has a dissipative form as follows:

$$\begin{cases} u_{tt}(x,t) + 2\alpha u_t(x,t) + \beta^2 u(x,t) = u_{xx}(x,t) + f(x,t), & x \in \Omega := (a,b),\ t > 0, \\ u(x,0) = f_1(x), \quad u_t(x,0) = f_2(x), \\ u(a,t) = g_1(t), \quad u(b,t) = g_2(t), \quad 0 < t < \infty, \end{cases} \tag{1}$$

where α and β are known real constants. In recent times, considerable attention has been given to the development of stable algorithms for numerically solving (1). In [12], Mehrdad Lakestani used interpolating scaling functions to solve the problem. It constructed an operational matrix for differentiation and used it to convert the problem into a set of algebraic equations. A similar idea is used in [7] by Mehdi Dehghan to solve the problem with the Chebyshev cardinal functions. K. E. Bicer et al. constructed the operational matrix of derivatives from Bernoulli polynomials in [5]. They have proposed a collocation method, and the problem's solution is expressed as a weighted sum of Bernoulli polynomials. J. Biazar et al. considered the Adomian decomposition method to solve the problem in [4], whereas S. Yousefi used the Galerkin method with Legendre multiwavelet approximation in [21]. In this fascinating paper, S. Yousefi expands the solution on the wavelet basis, and the coefficients are computed using the Galerkin method. Mehdi Dehghan expresses the solution in thin plate splines radial basis functions and determines the unknown coefficients in [6]. A polynomial differential quadrature method is used to solve the two-dimensional case of the problem in [9, 15]. In 2013, R.C. Mittal et al. proposed a cubic spline collocation method [14] for the numerical solution. A modified set of cubic b-spline basis functions are used to expand the solution, and the properties of B-splines convert the problem into a system of ordinary differential equations. Even though they have been solved by strong stability preserving Runge-Kutta methods, the results are not accurate near the boundaries. This is due to the lack of polynomial reproduction of the modification there. Hence a proper improvement of the boundary splines is needed to maintain the optimal accuracy order of four of the standard cubic B-splines in the entire domain. An efficient modification has been proposed in [2] to solve the problem using the differential quadrature technique. In this work, we use the collocation method as in [14] with the modification proposed in [2]. These newly defined basis functions could approximate the solution better when compared to other works [6, 8, 12, 14, 16, 17] in the literature. Since the newly modified cubic B-splines can produce the whole space of polynomials up to degree three, we call the functions to have the optimal accuracy order four. By using such basis functions for the collocation purpose, not only can we maintain the optimal accuracy of the standard cubic B-splines in the numerical solution, but also we could get an approximate solution with less numerical errors as compared with the works that used the old modification that found in [14]. The structure of the paper is as follows. In Sect. 2, the modified set of basis functions with accuracy order four is given. In Sect. 3, we discuss the way of computing the numerical solution with using defined modified set of cubic splines. The convergence of the solution has been analyzed in Sect. 4, and the stability of the scheme has been discussed in Sect. 5. In Sect. 6, we demonstrate the results of some test problems to ensure the computation's effectiveness and we conclude the article with Sect. 7 by adding main findings obtained.

2 Cubic Splines and Boundary Modification

Partition the domain $\Omega := (a, b)$ uniformly into N subintervals with the node points $a = x_0 < x_1 < x_2 < \cdots < x_N = b$, where $x_j = a + jh$, $h = \frac{b-a}{N}$ and $j = 0, 1, \ldots, N$. The standard cubic B-spline $B_j(x)$ is defined to be

$$B_j(x) := \frac{1}{h^3}\begin{cases} (x - x_{j-2})^3, & x \in \left[x_{j-2}, x_{j-1}\right), \\ (x - x_{j-2})^3 - 4(x - x_{j-1})^3, & x \in \left[x_{j-1}, x_j\right), \\ (x_{j+2} - x)^3 - 4(x_{j+1} - x)^3, & x \in \left[x_j, x_{j+1}\right), \\ (x_{j+2} - x)^3, & x \in \left[x_{j+1}, x_{j+2}\right), \\ 0, & \text{otherwise.} \end{cases} \tag{2}$$

Each B_j is compactly supported in the interval $[x_{j-2}, x_{j+2}]$, which is symmetric about x_j and twice continuously differentiable. The function values of B_j's and its derivatives at the node points are described in Table 1. By counting all possible B_j's that have nonzero contribution in the computational domain, we get the set of $N + 3$ B-splies $\{B_{-1}, B_0, B_1, \ldots, B_{N-1}, B_N, B_{N+1}\}$. Here the elements B_{-1} and B_{N+1} are the B-splines that are centered at the undefined node points x_{-1} and x_{N+1} and are supported in the interval $[x_{-3}, x_1]$ and $[x_{N-1}, x_{N+3}]$. To better handle the undefined node points and to reduce the number of functions to $N + 1$, the B-splines $\{B_{-1}, B_0, B_1, \ldots, B_N, B_{N+1}\}$ are arranged effectively without affecting its accuracy order four in [2] and is given as follows:

Table 1 Derivatives of cubic splines at each node

x	x_{j-2}	x_{j-1}	x_j	x_{j+1}	x_{j+2}
$B_j(x)$	0	1	4	1	0
$B_j'(x)$	0	$3/h$	0	$-3/h$	0
$B_j''(x)$	0	$6/h^2$	$-12/h^2$	$6/h^2$	0

$$\begin{aligned}
\tilde{B}_0 &:= B_0 + 4B_{-1}, & \tilde{B}_N &:= B_N + 4B_{N+1},\\
\tilde{B}_1 &:= B_1 - \tfrac{7}{2}B_{-1} + \tfrac{5}{8}B_0, & \tilde{B}_{N-1} &:= B_{N-1} - \tfrac{7}{2}B_{N+1} + \tfrac{5}{8}B_N,\\
\tilde{B}_2 &:= B_2 + \tfrac{88}{37}B_{-1} - \tfrac{21}{37}B_0 - \tfrac{4}{37}B_1, & \tilde{B}_{N-2} &:= B_{N-2} + \tfrac{88}{37}B_{N+1} - \tfrac{21}{37}B_N - \tfrac{4}{37}B_{N-1},\\
\tilde{B}_3 &:= B_3 - B_{-1} + \tfrac{1}{4}B_0 - \tfrac{1}{4}B_2, & \tilde{B}_{N-3} &:= B_{N-3} - B_{N+1} + \tfrac{1}{4}B_N - \tfrac{1}{4}B_{N-2},
\end{aligned} \tag{3}$$

where the set $\{\tilde{B}_0, \tilde{B}_1, \ldots, \tilde{B}_N\}$ is capable to produce the space of polynomials up to degree three and thus it has accuracy order four.

3 B-Spline Collocation

In this section, we describe the collocation method using the above set of basis functions $\tilde{B}_j$. We approximate the solution of Eq. (1) as

$$u(x,t) \approx \sum_{j=0}^{N} d_j(t)\tilde{B}_j(x), \tag{4}$$

where $d_j(t),\ j = 0, 1, \ldots, N$ are the unknown variable coefficients, which are expressed as a function of time t. Since we consider a second-order one-dimensional hyperbolic telegraph equation, in order to apply the collocation procedure, we should convert the second-order equation into a system of first-order ordinary differential equations via a transformation and discretization. Here we use the transformation $w(x,t) := u_t(x,t)$ and write Eq. (1) as a system

$$w = u_t, \quad w_t = -2\alpha w - \beta^2 u + u_{xx} + f(x,t). \tag{5}$$

By (4) we discretize the (5) as

$$\begin{aligned}
w(x_j,t) &= \textstyle\sum_{k=0}^{N} \dot{d}_k(t)\tilde{B}_k(x_j), \quad \text{for } j = 0, \ldots, N,\\
w_t(x_0,t) &= \ddot{g}_1(t),\\
w_t(x_j,t) &= -2\alpha w(x_j,t) - \beta^2 \textstyle\sum_{k=0}^{N} d_k(t)\tilde{B}_k(x_j) + \sum_{k=0}^{N} d_k(t)\tilde{B}_k''(x_j) + f(x_j,t),\\
w_t(x_N,t) &= \ddot{g}_2(t).
\end{aligned}$$

Denote $\dot{w}_j := w_t(x_j,t)$ and rewrite this system using Table 1 as follows

$$\begin{aligned}
\dot{g}_1(t) &= 8\dot{d}_0, & w_{N-3} &= \dot{d}_{N-4} + \tfrac{15}{4}\dot{d}_{N-3} + \dot{d}_{N-2},\\
w_1 &= \dot{d}_0 + \tfrac{37}{8}\dot{d}_1, & w_{N-2} &= \tfrac{144}{37}\dot{d}_{N-2} + \dot{d}_{N-1},\\
w_2 &= \dot{d}_1 + \tfrac{144}{37}\dot{d}_2, & w_{N-1} &= \tfrac{37}{8}\dot{d}_{N-1} + \dot{d}_N,\\
w_3 &= \dot{d}_2 + \tfrac{15}{4}\dot{d}_3 + \dot{d}_4, & \dot{g}_2(t) &= 8\dot{d}_N,\\
w_j &= \dot{d}_{j-1} + 4\dot{d}_j + \dot{d}_{j+1}, \quad j = 4, 5 \ldots N-4,
\end{aligned}$$

$$\begin{aligned}
\dot{w}_0 &= \ddot{g}_1(t),\\
\dot{w}_1 &= -2\alpha w_1 - \beta^2(d_0 + \tfrac{37}{8}d_1) + \tfrac{1}{h^2}(6d_0 - \tfrac{33}{4}d_1 + \tfrac{144}{37}d_2) + f(x_1, t),\\
\dot{w}_2 &= -2\alpha w_2 - \beta^2(d_1 + \tfrac{144}{37}d_2) + \tfrac{1}{h^2}(6d_1 - \tfrac{468}{37}d_2 + 9d_3) + f(x_2, t),\\
\dot{w}_3 &= -2\alpha w_3 - \beta^2(d_2 + \tfrac{15}{4}d_3 + d_4) + \tfrac{1}{h^2}(6d_2 - \tfrac{27}{2}d_3 + 6d_4) + f(x_3, t),\\
\dot{w}_j &= -2\alpha w_j - \beta^2(d_{j-1} + 4d_j + d_{j+1}) + \tfrac{6}{h^2}(d_{j-1} - 2d_j + d_{j+1}) + f(x_j, t),
\end{aligned}$$

$$\begin{aligned}
\dot{w}_{N-3} &= -2\alpha w_{N-3} - \beta^2(d_{N-4} + \tfrac{15}{4}d_{N-3} + d_{N-2}) + \tfrac{1}{h^2}(6d_{N-4} - \tfrac{27}{2}d_{N-3} + 6d_{N-2})\\
&\qquad + f(x_{N-3}, t),\\
\dot{w}_{N-2} &= -2\alpha w_{N-2} - \beta^2(\tfrac{144}{37}d_{N-2} + d_{N-1}) + \tfrac{1}{h^2}(9d_{N-3} - \tfrac{468}{37}d_{N-2} + 6d_{N-1})\\
&\qquad + f(x_{N-2}, t),\\
\dot{w}_{N-1} &= -2\alpha w_{N-1} - \beta^2(\tfrac{37}{8}d_{N-1} + d_N) + \tfrac{1}{h^2}(\tfrac{144}{37}d_{N-2} - \tfrac{33}{4}d_{N-1} + 6d_N)\\
&\qquad + f(x_{N-1}, t),\\
\dot{w}_N &= \ddot{g}_2(t).
\end{aligned}$$

The matrix equivalent of the above set of linear equations as a system of first-order ordinary differential equations can be written as

$$\begin{bmatrix} \dot{d}_0\\ \dot{d}_1\\ \dot{d}_2\\ \dot{d}_3\\ \dot{d}_4\\ \vdots\\ \dot{d}_{N-4}\\ \dot{d}_{N-3}\\ \dot{d}_{N-2}\\ \dot{d}_{N-1}\\ \dot{d}_N \end{bmatrix} = \begin{bmatrix} 8 & 0 & 0 & 0\;0\;0\;0 & \dots & \dots & 0\\ 1 & \frac{37}{8} & 0 & 0\;0\;0\;0 & \dots & \dots & 0\\ 0 & 1 & \frac{144}{37} & 0\;0\;0\;0 & \dots & \dots & 0\\ 0 & 0 & 1 & \frac{15}{4}\;1\;0\;0 & \dots & \dots & 0\\ 0 & 0 & 0 & 1\;4\;1\;0 & \dots & \dots & 0\\ & & \vdots & \vdots & \vdots & & \\ 0 & \dots & \dots & 0\;1\;4\;1 & 0 & 0 & 0\\ 0 & \dots & \dots & 0\;0\;1\;\frac{15}{4} & 1 & 0 & 0\\ 0 & \dots & \dots & 0\;0\;0\;0 & \frac{144}{37} & 1 & 0\\ 0 & \dots & \dots & 0\;0\;0\;0 & 0 & \frac{37}{8} & 1\\ 0 & \dots & \dots & 0\;0\;0\;0 & 0 & 0 & 8 \end{bmatrix}^{-1} \begin{bmatrix} F_0\\ F_1\\ F_2\\ F_3\\ F_4\\ \vdots\\ F_{N-4}\\ F_{N-3}\\ F_{N-2}\\ F_{N-1}\\ F_N \end{bmatrix}, \quad \begin{bmatrix} \dot{w}_0\\ \dot{w}_1\\ \dot{w}_2\\ \dot{w}_3\\ \dot{w}_4\\ \vdots\\ \dot{w}_{N-4}\\ \dot{w}_{N-3}\\ \dot{w}_{N-2}\\ \dot{w}_{N-1}\\ \dot{w}_N \end{bmatrix} = \begin{bmatrix} G_0\\ G_1\\ G_2\\ G_3\\ G_4\\ \vdots\\ G_{N-4}\\ G_{N-3}\\ G_{N-2}\\ G_{N-1}\\ G_N \end{bmatrix}, \tag{6}$$

where $F_0 := \dot{g}_1(t)$, $F_j := w_j$ for $j = 1, 2, \dots N-1$, $F_N := \dot{g}_2(t)$, and

$$\begin{aligned}
G_0 &= \ddot{g}_1(t),\\
G_1 &= -2\alpha w_1 - \beta^2(d_0 + \tfrac{37}{8}d_1) + \tfrac{1}{h^2}(6d_0 - \tfrac{33}{4}d_1 + \tfrac{144}{37}d_2) + f(x_1, t),\\
G_2 &= -2\alpha w_2 - \beta^2(d_1 + \tfrac{144}{37}d_2) + \tfrac{1}{h^2}(6d_1 - \tfrac{468}{37}d_2 + 9d_3) + f(x_2, t),\\
G_3 &= -2\alpha w_3 - \beta^2(d_2 + \tfrac{15}{4}d_3 + d_4) + \tfrac{1}{h^2}(6d_2 - \tfrac{27}{2}d_3 + 6d_4) + f(x_3, t),\\
G_j &= -2\alpha w_j - \beta^2(d_{j-1} + 4d_j + d_{j+1}) + \tfrac{6}{h^2}(d_{j-1} - 2d_j + d_{j+1}) + f(x_j, t),\\
G_{N-3} &= -2\alpha w_{N-3} - \beta^2(d_{N-4} + \tfrac{15}{4}d_{N-3} + d_{N-2}) + \tfrac{1}{h^2}(6d_{N-4} - \tfrac{27}{2}d_{N-3} + 6d_{N-2})\\
&\qquad + f(x_{N-3}, t),\\
G_{N-2} &= -2\alpha w_{N-2} - \beta^2(\tfrac{144}{37}d_{N-2} + d_{N-1}) + \tfrac{1}{h^2}(9d_{N-3} - \tfrac{468}{37}d_{N-2} + 6d_{N-1})\\
&\qquad + f(x_{N-2}, t),
\end{aligned}$$

$$G_{N-1} = -2\alpha w_{N-1} - \beta^2(\tfrac{37}{8}d_{N-1} + d_N) + \tfrac{1}{h^2}(\tfrac{144}{37}d_{N-2} - \tfrac{33}{4}d_{N-1} + 6d_N) + f(x_{N-1}, t),$$
$$G_N = \ddot{g}_1(t).$$

By inverting the tridiagonal matrix mentioned in Eq. (6), we will get a system of first-order ordinary differential equations which can be easily solved by Runge-Kutta methods. This is the primary purpose of splitting the Telegraph equation in Eq. (5). We can compute the solution at a time level n from this system if we know the coefficient vector d^n at that time level. First, we need to solve for $\dot{d}_j$, $j = 0, 1, 2, \ldots N$ by inverting the above diagonally dominant tridiagonal matrix using the Thomas algorithm. Then we will obtain a system of $2(N+1)$ first-order ordinary differential equations, which is then solved by the SSP-RK54 scheme mentioned in [18]. We initiate the procedure with the vectors d^0 and w^0 obtained from the initial conditions $u(x_j, 0) = f_1(x_j)$ and $u_t(x_j, 0) = f_2(x_j)$ for $j = 0, \ldots, N$. Equivalently we can write the corresponding initial coefficients as

$$\begin{bmatrix} d_0^0 \\ d_1^0 \\ d_2^0 \\ d_3^0 \\ d_4^0 \\ \vdots \\ d_{N-4}^0 \\ d_{N-3}^0 \\ d_{N-2}^0 \\ d_{N-1}^0 \\ d_N^0 \end{bmatrix} = \begin{bmatrix} 8 & 0 & 0 & 0\;0\;0\;0 & \ldots & \ldots & 0 \\ 1 & \frac{37}{8} & 0 & 0\;0\;0\;0 & \ldots & \ldots & 0 \\ 0 & 1 & \frac{144}{37} & 0\;0\;0\;0 & \ldots & \ldots & 0 \\ 0 & 0 & 1 & \frac{15}{4}\;1\;0\;0 & \ldots & \ldots & 0 \\ 0 & 0 & 0 & 1\;4\;1\;0 & \ldots & \ldots & 0 \\ & & \vdots & \vdots & \vdots & & \\ 0 & \ldots\ldots & 0\;1\;4\;1 & 0 & 0 & 0 \\ 0 & \ldots\ldots & 0\;0\;1\;\frac{15}{4} & 1 & 0 & 0 \\ 0 & \ldots\ldots & 0\;0\;0\;0 & \frac{144}{37} & 1 & 0 \\ 0 & \ldots\ldots & 0\;0\;0\;0 & 0 & \frac{37}{8} & 1 \\ 0 & \ldots\ldots & 0\;0\;0\;0 & 0 & 0 & 8 \end{bmatrix}^{-1} \begin{bmatrix} f_1(x_0) \\ f_1(x_1) \\ f_1(x_2) \\ f_1(x_3) \\ f_1(x_4) \\ \vdots \\ f_1(x_{N-4}) \\ f_1(x_{N-3}) \\ f_1(x_{N-2}) \\ f_1(x_{N-1}) \\ f_1(x_N) \end{bmatrix}, \quad \begin{bmatrix} w_0^0 \\ w_1^0 \\ \vdots \\ w_{N-1}^0 \\ w_N^0 \end{bmatrix} = \begin{bmatrix} f_2(x_0) \\ f_2(x_1) \\ \vdots \\ f_2(x_{N-1}) \\ f_2(x_N) \end{bmatrix}.$$

The progress of solution at each time jump can be computed using Eqs. (4) and (6).

4 Convergence Analysis of the Numerical Scheme

Let's consider the solution $u(x, t)$ of the telegraph equation four times continuously differentiable, and the fourth derivative is continuous. Using Taylor series representation centered at $x = a_0 \in \Omega$, we can express $u(x, t)|_{t=t_0}$ as

$$u(x, t_0) = u(a_0, t_0) + \frac{u^{(1)}(a_0, t_0)}{1!}(x - a_0)^1 + \frac{u^{(2)}(a_0, t_0)}{2!}(x - a_0)^2 + \frac{u^{(3)}(a_0, t_0)}{3!}(x - a_0)^3 + \frac{u^{(4)}(\zeta, t_0)}{4!}(x - a_0)^4,$$

where ζ is a point between a and x. Recall that the set of modified cubic B-splines $\{\tilde{B}_0, \tilde{B}_1, \ldots, \tilde{B}_N\}$ is of accuracy order four; thus, any polynomials up to degree three can be uniquely represented as a linear combination of these functions. Since the part

$$u(a_0,t_0)+\frac{u^{(1)}(a_0,t_0)}{1!}(x-a_0)^1+\frac{u^{(2)}(a_0,t_0)}{2!}(x-a_0)^2+\frac{u^{(3)}(a_0,t_0)}{3!}(x-a_0)^3=p(x)$$

being a cubic polynomial, thus we can find unique scalars $d_0(t_0), d_1(t_0), \ldots, d_N(t_0)$ such that

$$p(x)=\sum_{j=0}^{N} d_j(t_0)\tilde{B}_j(x).$$

Hence

$$\left|u(x,t_0)-\sum_{j=0}^{N} d_j(t_0)\tilde{B}_j(x)\right|=\frac{|u^{(4)}(\zeta,t_0)|}{4!}|x-a_0|^4=\mathcal{O}(|x-a_0|^4).$$

But in the case of second derivative of u with respect to x we have

$$\left|u_{xx}(x,t_0)-\sum_{j=0}^{N} d_j(t_0)\tilde{B}_j''(x)\right|=\frac{\left|u^{(4)}(\zeta,t_0)\right|\cdot 4\cdot 3}{4!}|x-a|^2=\mathcal{O}(|x-a|^2),$$

and thus we can conclude that the numerical method is of order two convergence.

5 Stability Analysis

We use the asymptotic stability notion of the proposed numerical scheme as defined in [1, 11, 13] for a discrete form of the problem

$$\frac{dU}{dt}=LU, \tag{7}$$

where L is a diagonalizable matrix. It is well known that the stability of the nonhomogeneous system

$$\frac{dU}{dt}=LU+W$$

with a free term W, is equivalent to the stability of the corresponding homogeneous system. In our case, the system is nonhomogeneous and thus it is enough to analyze the corresponding homogeneous system.

Definition 1 The region of absolute stability of a numerical method is defined for the scalar model problem (7) to be the set of all $\lambda\delta t$ such that $\|U_n\|$ is bounded as

$t \to \infty$. We say that a numerical method is asymptotically stable for a particular problem if for sufficiently small $\delta t > 0$, the product of δt times every eigenvalue of L lies within the region of absolute stability [1, 11, 13].

5.1 Stability of the Proposed Method

We consider the system of first-order ordinary differential equations (6), and by omitting the boundary, it can be written in the following form:

$$\frac{d}{dt}\begin{bmatrix} \dot{d} \\ \dot{w} \end{bmatrix} = \begin{bmatrix} O_{N-1} & [A^{-1}]_{i,j=1}^{N-1} \\ [-\beta^2 A + C]_{i,j=1}^{N-1} & -2\alpha I_{N-1} \end{bmatrix} \begin{bmatrix} d \\ w \end{bmatrix} + \begin{bmatrix} F \\ G \end{bmatrix}, \tag{8}$$

where O_{N-1} and I_{N-1} are respectively the zero matrix and the identity matrix both of order $N-1$,

$$\begin{aligned} F =& \big[-\big([A^{-1}]_{1,0}w_0 + [A^{-1}]_{1,N}w_N\big), \ldots, -\big([A^{-1}]_{N-1,0}w_0 + [A^{-1}]_{N-1,N}w_N\big)\big]^T \\ G =& \Big[-\big((-\beta^2[A]_{1,0} + [C]_{1,0})d_0 + (-\beta^2[A]_{1,N} + [C]_{1,N})d_N\big) + f(x_1, t_n), \cdots \\ & \cdots, -\big((-\beta^2[A]_{N-1,0} + [C]_{N-1,0})d_0 + (-\beta^2[A]_{N-1,N} + [C]_{N-1,N})d_N\big) \\ & + f(x_{N-1}, t_n)\Big], \end{aligned}$$

$$\text{with } A = \begin{bmatrix} [A]_{0,0} & \cdots & [A]_{0,N} \\ \vdots & \ddots & \vdots \\ [A]_{N,0} & \cdots & [A]_{N,N} \end{bmatrix} = \begin{bmatrix} 8 & 0 & 0 & 0\;0\;0\;0 & \ldots\ldots & 0 \\ 1 & \frac{37}{8} & 0 & 0\;0\;0\;0 & \ldots\ldots & 0 \\ 0 & 1 & \frac{144}{37} & 0\;0\;0\;0 & \ldots\ldots & 0 \\ 0 & 0 & 1 & \frac{15}{4}\;1\;0\;0 & \ldots\ldots & 0 \\ 0 & 0 & 0 & 1\;4\;1\;0 & \ldots\ldots & 0 \\ & & \vdots & \vdots & \vdots & \\ 0\ldots\ldots & & & 0\;1\;4\;1 & 0 & 0\;0 \\ 0\ldots\ldots & & & 0\;0\;1\;\frac{15}{4} & 1 & 0\;0 \\ 0\ldots\ldots & & & 0\;0\;0\;0 & \frac{144}{37} & 1\;0 \\ 0\ldots\ldots & & & 0\;0\;0\;0 & 0 & \frac{37}{8}\;1 \\ 0\ldots\ldots & & & 0\;0\;0\;0 & 0 & 0\;8 \end{bmatrix},$$

$$\text{and } C = \begin{bmatrix} [C]_{0,0} & \cdots & [C]_{0,N} \\ \vdots & \ddots & \vdots \\ [C]_{N,0} & \cdots & [C]_{N,N} \end{bmatrix} = \frac{1}{h^2} \begin{bmatrix} 12 & -\frac{45}{2} & \frac{756}{37} & -9 & 0 & 0 & 0 & \dots & \dots & 0 \\ 6 & -\frac{33}{4} & \frac{144}{37} & 0 & 0 & 0 & 0 & \dots & \dots & 0 \\ 0 & 6 & -\frac{468}{37} & 9 & 0 & 0 & 0 & \dots & \dots & 0 \\ 0 & 0 & 6 & -\frac{27}{2} & 6 & 0 & 0 & \dots & \dots & 0 \\ 0 & 0 & 0 & 6 & -12 & 6 & 0 & \dots & \dots & 0 \\ & & \vdots & & \vdots & & & \vdots & & \\ 0 & \dots & \dots & 0 & 6 & -12 & 6 & 0 & 0 & 0 \\ 0 & \dots & \dots & 0 & 0 & 6 & -\frac{27}{2} & 6 & 0 & 0 \\ 0 & \dots & \dots & 0 & 0 & 0 & 9 & -\frac{468}{37} & 6 & 0 \\ 0 & \dots & \dots & 0 & 0 & 0 & 0 & \frac{144}{37} & -\frac{33}{4} & 6 \\ 0 & \dots & \dots & 0 & 0 & 0 & -9 & \frac{756}{37} & -\frac{45}{2} & 12 \end{bmatrix}.$$

Here for all the problems, the system (8) is found to be absolutely stable, and the region of stability is the left half plane. That is, all the eigenvalues of the coefficient matrix $\begin{bmatrix} O_{N-1} & [A^{-1}]_{i,j=1}^{N-1} \\ [-\beta^2 A + C]_{i,j=1}^{N-1} & -2\alpha I_{N-1} \end{bmatrix}$ have a negative real part. The eigenvalues of the coefficient matrix are plotted in Fig. 1 for all the test problems considered in this paper.

6 Numerical Experiments

In this section, we demonstrate the numerical results of some test examples using the method described in Sect. 3. L_2, L_∞, and root mean square (RMS) errors are calculated by the formulae:

$$\|u_N(\cdot,t) - u(\cdot,t)\|_\infty = \max_{0 \le j \le N} |u_N(x_j,t) - u(x_j,t)|,$$

$$\|u_N(\cdot,t) - u(\cdot,t)\|_2 = \left(h \sum_{j=0}^{N} |u_N(x_j,t) - u(x_j,t)|^2\right)^{1/2},$$

$$\text{Root Mean Suare Error (RMS)} = \left(\frac{1}{N+1} \sum_{j=0}^{N} |u_N(x_j,t) - u(x_j,t)|^2\right)^{1/2},$$

where u_N is the numerical solution of the telegraph equation in (1) using uniform points $a = x_0 < x_1 < \cdots < x_{N-1} < x_N = b$ and u stands for the exact solution of (1).

Example 1 Consider the telegraph equation in (1) for $\alpha = 10$, $\beta = 5$ and $\Omega = (0, 2)$ with the initial and boundary conditions $u(x,0) = \tan\left(\frac{x}{2}\right)$, $u_t(x,0) = \frac{1}{2}\left(1 + \tan^2\left(\frac{x}{2}\right)\right)$ and $u(0,t) = \tan\left(\frac{t}{2}\right)$, $u(2,t) = \tan\left(\frac{2+t}{2}\right)$, respectively. The source term is $f(x,t) = \alpha\left(1 + \tan^2\left(\frac{x+t}{2}\right)\right) + \beta^2 \tan\left(\frac{x+t}{2}\right)$. The exact solution to this problem is $u(x,t) = \tan(\frac{x+t}{2})$. The implementation is done with $N = 100$. The computed solution is plotted by taking $\Delta t = 0.001$ for $t = 0.2, 0.4, 0.6, 0.8, 1.0$ and

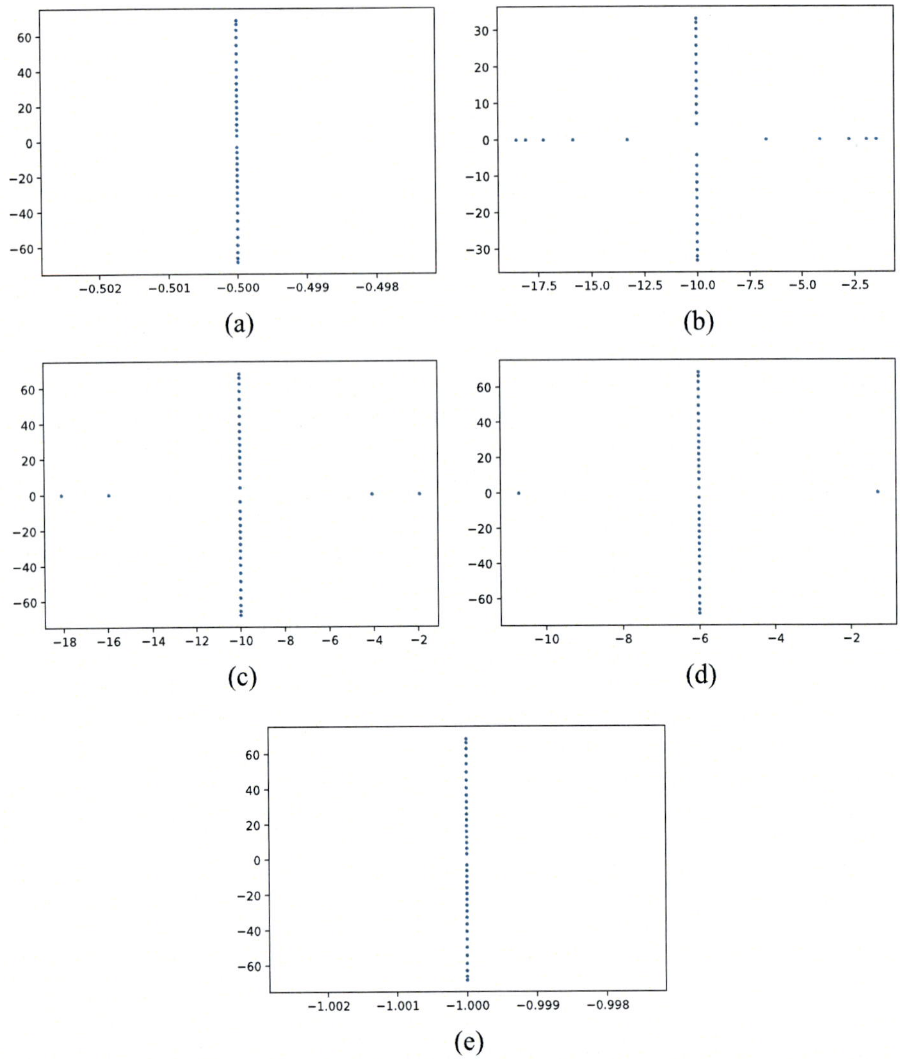

Fig. 1 The eigenvalue plots of the coefficient matrix for **a** Example 2 **b** Example 1 **c** Example 3 **d** Example 4 **e** Example 5 for $N = 20$

the surface plot is also given for $t = 0$ to $t = 1$ along with in Fig. 2. The L_2 and L_∞ errors are calculated and compared with results in [14] in Table 2. The order of convergence of the proposed numerical scheme is found to be two and is calculated in Table 3. The tabulated data shows that the approximation with the proposed modified basis functions is much more accurate. □

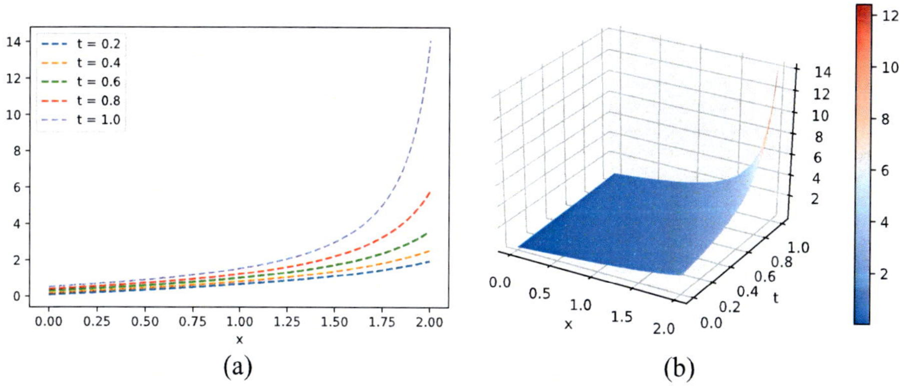

Fig. 2 Example 1. **a** is for approximate solutions with $N = 100$ and $\Delta t = 0.001$. **b** is for the surface plot

Table 2 Example 1: L_2 and L_∞ errors comparison for $N = 100$ and for various Δt

Methods	t	$\Delta t = 0.01$		$\Delta t = 0.001$	
		L_2	L_∞	L_2	L_∞
New	0.2	2.1763×10^{-6}	4.4785×10^{-6}	2.1694×10^{-6}	4.4651×10^{-6}
	0.4	7.0870×10^{-6}	1.4006×10^{-5}	7.0737×10^{-6}	1.3981×10^{-5}
	0.6	2.1328×10^{-5}	4.5082×10^{-5}	2.1296×10^{-5}	4.5018×10^{-5}
	0.8	8.8686×10^{-5}	2.1629×10^{-4}	8.8583×10^{-5}	2.1605×10^{-4}
	1.0	9.7829×10^{-4}	3.1307×10^{-3}	9.7702×10^{-4}	3.1293×10^{-3}
Method	t	$\Delta t = 0.0001$		$\Delta t = 0.001$	
		L_2	L_∞	L_2	L_∞
[14]	0.2	5.0305×10^{-5}	3.4797×10^{-5}	1.8782×10^{-4}	2.6332×10^{-4}
	0.4	9.5163×10^{-5}	5.3456×10^{-5}	4.8888×10^{-4}	6.9997×10^{-4}
	0.6	2.2041×10^{-4}	9.4725×10^{-5}	9.4864×10^{-4}	1.4860×10^{-3}
	0.8	7.8267×10^{-4}	1.8792×10^{-4}	1.8743×10^{-3}	3.4057×10^{-3}
	1.0	7.9277×10^{-3}	5.8720×10^{-4}	5.0981×10^{-3}	1.1148×10^{-2}
Method	t	$\Delta t = 0.0001$		$\Delta t = 0.001$	
		L_2	L_∞	L_2	L_∞
[16]	0.2	2.13×10^{-5}	3.35×10^{-5}	1.85×10^{-4}	2.79×10^{-4}
	0.4	5.25×10^{-5}	7.92×10^{-5}	4.43×10^{-4}	6.35×10^{-4}
	0.6	1.02×10^{-4}	1.63×10^{-4}	7.96×10^{-4}	1.17×10^{-3}
	0.8	2.39×10^{-4}	4.65×10^{-4}	1.47×10^{-3}	2.40×10^{-3}
	1.0	1.73×10^{-3}	5.59×10^{-4}	4.19×10^{-3}	9.51×10^{-3}
Method	t	$\Delta t = 0.001$			
		L_2	L_∞		
[17]	0.2	9.99×10^{-6}	6.83×10^{-5}		
	0.4	7.07×10^{-6}	4.28×10^{-5}		

Table 3 Example 1: Convergence rate of the problem for with $\Delta t = 0.001$ at $t = 1$ s

N	L_∞ error	CPU time (sec)	Convergence rate
2^5	2.97×10^{-2}	0.26	–
2^6	5.98×10^{-3}	0.31	2.31
2^7	2.17×10^{-3}	0.50	1.46
2^8	6.82×10^{-4}	0.63	1.67
2^9	1.85×10^{-4}	2.07	1.87
2^{10}	4.75×10^{-5}	9.30	1.96

Example 2 Consider the telegraph equation in (1) for $\alpha = \frac{1}{2}, \beta = 1$ and $\Omega = (0, 1)$ with the initial and boundary conditions

$$u(x, 0) = 0 = u_t(x, 0) \quad \text{and} \quad u(0, t) = 0 = u(1, t),$$

respectively. The source term is $f(x, t) = \frac{1}{e^t}\left(2t^2 + (x - x^2)(t^2 - 2t + 2)\right)$. The exact solution to this problem is $u(x, t) = \frac{(x-x^2)t^2}{e^t}$. Computed solution is calculated by using $N = 100$ and $\Delta t = 0.001$. Figure 3 interprets the behavior of the solution we obtained as time progresses from $t = 0$ to $t = 5$. The L_2, L_∞ and RMS errors tabulated and perform a comparison with results found in two works from literature shown in Tables 4 and 5. Second-order polynomial in the exact solution to this problem is effectively captured by our modified basis functions on the boundary and the better performance in the computational results as well. □

Example 3 Consider the telegraph equation in (1) on $\Omega = (0, 1)$ with the initial and boundary conditions $u(x, 0) = \sinh(x)$, $u_t(x, 0) = -2\sinh(x)$ and $u(0, t) = 0$, $u(1, t) = e^{-2t}\sinh(1)$, respectively. The source term is $f(x, t) = e^{-2t}\sinh(x)$

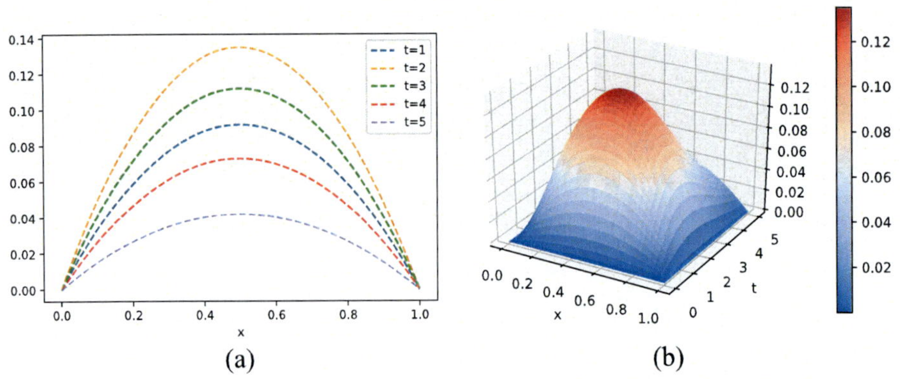

Fig. 3 Example 2. **a** is for approximate solutions with $N = 100$ and $\Delta = 0.001$. **b** is for the surface plot

Table 4 Example 2: L_2 and L_∞ errors comparison for $N = 100$, $\Delta t = 0.001$ and for various time

t	New Method		[14]		[6]	
	L_2	L_∞	L_2	L_∞	L_2	L_∞
1	3.8182×10^{-12}	5.3405×10^{-12}	4.5526×10^{-5}	5.9153×10^{-5}	1.4386×10^{-4}	1.8479×10^{-5}
2	3.1063×10^{-12}	4.3673×10^{-12}	1.4307×10^{-5}	1.7864×10^{-5}	8.0879×10^{-5}	1.0713×10^{-5}
3	5.0843×10^{-13}	7.2346×10^{-13}	6.4273×10^{-6}	1.4309×10^{-5}	1.2944×10^{-4}	1.8161×10^{-5}
4	1.0941×10^{-12}	1.5267×10^{-12}	8.9203×10^{-6}	1.3529×10^{-5}	1.1845×10^{-4}	1.6489×10^{-5}
5	1.5544×10^{-13}	2.2373×10^{-13}	3.0161×10^{-6}	5.2032×10^{-6}	7.5545×10^{-5}	1.0455×10^{-5}

Table 5 Example 2: L_2 and L_∞ errors comparison for $N = 100$, $\Delta t = 0.005$ and for various time

t	New Method			[8]		
	L_2	L_∞	RMS	L_2	L_∞	RMS
1	8.64×10^{-12}	1.04×10^{-11}	8.60×10^{-12}	7.54×10^{-6}	9.93×10^{-7}	7.58×10^{-7}
2	7.84×10^{-12}	8.97×10^{-12}	7.80×10^{-12}	1.57×10^{-5}	2.12×10^{-6}	2.58×10^{-6}
t	[16]			[17]		
	L_2	L_∞	RMS	L_2	L_∞	RMS
1	6.31×10^{-5}	8.76×10^{-5}	–	2.61×10^{-8}	1.67×10^{-7}	1.68×10^{-8}
2	2.34×10^{-5}	3.29×10^{-5}	–	6.10×10^{-8}	4.72×10^{-7}	4.74×10^{-8}

Table 6 Example 3: L_2, L_∞ and RMS errors comparison for $N = 51$, $\Delta t = 0.01$, $\alpha = 10$, $\beta = 5$ at $t = 2.0$ s

t	Method	L_2	L_∞	RMS
1	New	1.33×10^{-7}	1.93×10^{-7}	1.32×10^{-7}
	[8]	8.35×10^{-5}	1.70×10^{-5}	1.19×10^{-5}
2	New	3.78×10^{-8}	5.33×10^{-8}	3.75×10^{-8}
	[8]	2.37×10^{-5}	4.63×10^{-6}	3.38×10^{-6}

$(3 - 4\alpha + \beta^2)$. The exact solution to this problem is $u(x, t) = e^{-2t}\sinh(x)$. The numerical approximation using our method guarantees better accuracy. This can be observed from the error comparison Tables 6 and 7 and the surface plot in Fig. 4. The order of convergence is calculated in Table 8. The results here show better performance of the proposed scheme than the results shown in [8, 12]. □

Example 4 Consider the telegraph equation in (1) for $\Omega = (0, 1)$ with initial and boundary conditions $u(x, 0) = \sin(x)$, $u_t(x, 0) = 0$ and $u(0, t) = 0$, $u(1, t) = \cos(t)\sin(1)$. The source term is $f(x, t) = -2\alpha\sin(t)\sin(x) + \beta^2\cos(t)\sin(x)$.

Table 7 Example 3: Absolute error comparison for $N = 100$, $\Delta t = 0.001$

x	$\alpha = 20,\ \beta = 10$				$\alpha = 10,\ \beta = 5$			
	$t = 0.5$		$t = 1$		$t = 0.5$		$t = 1$	
	New	[12]	New	[12]	New	[12]	New	[12]
0.0	3.7×10^{-19}	7.2×10^{-21}	1.3×10^{-19}	1.4×10^{-15}	3.7×10^{-19}	2.3×10^{-13}	1.3×10^{-19}	1.4×10^{-10}
0.1	3.5×10^{-9}	1.1×10^{-5}	2.2×10^{-9}	2.8×10^{-7}	9.5×10^{-9}	4.4×10^{-6}	9.0×10^{-9}	2.1×10^{-7}
0.2	7.0×10^{-9}	8.7×10^{-7}	4.6×10^{-9}	7.6×10^{-7}	1.9×10^{-8}	1.1×10^{-6}	1.8×10^{-8}	8.5×10^{-7}
0.3	1.0×10^{-8}	1.0×10^{-6}	6.9×10^{-9}	6.5×10^{-7}	2.8×10^{-8}	1.1×10^{-6}	2.7×10^{-8}	7.3×10^{-7}
0.4	1.4×10^{-8}	2.1×10^{-7}	9.4×10^{-9}	5.0×10^{-7}	3.9×10^{-8}	1.3×10^{-7}	3.5×10^{-8}	4.5×10^{-7}
0.5	1.8×10^{-8}	7.5×10^{-8}	1.1×10^{-8}	1.0×10^{-7}	4.9×10^{-8}	2.3×10^{-7}	4.3×10^{-8}	8.0×10^{-8}
0.6	2.2×10^{-8}	3.5×10^{-7}	1.4×10^{-8}	1.2×10^{-7}	5.9×10^{-8}	7.9×10^{-7}	4.7×10^{-8}	2.3×10^{-7}
0.7	2.6×10^{-8}	3.6×10^{-7}	1.5×10^{-8}	8.0×10^{-8}	6.6×10^{-8}	8.9×10^{-7}	4.8×10^{-8}	3.0×10^{-7}
0.8	2.8×10^{-8}	2.2×10^{-7}	1.5×10^{-8}	2.3×10^{-8}	6.4×10^{-8}	8.4×10^{-7}	4.2×10^{-8}	2.9×10^{-7}
0.9	2.2×10^{-8}	1.6×10^{-8}	1.1×10^{-8}	8.2×10^{-8}	4.5×10^{-8}	7.8×10^{-7}	2.6×10^{-8}	2.7×10^{-7}
1.0	6.5×10^{-11}	4.8×10^{-7}	8.9×10^{-11}	2.6×10^{-7}	6.5×10^{-11}	5.9×10^{-7}	8.9×10^{-11}	2.6×10^{-7}

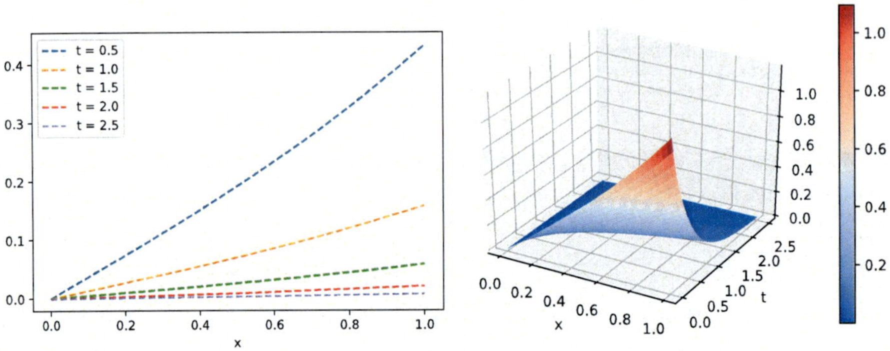

(a) (b)

Fig. 4 Example 3. **a** is for approximate solutions with $N = 33$ and $\Delta = 0.001$, $\alpha = 10$, $\beta = 5$. **b** is for the surface plot

Table 8 Example 3: Convergence rate of the problem for with $\Delta t = 0.01$ at $t = 1$ s

N	L_∞ error	CPU time (sec)	Convergence rate
2^3	5.47×10^{-6}	0.021	–
2^4	1.75×10^{-6}	0.020	1.64
2^5	4.67×10^{-7}	0.024	1.90
2^6	1.18×10^{-7}	0.025	1.98

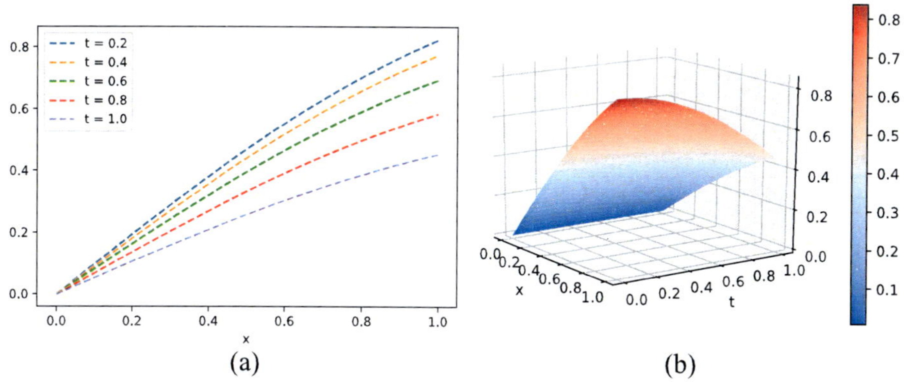

Fig. 5 Example 4. **a** is for approximate solutions with $N = 100$, $\Delta = 0.001$, $\alpha = 6$, $\beta = 2$. **b** is for the surface plot

The exact solution to this problem is $u(x, t) = \cos(t)\ \sin(x)$. An approximation is done by taking $N = 100$ here. Computed solution is plotted for various values of t up to 1 s shown in Fig. 5. The L_2 and L_∞ errors up to time 5 s are calculated and perform a comparison with two more works, which is tabulated in Tables 9 and 10. Comparison of root-mean-square error can be found in Table 9. Absolute error is compared in Table 11 and the order of convergence of the proposed method is calculated in Table 12. □

Example 5 Consider the telegraph equation in (1) for $\alpha = \beta = 1$ and $\Omega = (0, 1)$ with the initial and boundary conditions $u(x, 0) = x^2 = u_t(x, 0)$ and $u(0, t) = 0$, $u(1, t) = e^t$, respectively. The source term is $f(x, t) = e^t \left(\beta^2 x^2 + 2\alpha x^2 + x^2 - 2\right)$. In this problem exact solution is $u(x, t) = x^2 e^t$ and note that the function x^2, here is well captured by the proposed modified cubic B-splines and hence we get a better approximation as shown in Table 13. L_2, L_∞ and RMS errors are calculated with $N = 32$ grid points for various Δt and the numerical solution is drawn up to 12 s as in Fig. 6. □

Table 9 Example 4: L_2, L_∞ and RMS errors comparison for $N = 100$, $\Delta t = 0.0001$, $\alpha = 6$, $\beta = 2$ and for various time

t	New Method			[14]		
	L_2	L_∞	RMS	L_2	L_∞	RMS
0.2	3.9166×10^{-8}	6.1033×10^{-8}	3.8971×10^{-8}	2.6903×10^{-6}	5.2412×10^{-6}	2.6775×10^{-6}
0.4	8.6074×10^{-8}	1.2869×10^{-7}	8.5647×10^{-8}	5.6183×10^{-6}	8.6095×10^{-6}	5.5904×10^{-6}
0.6	1.1927×10^{-7}	1.7324×10^{-7}	1.1868×10^{-7}	9.7545×10^{-6}	1.2529×10^{-5}	9.7061×10^{-6}
0.8	1.3897×10^{-7}	1.9859×10^{-7}	1.3828×10^{-7}	1.3774×10^{-5}	2.0274×10^{-5}	1.3706×10^{-5}
1.0	1.4652×10^{-7}	2.0748×10^{-7}	1.4579×10^{-7}	1.7347×10^{-5}	2.7555×10^{-5}	1.7261×10^{-5}
t	[16]			[17]		
	L_2	L_∞	RMS	L_2	L_∞	RMS
0.2	2.96×10^{-6}	4.63×10^{-6}	2.94×10^{-6}	2.33×10^{-8}	2.10×10^{-7}	1.49×10^{-8}
0.4	6.77×10^{-6}	1.01×10^{-5}	6.73×10^{-6}	4.55×10^{-8}	4.30×10^{-7}	3.04×10^{-8}

Table 10 Example 4: L_2 and L_∞ errors comparison for various N where $\Delta t = 0.001$, $\alpha = 6$, $\beta = 2$

Methods	t	N=100		N=200	
		L_2	L_∞	L_2	L_∞
	0.2	3.9166×10^{-8}	6.1034×10^{-8}	9.8224×10^{-9}	1.5347×10^{-8}
	0.4	8.6075×10^{-8}	1.2869×10^{-7}	2.1560×10^{-8}	3.2243×10^{-8}
New	0.6	1.1927×10^{-7}	1.7324×10^{-7}	2.9864×10^{-8}	4.3376×10^{-8}
	0.8	1.3897×10^{-7}	1.9859×10^{-7}	3.4788×10^{-8}	4.9705×10^{-8}
	1.0	1.4652×10^{-7}	2.0748×10^{-7}	3.6671×10^{-8}	5.1922×10^{-8}
	0.2	3.6711×10^{-5}	7.9139×10^{-5}	3.4308×10^{-5}	6.8272×10^{-5}
	0.4	8.8989×10^{-5}	1.5967×10^{-4}	8.5756×10^{-5}	1.4935×10^{-4}
[14]	0.6	1.3733×10^{-4}	2.3362×10^{-4}	1.3367×10^{-4}	2.2410×10^{-4}
	0.8	1.7913×10^{-4}	2.9820×10^{-4}	1.7532×10^{-4}	2.8978×10^{-4}
	1.0	2.1302×10^{-4}	3.5085×10^{-4}	2.0929×10^{-4}	3.4386×10^{-4}

7 Discussion

This paper employs the modified cubic splines to numerically solve the 2nd− order one-dimensional hyperbolic telegraph equation in (1) with Dirichlet boundary conditions. By using an optimally accurate basis functions, the modified cubic B-splines, we could be able to approximate the solution of the problem with less computational error as comparing with some other known works in the literature. A discretized form of the differential equation is found against the modified set on a uniformly

Table 11 Example 4: Absolute error comparison for $N = 100$, $\Delta t = 0.001$

x	$\alpha = 20,\ \beta = 10$				$\alpha = 10,\ \beta = 5$			
	$t = 0.5$		$t = 1$		$t = 0.5$		$t = 1$	
	New	[12]	New	[12]	New	[12]	New	[12]
0.0	8.9×10^{-19}	1.9×10^{-11}	5.4×10^{-19}	3.2×10^{-12}	8.9×10^{-19}	2.4×10^{-13}	5.4×10^{-19}	1.1×10^{-11}
0.1	5.6×10^{-9}	5.0×10^{-5}	5.7×10^{-9}	1.7×10^{-5}	1.4×10^{-8}	6.0×10^{-6}	1.8×10^{-8}	6.1×10^{-7}
0.2	1.1×10^{-8}	1.6×10^{-6}	1.1×10^{-8}	4.3×10^{-6}	2.8×10^{-8}	4.3×10^{-6}	3.7×10^{-8}	1.6×10^{-6}
0.3	1.6×10^{-8}	4.4×10^{-6}	1.7×10^{-8}	3.6×10^{-6}	4.2×10^{-8}	2.0×10^{-6}	5.5×10^{-8}	1.3×10^{-6}
0.4	2.2×10^{-8}	1.7×10^{-6}	2.2×10^{-8}	2.6×10^{-6}	5.5×10^{-8}	6.5×10^{-7}	7.2×10^{-8}	8.0×10^{-7}
0.5	2.7×10^{-8}	5.0×10^{-7}	2.7×10^{-8}	5.6×10^{-7}	6.8×10^{-8}	2.9×10^{-7}	8.7×10^{-8}	5.1×10^{-7}
0.6	3.1×10^{-8}	1.4×10^{-6}	3.2×10^{-8}	6.8×10^{-7}	7.9×10^{-8}	7.6×10^{-7}	9.6×10^{-8}	2.9×10^{-7}
0.7	3.6×10^{-8}	1.4×10^{-6}	3.5×10^{-8}	4.4×10^{-7}	8.7×10^{-8}	6.9×10^{-7}	9.9×10^{-8}	9.4×10^{-8}
0.8	3.7×10^{-8}	1.0×10^{-6}	3.4×10^{-8}	4.2×10^{-8}	8.4×10^{-8}	4.1×10^{-7}	8.8×10^{-8}	3.3×10^{-8}
0.9	3.1×10^{-8}	6.8×10^{-7}	2.6×10^{-8}	2.8×10^{-8}	6.0×10^{-8}	5.7×10^{-8}	5.8×10^{-8}	1.4×10^{-7}
1.0	9.0×10^{-12}	2.2×10^{-7}	3.3×10^{-11}	3.0×10^{-7}	9.0×10^{-12}	4.0×10^{-7}	3.3×10^{-11}	3.0×10^{-7}

Table 12 Example 4: Convergence rate of the problem for with $\Delta t = 0.01$ at $t = 1$ s

N	L_∞ error	CPU time (sec)	Convergence rate
2^3	2.44×10^{-5}	0.17	–
2^4	7.55×10^{-6}	0.17	1.69
2^5	1.99×10^{-6}	0.21	1.99
2^6	5.05×10^{-7}	0.25	1.97
2^7	1.26×10^{-7}	0.39	2.00

discretized domain to form a system of first-order ordinary differential equations. This ODE system is solved by the SSP-RK54 scheme, which ensures the strong stability property of the procedure. The novelty of the approximation lies in the modification of cubic B-splines. Since these basis functions could reproduce the polynomials having degrees up to three on the whole domain, the approximate solution will possess that much accuracy. Zhimin Hong et al. [8] solved the telegraph equation in (1) using finite difference schemes to discretize time-space derivatives,

Table 13 Example 5: L_2, L_∞ and RMS errors comparison for various $\Delta\ t$ where $N = 32$

t	$\Delta\ t = 0.01$			$\Delta\ t = 0.001$		
	L_2	L_∞	RMS	L_2	L_∞	RMS
2	5.4234×10^{-10}	8.8826×10^{-10}	5.3406×10^{-10}	3.0167×10^{-10}	5.6077×10^{-10}	2.9707×10^{-10}
4	4.0638×10^{-9}	7.6402×10^{-9}	4.0018×10^{-9}	2.5398×10^{-9}	4.7043×10^{-9}	2.5010×10^{-9}
6	3.0037×10^{-8}	5.6985×10^{-8}	2.9579×10^{-8}	1.9078×10^{-8}	3.5321×10^{-8}	1.8787×10^{-8}
8	2.2195×10^{-7}	4.2157×10^{-7}	2.1856×10^{-7}	1.4128×10^{-7}	2.6155×10^{-7}	1.3913×10^{-7}
10	1.6400×10^{-6}	3.1155×10^{-6}	1.6149×10^{-6}	1.0443×10^{-6}	1.9332×10^{-6}	1.0283×10^{-6}

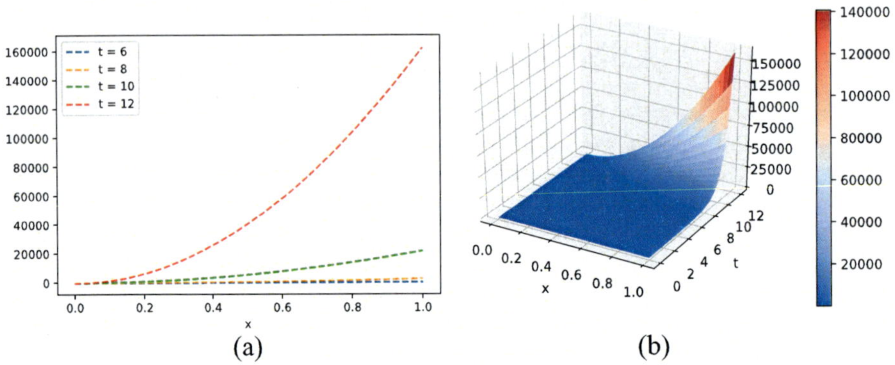

Fig. 6 Example 5. **a** is for approximate solutions with $N = 33$ and $\Delta = 0.01$. **b** is for the surface plot

and they obtained an extensive sparse system which is then solved by adaptive Monte Carlo methods in [8]. The numerical experiments Examples 2, 3 demonstrate that our proposed method has a significant reduction in L_2 and L_∞ errors up to the order of 10^{-2} to 10^{-5} in comparison with the method in [8] (see Tables 5 and 6 for details). In [16], Tahir Nazir et al. discretized the telegraph equation in (1) on a uniform mesh using the cubic trigonometric B-spline collocation method, while Shokofeh Sharifi et al. [17] used collocation with extended cubic B-spline instead. The excellent results in [16, 17] are compared with other known methods in [6, 14]. The errors in the numerical approximation using our proposed method seem to be much smaller for Examples 1, 2, 4 in comparison with the results found in [16, 17] as well. Similarly in the case of the results found in [12] (see Tables 7 and 11). Generally speaking, the numerical experiments illustrate that our proposed method performs better than many other related methods reported in the literature such as [6, 8, 12, 14, 16, 17]. The stability and convergence of the proposed numerical scheme have been systematically analyzed in this article. The scheme is found to be absolutely stable, and the region of stability is determined to be the left half plane. Moreover, the scheme has second-order convergence and is verified through the problems reported in this paper.

Acknowledgements Research of Noufal Asharaf supported with grant PL(UGC)1/SPG/SMNRI/ 2018–19 by Cochin University of Science and Technology, and Athira Babu under the Senior Research Fellowship of University Grants Commission.

References

1. Ashpazzadeh, E., Han, B., Lakestani, M.: Biorthogonal multiwavelets on the interval for numerical solutions of Burgersâ equation. J. Comput. Appl. Math. **317**, 510–534 (2017)
2. Babu, A., Han, B., Asharaf, N.: Numerical solution of the hyperbolic telegraph equation using cubic b-spline-based differential quadrature of high accuracy. In: Computational Methods for Differential Equations (2022)
3. Banasiak, J., Mika, J.R.: Singularly perturbed telegraph equations with applications in the random walk theory. Int. J. Stoch. Anal. **11**(1), 9–28 (1998)
4. Biazar, J., Ebrahimi, H.: An approximation to the solution of telegraph equation by Adomian decomposition method. Int. Math. Forum 2231–2236 (2007)
5. Biçer, K.E., Yalçinbaş, S.: Numerical solution of telegraph equation using Bernoulli collocation method. Proc. Natl. Acad. Sci., India Sect. A: Phys. Sci. **89**(4)(209), 769–775
6. Dehghan, M., Shokri, A.: A numerical method for solving the hyperbolic telegraph equation. Numer. Methods Part. Differ. Equ.: Int. J. **24**(4), 1080–1093 (2008)
7. Dehghan, M., Lakestani, M.: The use of Chebyshev cardinal functions for solution of the second-order one-dimensional telegraph equation. Numer. Methods Part. Differ. Equ. **25**(4), 931–938 (2009)
8. Hong, Z., Wang, Y., Hao, H.: Adaptive Monte Carlo methods for solving hyperbolic telegraph equation. J. Comput. Appl. Math. **345**, 405–415 (2019)
9. Jiwari, R., Pandit, S., Mittal, R.: A differential quadrature algorithm to solve the two dimensional linear hyperbolic telegraph equation with Dirichlet and Neumann boundary conditions. Appl. Math. Comput. **218**(13), 7279–7294 (2012)
10. Jordan, P., Puri, A.: Digital signal propagation in dispersive media. J. Appl. Phys. **85**(3), 1273–1282 (1999)
11. Kumar, B.V.R., Mehra, M.: A Wavelet-Taylor Galerkin method for parabolic and hyperbolic partial differential equations. Int. J. Comput. Methods **2**(1), 75–97 (2005)
12. Lakestani, M., Saray, B.N.: Numerical solution of telegraph equation using interpolating scaling functions. Comput. Math. Appl. **60**(7), 1964–1972 (2010)
13. Lakestani, M., Dehghan, M.: Numerical solutions of the generalized Kuramoto-Sivashinsky equation using B-spline functions. Appl. Math. Model. **36**(2), 605–617 (2012)
14. Mittal, R., Bhatia, R.: Numerical solution of second order one dimensional hyperbolic telegraph equation by cubic b-spline collocation method. Appl. Math. Comput. **220**, 496–506 (2013)
15. Mittal, R., Bhatia, R.: A numerical study of two dimensional hyperbolic telegraph equation by modified b-spline differential quadrature method. Appl. Math. Comput. **244**, 976–997 (2014)
16. Nazir, T., Abbas, M., Yaseen, M.: Numerical solution of second-order hyperbolic telegraph equation via new cubic trigonometric b-splines approach. Cogent Math. Stat. **4**(1), 1382061 (2017)
17. Sharifi, S., Rashidinia, J.: Numerical solution of hyperbolic telegraph equation by cubic b-spline collocation method. Appl. Math. Comput. **281**, 28–38 (2016)
18. Spiteri, R.J., Ruuth, S.J.: A new class of optimal high-order strong-stability-preserving time discretization methods. SIAM J. Numer. Anal. **40**(2), 469–491 (2002)

19. Wallace, P.R.: Mathematical Analysis of Physical Problems. Courier Corporation (1984)
20. Weston, V., He, S.: Wave splitting of the telegraph equation in R3 and its application to inverse scattering. Inverse Prob. **9**(6), 789 (1993)
21. Yousefi, S.: Legendre multiwavelet Galerkin method for solving the hyperbolic telegraph equation. Numer. Methods Part. Differ. Equ.: Int. J. **26**(3), 535–543 (2010)